Recent Progress in
Environmental Engineering

Recent Progress in Environmental Engineering

Edited by Marrianne Fox

SYRAWOOD
PUBLISHING HOUSE

New York

Published by Syrawood Publishing House,
750 Third Avenue, 9th Floor,
New York, NY 10017, USA
www.syrawoodpublishinghouse.com

Recent Progress in Environmental Engineering
Edited by Marrianne Fox

International Standard Book Number: 978-1-68286-542-2 (Hardback)

Cataloging-in-Publication Data

Recent progress in environmental engineering / edited by Marrianne Fox.
    p. cm.
Includes bibliographical references and index.
ISBN 978-1-68286-542-2
1. Environmental engineering. 2. Environmental protection.
3. Environmental sciences. I. Fox, Marrianne.
TD145 .R43 2018
628--dc23

TABLE OF CONTENTS

Permissions

List of Contributors

Index

PREFACE

The branch of engineering which focuses on the environment and the proper utilization of resources is known as environmental engineering. The increase in population as well as the use of technology that emits harmful gases results in depletion of natural resources. These issues are tackled by environmental engineering. It involves controlling air pollution, managing wastewater, a sustainable use of resources, recycling, etc. Coherent flow of topics, student-friendly language and extensive use of examples make this book an invaluable source of knowledge. While understanding the long-term perspectives of the topics, the book makes an effort in highlighting their impact as a modern tool for the growth of discipline.

All of the data presented henceforth, was collaborated in the wake of recent advancements in the field. The aim of this book is to present the diversified developments from across the globe in a comprehensible manner. The opinions expressed in each chapter belong solely to the contributing authors. Their interpretations of the topics are the integral part of this book, which I have carefully compiled for a better understanding of the readers.

At the end, I would like to thank all those who dedicated their time and efforts for the successful completion of this book. I also wish to convey my gratitude towards my friends and family who supported me at every step.

Editor

A hybrid solar photovoltaic-wind turbine-Rankine cycle for electricity generation in Turkish Republic of Northern Cyprus

Samuel Asumadu-Sarkodie[1]*, Çağlan Sevinç[1] and Herath M.P.C. Jayaweera[1]

*Corresponding author: Samuel Asumadu-Sarkodie, Sustainable Environment and Energy Systems, Middle East Technical University Northern Cyprus Campus, Kalkanli-Guzelyurt, Turkey.
E-mail: samuelsarkodie@yahoo.com, samuel.sarkodie@metu.edu.tr
Reviewing editor: Shashi Dubey, Hindustan College of Engineering, India

Abstract: This paper presents an energy demand model by designing a hybrid solar-wind-thermal power generation system of the Turkish Republic of Northern Cyprus, a promising substitute for the expensive battery banks. The study models the future energy demand of Turkish Republic of Northern Cyprus based on the IPCC emissions scenario A1B and A2 by designing a new hybrid solar-wind-thermal power system that satisfies the current and future requirements of firm capacity during peak periods. The study suggests an improvement in a hybrid solar-wind-thermal power system performance by predicting reliable outputs that can integrate renewable energy technologies to conventional power generation. The energy consumption prediction model emphasizes the energy requirement that has a growing demand from 300 to 400 GWh in scenario A1B and 150–450 GWh in scenario A2 from 2010 to 2050. The proposed design can meet 400 GWh of electricity demand in TRNC based on IPCC scenario A1B and 450 GWh of electricity demand in TRNC based on IPCC scenario A2. The percentage contribution of solar, wind and thermal energy for 2010, 2020, 2030, 2040 and 2050 are presented along with CO_2 emissions and water consumption for each of the years.

Subjects: Energy & Fuels; Power & Energy; Renewable Energy

ABOUT THE AUTHORS

Samuel Asumadu-Sarkodie is a multidisciplinary researcher who currently studies Masters in Sustainable Environment and Energy Systems at Middle East Technical University, Northern Cyprus Campus where he is also a graduate assistant in the Chemistry department. His research interest includes but not limited to: renewable energy, econometrics, energy economics, climate change and sustainable development.

Herath M.P.C. Jayaweera studies masters in Sustainable Environment and Energy Systems at Middle East Technical University, Northern Cyprus Campus where he is a research assistant and a graduate assistant in the Physics department. His research interest includes: Integrated circuit design and Energy Harvesting.

PUBLIC INTEREST STATEMENT

Energy demand modelling is of much interest among scientist and engineers concerned with the issues of energy production and utilization due to its usefulness in planning and formulating energy policies. Future energy consumption in Turkish Republic of Northern Cyprus is modelled with regression analysis based on GDP and population. Due to uncertainties about the future, the study employs the Intergovernmental Panel on Climate Change scenarios to forecast the future energy demand. The output of energy demand is used to design a hybrid solar-wind-thermal power generation for Turkish Republic of Northern Cyprus. Water consumption by the proposed system and carbon dioxide emission reduction contribution are analysed, respectively. The study will provide an alternative solution to the existing battery banks used as storage for a hybrid PV-Wind energy systems and will serve as an informative tool for local and foreign investors in Turkish Republic of Northern Cyprus and also serve as a foundation for future research.

Keywords: energy conversion/recovery; IPCC scenarios, Northern Cyprus; hybrid solar-wind-thermal; power generation

1. Introduction

Clean combustion, high thermodynamic efficiency and renewable energy technologies are gaining world attention, due to their necessity to satisfy demands for CO_2 emission mitigation *via* the reduction of combustion generated pollution (Beér, 2000). CO_2 emissions can be mitigated if there are alternatives that are accessible, available and economical viable to replace the conventional fossil fuels (Asumadu-Sarkodie & Owusu, 2016d). Renewable energy is the cleanest and promising technology that can help mitigate CO_2 emissions (Asumadu-Sarkodie & Owusu, 2016g). The concept of sustainability demands a balance between energy availability, accessibility, economic viability, environmental impacts and social interactions (Asumadu-Sarkodie & Owusu, 2016b; Owusu & Asumadu-Sarkodie, 2016; Owusu, Asumadu-Sarkodie, & Ameyo, 2016). Security of energy supplies is of vital importance to our way of life and economic endeavours (Nakata, 2004). Solar and wind energy are some examples of renewable energy technologies that are widespread, environmental friendly and freely available throughout the year. Wind power plants may not be viable at all sites due to low wind speed and its huge uneven variation as compared to solar energy (Nema, Nema, & Rangnekar, 2009). Likewise, due to fluctuating weather conditions, cost to integrate of solar systems may be drastically high (Dersch et al., 2004). However, combined solar wind systems are very complementary (Yang, Lu, & Burnett, 2003) in a way that, during the times and seasons where solar resources do not exist or are low in capacity, wind resources take over. Nonetheless, because of dispatchability issues (Gyuk et al., 2005) and under-development of storage systems (Mohd, Ortjohann, Schmelter, Hamsic, & Morton, 2008), a heat engine should also be combined with these systems in order to supply the base load (Larson & Williams, 1987), which increases global warming due to greenhouse gases emissions. As the effects of global warming are seen in a long-term (Solomon, Plattner, Knutti, & Friedlingstein, 2009), well-defined scenarios that predict future should be investigated in order to estimate related emissions of hybrid solar-wind-thermal power plants.

Over increasing energy demand due to increasing population (Asumadu-Sarkodie & Owusu, 2016a, 2016e, 2016f; Dincer, 2000; Ehrlich & Holdren, 1971), decline of fossil fuels and global warming due to GHG emissions have led to the demand for clean, inexhaustible, environmentally friendly and renewable energy resources (Asumadu-Sarkodie & Owusu, 2016c; Demirbas, 2005; Hepbasli, 2008). The estimation of the future energy demand (Akay & Atak, 2007; Berndt & Wood, 1975; Ceylan & Ozturk, 2004; Crompton & Wu, 2005; Dunkerley, 1982; Ediger & Akar, 2007; Ediger & Tatlıdil, 2002; Ersel Canyurt, Ceylan, Kemal Ozturk, & Hepbasli, 2004; Haldenbilen & Ceylan, 2005; Maddala, Trost, Li, & Joutz, 1997; Murat & Ceylan, 2006; Robinson, 1988; Taal, Bulatov, Klemeš, & Stehlík,, 2003; Ünler, 2008) has led to the development of new energy portfolio which are more efficient, sustainable and reliable. The implementation of a hybrid solar-wind system has received a considerable attention due to its zero-emission generating capacity.

Against the backdrop, it is essential to model the future energy demand based on a well-defined scenarios from IPCC and design a hybrid solar-wind-thermal power generation for Turkish Republic of Northern Cyprus (TRNC). To the best of our knowledge, scientific studies on modelling energy demand is limited in TRNC. As a contribution to literature, our propose study will examine the case of TRNC while providing an alternative solution to the existing battery banks used as storage for a hybrid PV-Wind energy systems.

1.1. Energy demand model

Access to energy is linked to the human development (industrial production, health, agricultural output, access to water, population, education, quality of life, economic output, etc.) of every nation (Asumadu-Sarkodie & Owusu, 2016b, 2016c, 2016g). Energy forecasting is a vital input required in the conceptual framework of sustainable energy policies of every nation. Presently, energy demand modelling is of much interest among scientist and engineers concerned with the issues of energy production and utilization (Ozturk, Canyurt, Hepbasli, & Utlu, 2004). Modelling is very useful in planning and formulating energy policies (Dincer & Dost, 1996). Modelling of energy demand is typically based on historical consumptions and the link of this consumption with altered variables such as population, gross domestic product, employment, export amount, and so on (Ceylan, Ozturk, Hepbasli, & Utlu, 2005; Kankal, Akpınar, Kömürcü, & Özşahin, 2011; Kavaklioglu, 2011). Furtado and Suslick (1993) employed the learning and translog models which indicated in their study that, GDP was the main determinant for petroleum consumption evolution in the future. Yang (2000) investigated the causality between energy consumption and GDP by using the Granger's technique. Yoo (2006) examined the causal relationship between electricity consumption and economic growth among the Association of South East Asian Nations.

1.2. Regression analysis model

Regression analysis model is one of the numerous energy models that are developed for the sustainable progress of any country (Suganthi & Samuel, 2012). O'Neill and Desai (2005) analysed the accuracy in the projections of United States energy consumption produced by the Energy Information Administration over the period from 1982 to 2000. It was evident in their study that, GDP projections were consistently too high, while energy intensity projections were consistently too low. Suganthi and Samuel (2012) reviewed on the energy models for demand forecasting and concluded that, energy demand management is a requirement for proper allocation of the available resources of any country. A study by Kankal et al. (2011) introduced the multiple linear regression analysis models to calculate the energy consumption using GDP and population, respectively. A regression-based daily peak load forecasting method with a transformation function and reflection method was proposed by Haida and Muto (1994). Lee and Chang (2007) studied on the impact of energy consumption on economic growth using evidence from linear and non-linear models in Taiwan. They concluded in their study that, a threshold regression provides a better empirical model than the standard linear model. Al-Ghandoor, Al-Hinti, Jaber, and Sawalha (2008) studied on the electricity consumption and its associated GHG emissions of the Jordan industrial sector using a multivariate linear regression analysis to identify the main drivers behind electricity consumption and to project future electricity consumption. Jónsson, Pinson, and Madsen (2010) studied on the impact of average market price behaviour on wind energy by using a non-parametric regression model. They concluded that, the effects of wind power forecast on the average behaviour, the non-linearity and time variations in the relationship are quite substantial. Lam, Tang, and Li (2008) analysed the seasonal variations in residential and commercial sector electricity consumption in Hong Kong using a multiple regression technique. They concluded that, regression models could be a good indicator for the annual electricity consumption. Gorucu (2004) evaluated and forecast the gas consumption in Ankara, Turkey using a multivariable regression analysis. The study included an approach that understood the factors affecting gas demand in Ankara, Turkey.

1.3. Previous studies on hybrid wind-solar energy systems

A lot of studies have been done on hybrid solar PV-wind power systems (Petrakopoulou, Robinson, & Loizidou, 2016; Ribeiro, Arouca, & Coelho, 2016; Romero Rodríguez, Salmerón Lissén, Sánchez Ramos, Rodríguez Jara, & Álvarez Domínguez, 2016). Bekele and Palm (2010) investigated on the feasibility and economic viability of a hybrid solar PV-wind-diesel battery power system to meet the load requirement of 200 families in an Ethiopian community. Their research contributes to improving electricity coverage in a community in Ethiopia, compares many scenarios to enable investors make a selection and proposes diesel generator and a battery as a storage facility to ensure uninterrupted power supply in the selected community. Nonetheless, the authors doubted the outcome of their study due to the limited coverage of their study. In the same vein, their results are not certain if it can be applicable to other regions in Ethiopia with similar or different weather patterns. Their proposed hybrid system was claimed to contribute to a reduction in environmental emissions without justification from their study.

Yang, Wei, and Chengzhi (2009) investigated on the optimal design of a hybrid solar-wind power system by employing battery banks as system storage. Importantly, their study improves on battery over-discharge situations which are associated with battery banks. Their study employs the concept of loss of power probability which eventually enables system optimum configuration. Nonetheless, battery bank storage systems are not ideal due to the occurrence of losses in storing periods, charging and discharging cycles. Their study measures a one-year field data which is somewhat difficult to be conclusive in the subject matter since the energy contribution from the various sources employed in their study varies.

In the same vein, Diaf, Belhamel, Haddadi, and Louche (2008) investigated on the assessment of a hybrid solar-wind power system by employing battery banks as system storage. Importantly, their study improves on the development of an optimum size of a system at a given load distribution that can fulfil a given energy requirement. Improving the optimal sizing of a hybrid solar-wind power system is critical to improving efficiency and economic viability of the proposed system. Their study also employs the concept of loss of power probability which enables system optimum configuration through an un-interruptible power supply from a wind generator. Nonetheless, battery bank storage systems have issues with reliability and are not ideal due to the occurrence of losses in charging, storing periods and discharging cycles. Also, assessing the optimal sizing of a hybrid solar-wind power system is location dependent, system reliability dependent and depends on the potentiality of wind and solar energy at a particular location.

Zhou, Lou, Li, Lu, and Yang (2010) investigated the control technologies for a hybrid solar-wind power system with battery storage. Their study suggested that a design for optimum resource allocation based on load demand is beneficial to reducing the initial and operation cost of a hybrid solar-wind power system. Nonetheless, probabilistic approaches are difficult to represent a changing performance of a hybrid solar-wind power system.

Cavalcanti and Motta (2015) investigated on a solar-powered-fuel-assisted Rankine cycle for power generation using an exergoeconomic analysis. Even though an economic analysis was performed in their study yet not feasible economically for investment since the cost of electricity is higher than conventional forms of power generation. Moreover, issues pertaining to water consumption and carbon dioxide emission as a result of the Rankine cycle were not assessed.

Nevertheless, our proposed study is a build-up on previous studies (Pehlivantürk, Özkan, & Baker, 2014) that mathematically modelled and simulated a concentrating solar power system in Middle East Technical University, Northern Cyprus Campus. Their micro solar power consisted of a parabolic trough collector (PTC), propane boiler, an organic Rankine cycle and a wet cooling tower. Their study was able to better the performance of the system by expanding the PTC field to burn little propane. Moreover, their study identified some non-concentrating evacuated tube collectors that possess improved thermal performance than PTCs. Nevertheless, little analysis was presented on water consumption by the Rankine cycle, their study was on a small-scale and the role of renewable energy was limited.

Majority of literature reviewed employs storage mechanisms such as battery banks that have issues with reliability and are not ideal due to the occurrence of losses in charging, storing periods and discharging cycles for hybrid solar-wind power system. Moreover, literature is limited in the scope of assessing a hybrid solar-wind-Rankine cycle power generation system that models energy demand and addresses environmental impacts by assessing water consumption and carbon dioxide emissions. Building on their research, our study models the future energy demand of TRNC based on the IPCC emissions scenario A1B and A2 by designing a new hybrid solar-wind-thermal power system that satisfies the current and future requirement of firm capacity during peak periods. Our study suggests an improvement in a hybrid solar-wind thermal power system performance by predicting reliable outputs that can integrate renewable energy technologies (solar and wind energy) to conventional power generation (Rankine cycle based on propane) sources. To the best of our knowledge, this is the first of its kind a study combines both hybrid-solar PV-wind-Rankine cycle to meet future energy demand based on predicted IPCC scenarios. Our research will serve as an informative tool for local and foreign investors in TRNC while contributing to future research. Finally, we are aware of the cost intensive nature and environmental unfriendliness of driving a Rankine cycle based on propane however, we opted for propane in order to accentuate the positives of renewable energy sources. In the long-run, the CO_2 emissions and water consumption for the Rankine cycle are estimated for the future electricity power demand based on the IPCC scenarios.

2. Methodology

2.1. Energy consumption and demand model

The energy consumption and demand model for the year 2004–2013 are the measured energy consumption values for TRNC from the Population Geography of the TRNC (Atasoy, 2012). However, the energy demand for year 2014–2050 was modelled using:

$$y = a_1 x_1 + b_1 x_2 + f_1 \qquad (1)$$

where y is the energy demand or consumption, x_1 is the population and x_2 is the GDP. The a_1 is the population coefficient, b_1 is the GDP coefficient (for $\$10^9$) and f_1 is the intercept and the calculated values of these coefficients for TRNC are 2.18, 66.47 and 302,265, respectively. Appendix A show the analysis of future renewable energy demand, population and GDP based on IPCC scenario A1B.

IPCC scenario AIB describes a "future world of very rapid economic growth, global population that peaks in mid-century and declines thereafter, and the rapid introduction of new and more efficient technologies as a result of not relying too heavily on one particular energy source, on the assumption that similar improvement rates apply to all energy supply and end-use technologies" (IPCC Fourth Assessment Report on Climate Change, 2015). Based on the IPCC scenario AIB, the current study assumes a 1.01% population rise until 2030 in accordance with Atasoy (2012) and assumes a 1.005% population declines during 2030–2050. GDP increase is assumed to be 10% until 2017, in accordance to the rate of increase in GDP in the fastest increasing economies such as Brazil (Bank, http://data.worldbank.org/country/brazil), and 1% thereafter. Moreover, the heat engine efficiency is increased by 5% for every 10 years to include efficiency assumption of Scenario A1B. Finally, renewable energy contribution is assumed to immediately increase to 20%, which is stated in EU 20-20-20 plan (European Commission, 2015) and keeping the same ratio until 2050.

IPCC scenario A2 assumes that economic growth will be more evenly distributed, i.e. none of the economic factors will be rapid but stable, and so GDP increase is taken as 2.3% per annum in accordance with Intergovernmental Panel on Climate Change Report on Economic Development on Emissions Scenarios (SRES) (Change IPoC, 2015a). As the population is assumed to be rising in the same way it does at the moment, its rise is always taken as 1.01%. Moreover, in the current scenario, future is described as a delayed renewable energy sources, so renewable energy contribution is modified to be 10, 12, 15, 17.5 and 20% during 2010–2020, 2020–2030, 2030–2040, 2040–2050, respectively. Finally, as technological advances are slower than Scenario A1B, heat engine efficiency is taken as 20, 22.5, 25, 27.5 and 30%, respectively, for the periods mentioned above.

2.2. Solar resources model

Using the TMY2 data, the terrestrial solar resources model for fixed surfaces is used to calculate the maximum solar resources and hence the PV power is expressed as:

$$E_{pv} = \eta_{sys} \eta_{pv} I_{pv} A_{pv} \qquad (2)$$

where E_{pv} is the energy output from the PV panel, η_{sys} is the system performances ratio, η_{pv} is the conversion efficiency of an individual PV module, I_{pv} is the hourly insolation in the plane of the PV modules and A_{pv} is the total area of PV modules (Duffie & Beckman, 1980).

$$\eta_{pv} = \eta_{ref} \left\{ 1 - \beta_{ref} \left[T_a - T_{ref} + (T_{noct} - T_a) \frac{I_{pv}}{I_{noct}} \right] \right\} \qquad (3)$$

where η_{ref} is the PV module efficiency at reference conditions, T_{ref} (25°C) and I_{noct} (800 Wh.m^{-2}) and T_a is the ambient temperature. I_{pv} is the insolation for the PV panel. In the implemented model, different solar areas were used to keep constant the solar power generation from year to year.

2.3. Wind resources model

The TMY wind resources are used to calculate the hub height wind speed for the wind turbine model. The hub height wind speed (U_h) can calculate from 10-m height wind speed (U_{tmy}) data as,

$$U_h = U_{tmy}\left(\frac{z_h}{z_{tmy}}\right)^\alpha \tag{4}$$

where Z_h is the wind turbine hub height and Z_{tmy} is the height for the TMY wind data, wind shear constant α is 0.147. The electric power generated from the corresponding wind resources was calculated using the WGT model (Johnson, 2005) (Appendix B).

2.4. Thermal energy model

Moreover, Pehlivanturk's work (Pehlivantürk et al., 2014) was used in order to model heat engine, hence CO_2 emission and water consumption. Although mathematical details are not given in this report for brevity, it is mandatory to state that the model flowchart is modified in a way that backup boiler is used when total electricity produced that would be generated with only solar and wind sources is not enough to supply the demand. So,

$$\dot{Q}_{boiler} = \left[\max\left(\dot{W}_{demand} - \dot{W}_{solar} - \dot{W}_{wind}, 0\right)\right] / (\text{cycle thermal efficiency}) \tag{5}$$

where $\dot{Q}_{boiler}$ is the rate of heat transfer to the heat transfer fluid in the boiler, $\dot{W}_{demand}$ rate of demand, $\dot{W}_{solar}$ rate of work from solar and $\dot{W}_{wind}$ is the rate of work from wind.

Using the work of Pehlivantürk et al. (2014), the analysis of water consumption of the heat engine is calculated as;

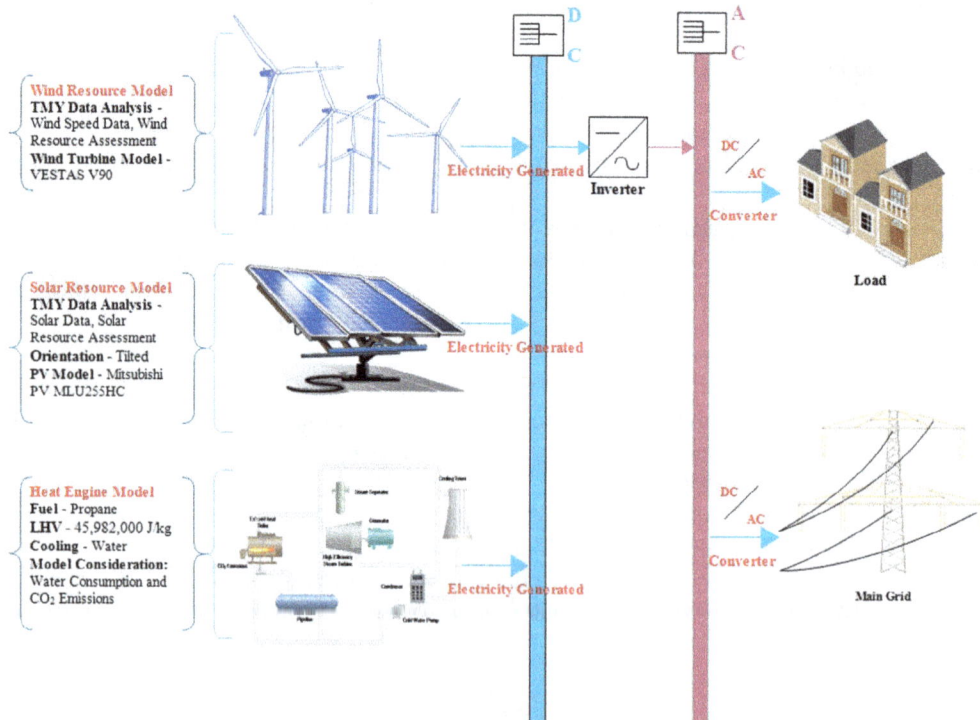

Figure 1. Schematic representation of the proposed hybrid solar photovoltaic-wind turbine-Rankine cycle.

$$\dot{Q}_c = \rho_{H_2O} V_{H_2O} h_{fg} \tag{6}$$

where the low temperature heat transfer from the Rankine cycle ($\dot{Q}_c$) is used to evaporate water at 25°C (isothermal latent cooling), ρ_{H2O} is the density of water, V_{H2O} is the volumetric rate of water consumption by the cooling heat engine and h_{fg} is the enthalpy of the evaporation of water. Figure 1 shows the schematic representation of the proposed hybrid solar photovoltaic-wind turbine-Rankine cycle.

3. Results and discussion

In Table 1, the regression analysis of future renewable energy demand, population and GDP for the years 2010–2050 is given. In Table 1 (Scenario A1B), a multiple R value of 0.95 shows that there is a strong relationship between the future renewable energy generated (demand), population and GDP for the corresponding years.

In Figure 2, the electricity generation by renewable energy sources for TRNC is given. In Figure 2, the energy demand increases gradually with increasing population growth which is in line with a multitude of literature. Between the years 2015–2050 based on IPCC scenario A1B, the renewable energy demand increases from 300 to 400 GWh whilst energy consumption gradually increases from 150 to 450 GWh in IPCC scenario A2. In Figure 2(a), as it can be seen, slope changes at year 2017 and 2030 because of the storyline and corresponding assumptions mentioned in methodology. It was stated that GDP tends to increase by 10% per annum until 2017, and tends to rise by 1% thereafter which explains the change in slope at 2017. Moreover, after 2030, population is assumed to drop, explaining the change in slope at 2030. Likewise, in Figure 2(b), there are jumps in electricity generation, which is because of increasing renewable energy contribution instantly during the corresponding years. The energy demand in these two scenarios cannot be supplied by the hybrid solar photovoltaic-wind power system alone. Therefore, to compensate for future energy demand, a Rankine cycle electricity generating system was added.

In Figure 3, the contribution percentage of hybrid-solar photovoltaic-wind-Rankine cycle for TRNC is given. TRNC reaches its maximum amount of solar radiation in mid-year periods during summer months and minimum solar radiation during winter months. As a result of that the maximum PV solar energy output can be achieved during the summer period as shown in Figure 3. In this regard, the contribution of Rankine cycle power generation is minimum in mid-year periods and maximum in winter and spring season. In the presented model, the energy demand that cannot be supplied by the hybrid solar photovoltaic-wind power system during peak hours is compensated using the Rankine cycle power generation as shown in Figure 3. Nonetheless, the wind power energy output for the entire year seems constant due to the non-variability of wind speed.

The model is designed to keep a constant contribution of the solar, wind and RC for each year during the period 2010–2050 as shown in Figure 4. The contribution of solar for the energy consumption is kept constant using the corresponding different solar areas of the PV panel and wind energy is kept constant by increasing the wind turbines. The Rankine cycle contribution to the energy generation is kept at 64% for each year instead of the battery bank.

Table 1. Regression analysis of future renewable energy demand, population and GDP for the years 2010–2050 for scenario A1B	
Regression statistics	*p < 0.05*
Multiple R	0.953
R^2	0.908
Adjusted R^2	0.882
Standard error	58,796.299
Observations	10

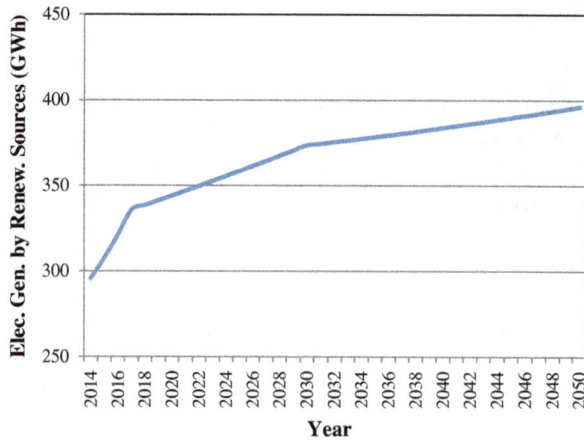

(a)   Scenario A1B model

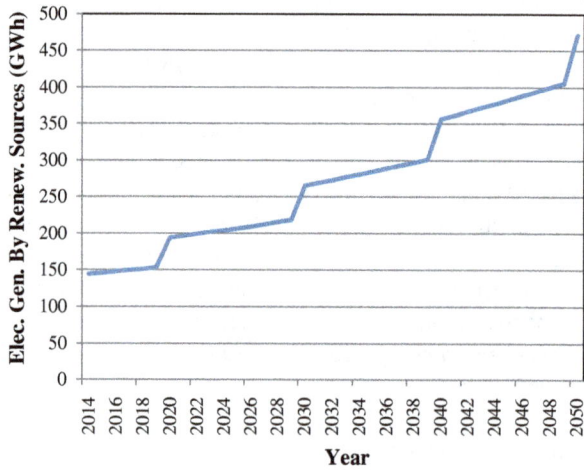

(b)   Scenario A2 model

Figure 2. Electricity generation by renewable energy sources for a period of 2014–2050 based on IPCC emission scenario A1B and A2.

In Figure 5, the hourly CO_2 emission from hybrid-solar photovoltaic-wind-Rankine cycle for TRNC is given. CO_2 emission is minimum at summer and maximum in winter and spring, which is due to increasing demand of electricity for heating purposes during the cold season and the high insolation of solar energy led to high contribution of solar PV energy for the system. In Figure 5(a), CO_2 emissions increase on a yearly basis due to increasing yearly energy demands.

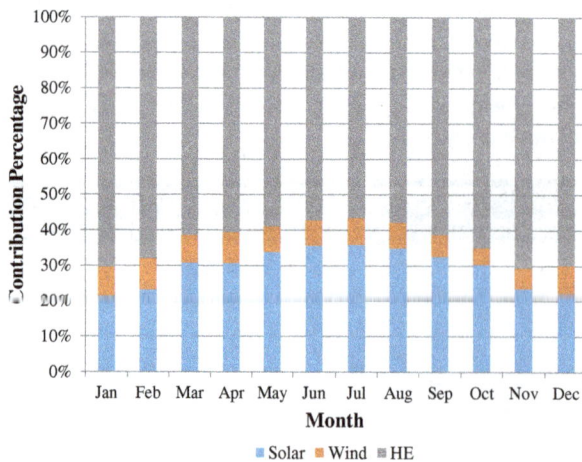

Figure 3. Contribution percentage of hybrid-solar photovoltaic-Rankine cycle model for 2010.

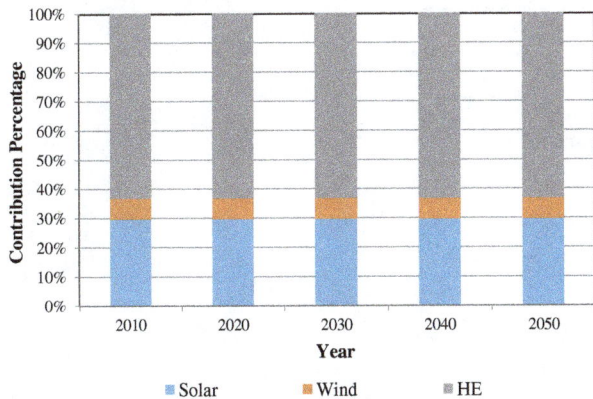

Figure 4. Contribution percentage of hybrid-solar photovoltaic-Rankine cycle model for period 2014–2050, according to the IPCC emission scenario A1B.

Nonetheless, the gap between the yearly CO_2 decreases with respect to the increasing years as depicted in Figure 6. As the rate of CO_2 emissions decreases with increasing CO_2 emissions and increasing years which is in line with IPCC emission scenario A1B (Change IPoC, 2015b) (Appendix B). In Figure 5(b), the increasing rate of the CO_2 emissions increases with increasing years for scenario A2 which is confirmed by Figure 6. However, at the beginning of year 2014, CO_2 emissions are comparably higher in scenario A1B as compared to scenario A2, but the opposite occurs at the end of 2050 as shown in Figure 6. Another reason for having different trends of emissions in Figure 5 is the RE contribution and heat engine efficiencies. Even though both scenarios have increasing demand curves as shown in Figure 2, RE contribution is always increasing in Scenario A2 with less efficiency

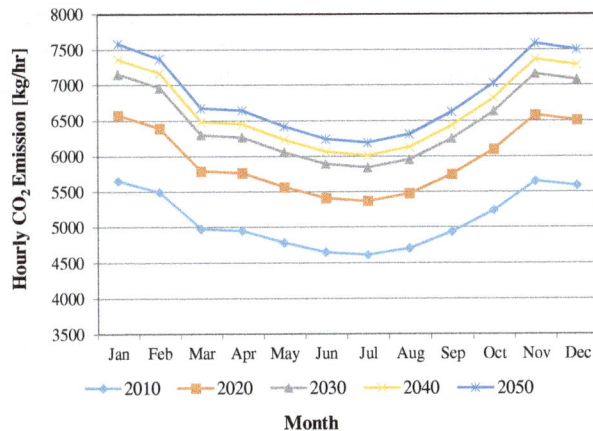

(a)　Scenario A1B model

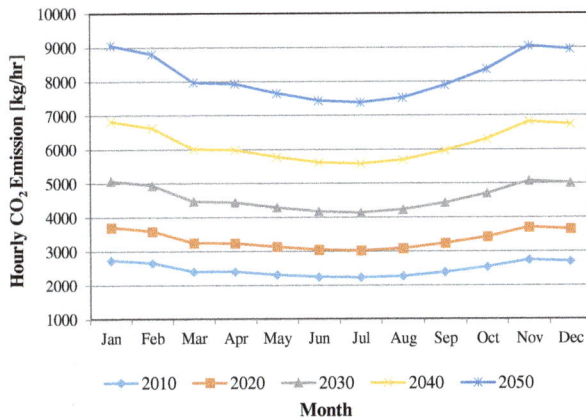

(b)　Scenario A2 model

Figure 5. The hourly CO_2 emission for the months in year for proposed model.

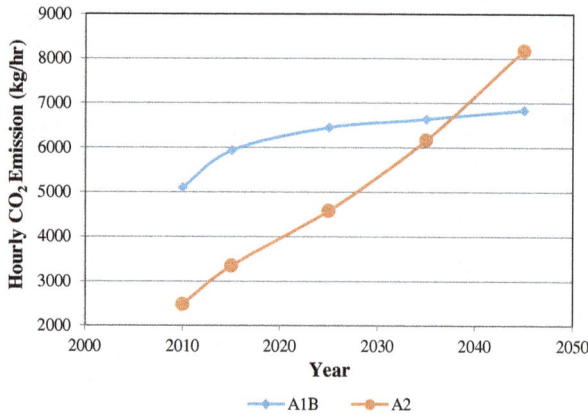

Figure 6. The hourly CO$_2$ emission for the proposed model for period 2011–2050.

increase compared to Scenario A1B, making efficiency change unable to compensate electricity generation.

In Figure 7, the annual increase in hourly CO$_2$ emission is given. For scenario A1B model, the annual increasing rate of CO$_2$ emissions decreases until year 2040 as shown in Figure 7(a). But for scenario A2, the annual increasing rate of CO$_2$ emissions increases until year 2050 as shown in Figure 7(b). This outcome is due to IPCC scenario A2 which describes a future with delayed renewable energy.

In Figure 8, the hourly water consumption for the hybrid-solar photovoltaic-wind-Rankine cycle for TRNC is given. A substantial amount of water is a requirement for the proposed model as shown in Figure 8. The minimum amount of water is required in summer periods due to the less contribution of Rankine cycle power generating system. Same as CO$_2$, water consumption increases yearly.

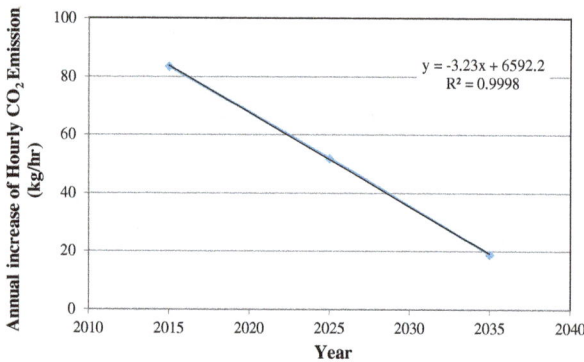

(a) Scenario A1B model

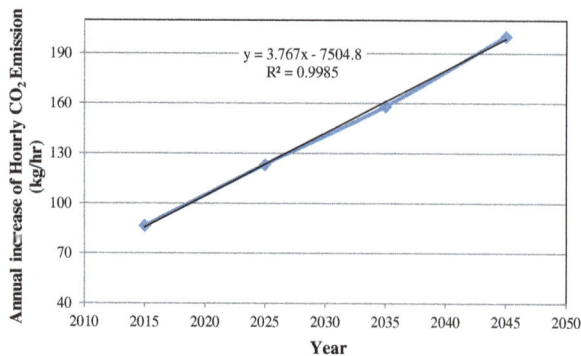

(b) Scenario A2 model

Figure 7. The annual increase in hourly CO$_2$ emission for the proposed model for period 2011–2050.

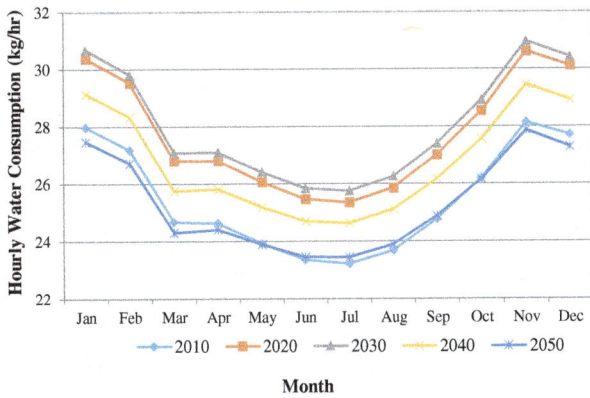

(a) Scenario A1B model

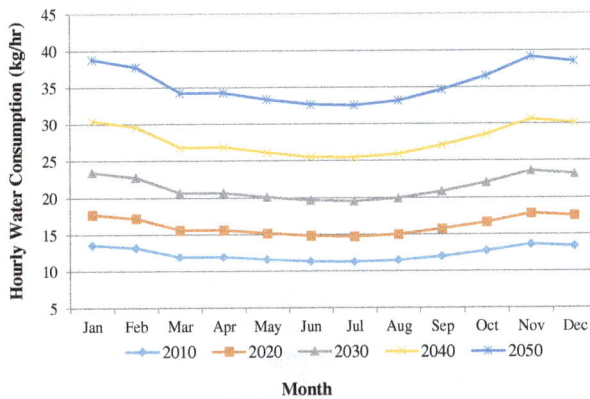

(b) Scenario A2 model

Figure 8. The hourly water consumption for the months in year for proposed model.

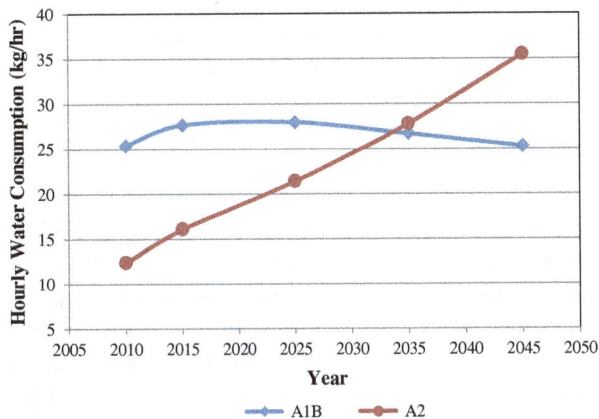

Figure 9. The hourly water consumption for the proposed model for period 2011–2050.

With reference to scenario A1B, water consumption for power generation increases until year 2030 and begins to decrease thereafter as shown in Figure 8(a) and Figure 9. The reason is that water consumption is directly related with heat engine efficiency, and after 30% of efficiency (year 2030), even though the demand is increasing, effect of efficiency increases compensates the increase in demand, resulting in less water to be consumed at cooling tower. Notwithstanding, water consumption gradually increases in scenario A2 as shown in Figure 8(b). The water consumption at the beginning of year 2010 is comparably low in scenario A2 compared to scenario A1B.

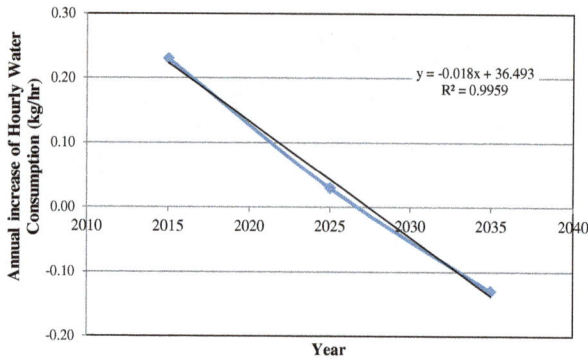

(a) Scenario A1B model

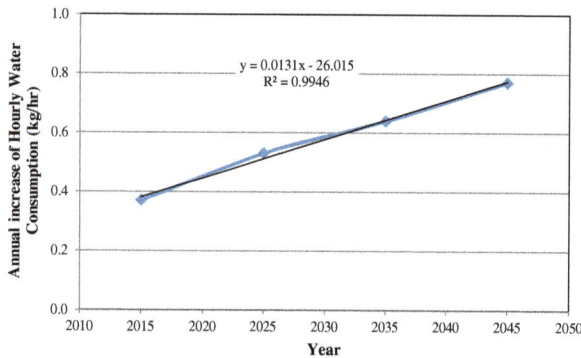

(b) Scenario A2 model

Figure 10. The annual increase in hourly CO_2 emission for the proposed model for period 2011–2050.

In Figure 10, the annual increase in hourly water consumption is given. In scenario A1B, water consumption gradually decreases at an increasing rate as shown in Figure 10(a). However, water consumption gradually increases at an increasing rate in scenario A2 as shown in Figure 10(b).

4. Conclusions and future work

In this study, a multiple linear regression analysis was used to calculate the future renewable energy consumption for TRNC through a linear interpolation of GDP and population of TRNC based on the IPCC emission scenarios A1B and A2. A mathematical model for an autonomous hybrid solar-wind-heat engine was developed and CO_2 emissions and water consumption were calculated for years 2010–2050.

The energy consumption prediction model clearly emphasizes that the energy requirement for electricity has a growing demand from 300 to 400 GWh in scenario A1B and 150 to 450 GWh in scenario A2 during the period 2010–2050, which is impossible to meet using an autonomous hybrid-solar photovoltaic-wind turbine. This led to a design of a hybrid-solar photovoltaic-wind turbine-Rankine cycle to compensate the future electricity demand of TRNC during peak hours. The results of the study are summarized as follows:

- The hybrid-solar photovoltaic-wind turbine-Rankine cycle can meet 400 GWh of electricity demand in TRNC based on scenario A1B.

- The hybrid-solar photovoltaic-wind turbine-Rankine cycle can meet 450 GWh of electricity demand in TRNC based on scenario A2.

- The annual increasing rate of CO_2 emissions decreases until year 2040 in IPCC scenario A1B.

- The annual increasing rate of CO_2 emissions increases until year 2050 in IPCC scenario A2.

The study can further be developed by following a probabilistic approach instead of a deterministic one. Instead of finding definite amounts of CO_2 emissions and water consumption, possible range of such values with a satisfying level of confidence interval would be handier to plan for the future. Moreover, the other IPCC scenarios such as B1 and B2 which were absent in the current study can be modelled to increase the number of strategic planning. Finally, a storage system such as a pumped hydro power system can be hybridized with the current system to store the excess amount of energy generated.

Nomenclature

E_{pv}	Energy output from the PV panel
η_{sys}	System performances ratio
η_{pv}	Conversion efficiency
I_{pv}	Hourly insolation in the plane of the PV modules
A_{pv}	Total area of PV modules
η_{ref}	PV module efficiency at reference conditions
T_a	Ambient temperature
U_h	Hub height wind speed
Z_h	Wind turbine hub height
Z_{tmy}	Height for the TMY wind data
α	Wind shear constant
$\dot{Q}_{boiler}$	Rate of heat transfer in the boiler

Acronyms

RE	Renewable Energy
TRNC	Turkish Republic of Northern Cyprus
GDP	Gross Domestic Product
IPCC	Intergovernmental Panel on Climate Change
TMY2	Typical Meteorological Year 2
HE	Heat Engine

Acknowledgements
We would like to thank Prof. Derek Baker, Department of Mechanical Engineering, Middle East Technical University, Ankara campus who provided the seed and earlier guidance for many of these ideas.

Funding
The authors received no direct funding for this research.

Author details
Samuel Asumadu-Sarkodie[1]
E-mail: samuelsarkodie@yahoo.com
ORCID ID: http://orcid.org/0000-0001-5035-5983
Çağlan Sevinç[1]
E-mail: caglan.sevinc@gmail.com
Herath M.P.C. Jayaweera[1]
E-mail: pradeep.jayaweera@metu.edu.tr

[1] Sustainable Environment and Energy Systems, Middle East Technical University Northern Cyprus Campus, Kalkanli-Guzelyurt, Turkey.

References
Akay, D., & Atak, M. (2007). Grey prediction with rolling mechanism for electricity demand forecasting of Turkey. *Energy, 32*, 1670–1675.

http://dx.doi.org/10.1016/j.energy.2006.11.014

Al-Ghandoor, A., Al-Hinti, I., Jaber, J., & Sawalha, S. (2008). Electricity consumption and associated GHG emissions of the Jordanian industrial sector: Empirical analysis and future projection. *Energy Policy, 36,* 258–267.

Asumadu-Sarkodie, S., & Owusu, P. A. (2016a). Feasibility of biomass heating system in Middle East Technical University, Northern Cyprus Campus. *Cogent Engineering.* doi:10.1080/23311916.2015.1134304

Asumadu-Sarkodie, S., & Owusu, P. A. (2016b). A review of Ghana's energy sector national energy statistics and policy framework. *Cogent Engineering.* doi:10.1080/23311916.2016.1155274

Asumadu-Sarkodie, S., & Owusu, P. A. (2016a). Carbon dioxide emissions, GDP, energy use and population growth: A multivariate and causality analysis for Ghana, 1971-2013. *Environmental Science and Pollution Research.* doi:10.1007/s11356-016-6511-x

Asumadu-Sarkodie, S., & Owusu, P. A. (2016d). Multivariate co-integration analysis of the Kaya factors in Ghana. *Environmental Science and Pollution Research.* doi:10.1007/s11356-016-6245-9

Asumadu-Sarkodie, S., & Owusu, P. A. (2016e). The potential and economic viability of solar photovoltaic in Ghana. *Energy sources, Part A: Recovery, utilization, and environmental effects.* doi:10.1080/15567036.2015.1122682

Asumadu-Sarkodie, S., & Owusu, P. A. (2016f). The potential and economic viability of wind farms in Ghana *Energy sources, Part A: Recovery, utilization, and environmental effects.* doi:10.1080/15567036.2015.1122680

Asumadu-Sarkodie, S., & Owusu, P. A. (2016g). The relationship between carbon dioxide and agriculture in Ghana, A comparison of VECM and ARDL Model. *Environmental Science and Pollution Research.* doi:10.1007/s11356-016-6252-x

Atasoy, A. (2012). *Kuzey Kıbrıs Türk Cumhuriyeti'nin Nüfus Coğrafyasi* [The population of the Turkish Republic of Northern Cyprus].

Beér, J. M. (2000). Combustion technology developments in power generation in response to environmental challenges. *Progress in Energy and Combustion Science, 26,* 301–327. http://dx.doi.org/10.1016/S0360-1285(00)00007-1

Bekele, G., & Palm, B. (2010). Feasibility study for a standalone solar–wind-based hybrid energy system for application in Ethiopia. *Applied Energy, 87,* 487–495. http://dx.doi.org/10.1016/j.apenergy.2009.06.006

Berndt, E. R., & Wood, D. O. (1975). Technology prices and the derived demand for energy. *The review of Economics and Statistics, 57,* 259–268.

Cavalcanti, E. J. C., & Motta, H. P. (2015). Exergoeconomic analysis of a solar-powered/fuel assisted Rankine cycle for power generation. *Energy, 88,* 555–562. doi:10.1016/j.energy.2015.05.081

Ceylan, H., & Ozturk, H. K. (2004). Estimating energy demand of Turkey based on economic indicators using genetic algorithm approach. *Energy Conversion and Management, 45,* 2525–2537.

Ceylan, H., Ozturk, H. K., Hepbasli, A., & Utlu, Z. (2005). Estimating energy and exergy production and consumption values using three different genetic algorithm approaches. Part 2: Application and scenarios. *Energy Sources, 27,* 629–639. http://dx.doi.org/10.1080/00908310490448631

Change IPoC. (2015a). *Report on economic development on emissions scenarios (SRES).* Retrieved from http://www.ipcc.ch/ipccreports/sres/emission/index.php?idp=100

Change IPoC. (2015b). *Special report on emissions scenarios (SRES).* Retrieved from http://www.ipcc.ch/ipccreports/tar/wg1/029.htm#f1

European Commission. (2015). *The 2020 climate and energy package.* Retrieved from http://ec.europa.eu/clima/policies/package/index_en.htm

Crompton, P., & Wu, Y. (2005). Energy consumption in China: Past trends and future directions. *Energy economics, 27,* 195–208.

Demirbas, A. (2005). Potential applications of renewable energy sources, biomass combustion problems in boiler power systems and combustion related environmental issues. *Progress in Energy and Combustion Science, 31,* 171–192. http://dx.doi.org/10.1016/j.pecs.2005.02.002

Dersch, J., Geyer, M., Herrmann, U. L. F., Jones, S. A., Kelly, B., Kistner, R., ... Price, H. (2004). Trough integration into power plants—a study on the performance and economy of integrated solar combined cycle systems. *Energy, 29,* 947–959. http://dx.doi.org/10.1016/S0360-5442(03)00199-3

Diaf, S., Belhamel, M., Haddadi, M., & Louche, A. (2008). Technical and economic assessment of hybrid photovoltaic/wind system with battery storage in Corsica island. *Energy Policy, 36,* 743–754. doi:10.1016/j.enpol.2007.10.028

Dincer, I. (2000). Renewable energy and sustainable development: A crucial review. *Renewable and Sustainable Energy Reviews, 4,* 157–175. http://dx.doi.org/10.1016/S1364-0321(99)00011-8

Dincer, I., & Dost, S. (1996). Energy intensities for Canada. *Applied Energy, 53,* 283–298. http://dx.doi.org/10.1016/0306-2619(95)00023-2

Duffie, J. A., & Beckman, W. A. (1980). *Solar engineering of thermal processes* (Vol. 3). New York, NY: Wiley.

Dunkerley, J. (1982). Estimation of energy demand: The developing countries. *The Energy Journal, 3,* 79–99.

Ediger, V. Ş., & Akar, S. (2007). ARIMA forecasting of primary energy demand by fuel in Turkey. *Energy Policy, 35,* 1701–1708.

Ediger, V. Ş., & Tatlıdil, H. (2002). Forecasting the primary energy demand in Turkey and analysis of cyclic patterns. *Energy Conversion and Management, 43,* 473–487.

Ehrlich, P. R., & Holdren, J. P. (1971). *Impact of population growth.*

Ersel Canyurt, O., Ceylan, H., Kemal Ozturk, H., & Hepbasli, A. (2004). Energy demand estimation based on two-different genetic algorithm approaches. *Energy Sources, 26,* 1313–1320.

Furtado, A. T., & Suslick, S. B. (1993). Forecasting of petroleum consumption in Brazil using the intensity of energy technique. *Energy Policy, 21,* 958–968.

Gorucu, F. (2004). Evaluation and forecasting of gas consumption by statistical analysis. *Energy Sources, 26,* 267–276. http://dx.doi.org/10.1080/00908310490256617

Gyuk, I., Kulkarni, P., Sayer, J. H., Boyes, J. D., Corey, G. P., & Peek, G. H. (2005). The united states of storage [electric energy storage]. *Power and Energy Magazine, IEEE, 3,* 31–39.

Haida, T., & Muto, S. (1994). Regression based peak load forecasting using a transformation technique. *IEEE Transactions on Power Systems, 9,* 1788–1794.

Haldenbilen, S., & Ceylan, H. (2005). Genetic algorithm approach to estimate transport energy demand in Turkey. *Energy Policy, 33,* 89–98.

Hepbasli, A. (2008). A key review on exergetic analysis and assessment of renewable energy resources for a sustainable future. *Renewable and Sustainable Energy Reviews, 12,* 593–661.

IPCC Fourth Assessment Report on Climate Change (2015). *Projections of future changes in climate.* Retrieved November 13, 2015, from https://www.ipcc.ch/publications_and_data/ar4/wg1/en/spmsspm-projections-of.html.

Johnson, D. (2005). Wind turbine power. In L. Gary (Ed.), *Wind turbine power, energy and torque* (Chapter 4).

Jónsson, T., Pinson, P., & Madsen, H. (2010). On the market impact of wind energy forecasts. *Energy Economics, 32,* 313–320. http://dx.doi.org/10.1016/j.eneco.2009.10.018

Kankal, M., Akpınar, A., Kömürcü, M. İ., & Özşahin, T. Ş. (2011). Modeling and forecasting of Turkey's energy consumption using socio-economic and demographic variables. *Applied Energy, 88,* 1927–1939. doi:10.1016/j.apenergy.2010.12.005

Kavaklioglu, K. (2011). Modeling and prediction of Turkey's electricity consumption using support vector regression. *Applied Energy, 88,* 368–375. doi:10.1016/j.apenergy.2010.07.021

Lam, J. C., Tang, H. L., & Li, D. H. (2008). Seasonal variations in residential and commercial sector electricity consumption in Hong Kong. *Energy, 33,* 513–523. http://dx.doi.org/10.1016/j.energy.2007.10.002

Larson, E. D., & Williams, R. H. (1987). Steam-injected gas turbines. *Journal of Engineering for Gas Turbines and Power, 109,* 55–63. http://dx.doi.org/10.1115/1.3240006

Lee, C. C., & Chang, C. P. (2007). The impact of energy consumption on economic growth: Evidence from linear and nonlinear models in Taiwan. *Energy, 32,* 2282–2294. http://dx.doi.org/10.1016/j.energy.2006.01.017

Maddala, G. S., Trost, R. P., Li, H., & Joutz, F. (1997). Estimation of short-run and long-run elasticities of energy demand from panel data using shrinkage estimators. *Journal of Business & Economic Statistics, 15,* 90–100.

Mohd, A., Ortjohann, E., Schmelter, A., Hamsic, N., & Morton, D. (2008). Challenges in integrating distributed energy storage systems into future smart grid. In *Industrial electronics, IEEE International Symposium on ISIE 2008* (pp. 1627–1632). IEEE.

Murat, Y. S., & Ceylan, H. (2006). Use of artificial neural networks for transport energy demand modeling. *Energy Policy, 34,* 3165–3172.

Nakata, T. (2004). Energy-economic models and the environment. *Progress in Energy and Combustion Science, 30,* 417–475. doi:10.1016/j.pecs.2004.03.001

Nema, P., Nema, R., & Rangnekar, S. (2009). A current and future state of art development of hybrid energy system using wind and PV-solar: A review. *Renewable and Sustainable Energy Reviews, 13,* 2096–2103.

O'Neill, B. C., & Desai, M. (2005). Accuracy of past projections of US energy consumption. *Energy Policy, 33,* 979–993.

Owusu, P., & Asumadu-Sarkodie, S. (2016). A review of renewable energy sources, sustainability issues and climate change mitigation. *Cogent Engineering.* doi:10.1080/23311916.2016.1167990

Owusu, P. A., Asumadu-Sarkodie, S., & Ameyo, P. (2016). A review of Ghana's water resource management and the future prospect. *Cogent Engineering.* doi:10.1080/23311916.2016.1164275

Ozturk, H. K., Canyurt, O. E., Hepbasli, A., & Utlu, Z. (2004). Residential-commercial energy input estimation based on genetic algorithm (GA) approaches: An application of Turkey. *Energy and Buildings, 36,* 175–183. http://dx.doi.org/10.1016/j.enbuild.2003.11.001

Pehlivantürk, C., Özkan, O., & Baker, D. K. (2014). Modeling and simulations of a micro solar power system. *International Journal of Energy Research, 38,* 1129–1144. http://dx.doi.org/10.1002/er.v38.9

Petrakopoulou, F., Robinson, A., & Loizidou, M. (2016). Simulation and analysis of a stand-alone solar-wind and pumped-storage hydropower plant. *Energy, 96,* 676–683. doi:10.1016/j.energy.2015.12.049

Ribeiro, A. E. D., Arouca, M. C., & Coelho, D. M. (2016). Electric energy generation from small-scale solar and wind power in Brazil: The influence of location, area and shape. *Renewable Energy, 85,* 554–563. doi:10.1016/j.renene.2015.06.071

Robinson, J. B. (1988). Unlearning and backcasting: Rethinking some of the questions we ask about the future. *Technological Forecasting and Social Change, 33,* 325–338. http://dx.doi.org/10.1016/0040-1625(88)90029-7

Romero Rodríguez, L., Salmerón Lissén, J. M., Sánchez Ramos, J., Rodríguez Jara, E. Á., Álvarez Domínguez, S. (2016). Analysis of the economic feasibility and reduction of a building's energy consumption and emissions when integrating hybrid solar thermal/PV/micro CHP systems. *Applied Energy, 165,* 828–838. doi:10.1016/j.apenergy.2015.12.080

Solomon, S., Plattner, G. K., Knutti, R., & Friedlingstein, P. (2009). Irreversible climate change due to carbon dioxide emissions. *Proceedings of the National Academy of Sciences, 106,* 1704–1709.

Suganthi, L., & Samuel, A. A. (2012). Energy models for demand forecasting—A review. *Renewable and Sustainable Energy Reviews, 16,* 1223–1240. doi: http://dx.doi.org/10.1016/j.rser.2011.08.014

Taal, M., Bulatov, I., Klemeš, J., & Stehlík, P. (2003). Cost estimation and energy price forecasts for economic evaluation of retrofit projects. *Applied Thermal Engineering, 23,* 1819–1835.

Ünler, A. (2008). Improvement of energy demand forecasts using swarm intelligence: The case of Turkey with projections to 2025. *Energy Policy, 36,* 1937–1944.

Yang, H. Y. (2000). A note on the causal relationship between energy and GDP in Taiwan. *Energy Economics, 22,* 309–317.

Yang, H., Lu, L., & Burnett, J. (2003). Weather data and probability analysis of hybrid photovoltaic-wind power generation systems in Hong Kong. *Renewable Energy, 28,* 1813–1824.

Yang, H., Wei, Z., & Chengzhi, L. (2009). Optimal design and techno-economic analysis of a hybrid solar–wind power generation system. *Applied Energy, 86,* 163–169. http://dx.doi.org/10.1016/j.apenergy.2008.03.008

Yoo, S. H. (2006). The causal relationship between electricity consumption and economic growth in the ASEAN countries. *Energy Policy, 34,* 3573–3582.

Zhou, W., Lou, C., Li, Z., Lu, L., & Yang, H. (2010). Current status of research on optimum sizing of stand-alone hybrid solar–wind power generation systems. *Applied Energy, 87,* 380–389. http://dx.doi.org/10.1016/j.apenergy.2009.08.012

Appendix A

Analysis of future renewable energy demand, population and GDP based on IPCC scenario A1B.

Year	Energy consumption (MWh)	Population	GDP (×10⁶ TL)	Demand for RE (MWh)
2004	900,000	218,066	2,456	
2005	1,025,000	220,289	3,070	
2006	1,125,000	256,644	3,988	
2007	1,200,000	268,011	4,604	
2008	1,245,000	270,691	5,080	
2009	1,230,000	273,398	5,376	
2010	1,245,000	276,132	5,614	
2011	1,475,000	278,893	6,612	
2012	1,375,000	281,682	7,425	
2013	1,375,000	284,499	7,507	
2014	1,478,013	287,344	8,258	295,603
2015	1,539,173	290,218	9,083	307,835
2016	1,605,886	293,120	9,992	321,177
2017	1,678,699	296,051	10,991	335,740
2018	1,692,463	299,011	11,101	338,493
2019	1,706,365	302,002	11,212	341,273
2020	1,720,406	305,022	11,324	344,081
2021	1,734,588	308,072	11,437	346,918
2022	1,748,911	311,152	11,552	349,782
2023	1,763,378	314,264	11,667	352,676
2024	1,777,989	317,407	11,784	355,598
2025	1,792,746	320,581	11,902	358,549
2026	1,807,651	323,786	12,021	361,530
2027	1,822,705	327,024	12,141	364,541
2028	1,837,909	330,295	12,262	367,582
2029	1,853,265	333,598	12,385	370,653
2030	1,868,775	336,934	12,509	373,755
2031	1,873,415	335,249	12,634	374,683
2032	1,878,157	333,573	12,760	375,631
2033	1,883,001	331,905	12,888	376,600
2034	1,887,948	330,245	13,017	377,590
2035	1,892,998	328,594	13,147	378,600
2036	1,898,153	326,951	13,278	379,631
2037	1,903,414	325,316	13,411	380,683
2038	1,908,780	323,690	13,545	381,756
2039	1,914,254	322,071	13,681	382,851
2040	1,919,835	320,461	13,817	383,967
2041	1,925,525	318,859	13,956	385,105
2042	1,931,323	317,264	14,095	386,265

Year	Energy consumption (MWh)	Population	GDP (×10⁶ TL)	Demand for RE (MWh)
2043	1,937,233	315,678	14,236	387,447
2044	1,943,253	314,100	14,379	388,651
2045	1,949,385	312,529	14,522	389,877
2046	1,955,629	310,966	14,668	391,126
2047	1,961,988	309,412	14,814	392,398
2048	1,968,460	307,865	14,962	393,692
2049	1,975,048	306,325	15,112	395,010
2050	1,981,753	304,794	15,263	396,351

Appendix B

Model Validation and Verification

The wind model employed in this study is verified with third-party software RETScreen by National Resources Canada and validated with VESTAS power model as shown in Figure 11(a). The CO_2 emission analysis model (observed) (Figure 6) fits into IPCC's Multi-Model Averages and Assessed Ranges for Surface Warming (expected) as shown in Figure 11(b).

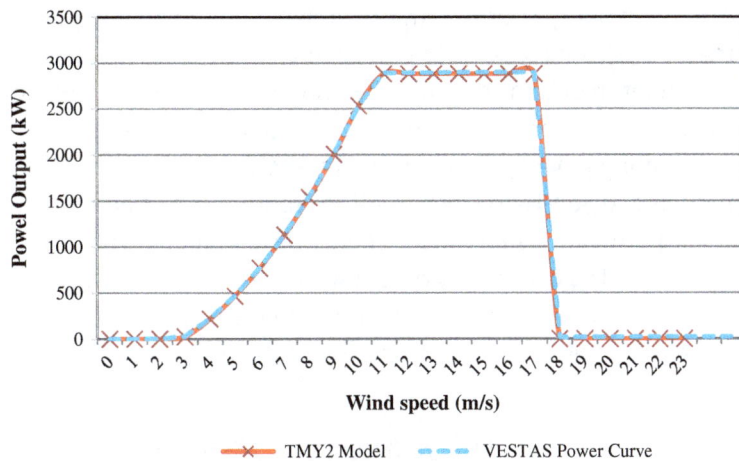

Figure 11a. Validation of model with vestas power curve.

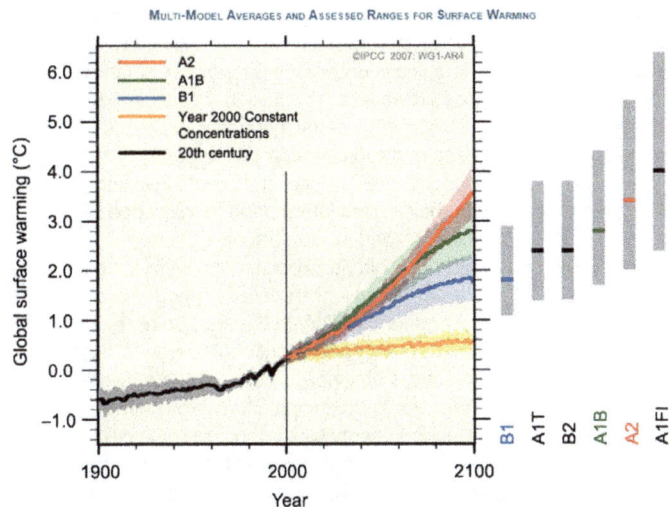

Figure 11b. Validation of model with IPCC multi-model averages and assessed ranges for surface warming (IPCC Fourth Assessment Report on Climate Change, 2015).

2

A review of Ghana's energy sector national energy statistics and policy framework

Samuel Asumadu-Sarkodie[1]* and Phebe Asantewaa Owusu[1]

*Corresponding author: Samuel Asumadu-Sarkodie, Sustainable Environment and Energy Systems, Middle East Technical University, Northern Cyprus Campus, Guzelyurt, Turkey
E-mail: samuel.sarkodie@metu.edu.tr
Reviewing editor: Shashi Dubey, Hindustan College of Engineering, India

Abstract: In this study, a review of Ghana's energy sector national energy statistics and policy framework is done to create awareness of the strategic planning and energy policies of Ghana's energy sector that will serve as an informative tool for both local and foreign investors, help in national decision-making for the efficient development and utilization of energy resources. The review of Ghana's energy sector policy is to answer the question, what has been done so far? And what is the way forward? The future research in Ghana cannot progress without consulting the past. In order to ensure access to affordable, reliable, sustainable, and modern energy for all, Ghana has begun expanding her economy with the growing Ghanaian population as a way to meet the SDG (1), which seeks to end poverty and improve well-being. There are a number of intervention strategies by Ghana's Energy sector which provides new, high-quality, and cost-competitive energy services to poor people and communities, thus alleviating poverty. Ghana's Energy sector has initiated the National Electrification Scheme, a Self-Help Electrification Program, a National Off-grid Rural Electrification Program, and a Renewable Energy Development Program (REDP). The REDP aims to: assess the availability of renewable energy resources, examine the technical feasibility and cost-effectiveness of promising renewable energy technologies, ensure the efficient production and use of the Ghana's renewable

ABOUT THE AUTHORS

Samuel Asumadu-Sarkodie is a multidisciplinary researcher who currently studies Masters in Sustainable Environment and Energy Systems at Middle East Technical University, Northern Cyprus Campus where he's also a Graduate Assistant in the Chemistry department. His research interest includes but not limited to: renewable energy, econometrics, energy economics, climate change, and sustainable development.

Phebe Asantewaa Owusu studies Masters in Sustainable Environment and Energy Systems at Middle East Technical University, Northern Cyprus Campus where she's also a Graduate Assistant in the Chemistry Department.

PUBLIC INTEREST STATEMENT

The main backbone of industrialized countries is the extensive nature of technological advancement through innovation and scientific research toward nation building. However, developing countries like Ghana have suffered many drawbacks in the field of health, water management, energy management, etc., due to limited and sporadic scientific research in these areas that provide local and private investors with the required information to make decisive choices toward its investment. Therefore, a multidisciplinary approach that unveils the core issues in Ghana in the scientific space would boost nation building. In this study, a review of Ghana's energy sector national energy statistics and policy framework is done to create awareness of the strategic planning and energy policies of Ghana's energy sector that will serve as a useful tool for both local and foreign investors and future research, thus facilitating in national decision-making for the efficient development and utilization of energy resources.

energy resources, and develop an information base that facilitates the establishment of a planning framework for the rational development and the use of the Ghana's renewable energy resources.

Subjects: Bio Energy; Clean Tech; Clean Technologies; Environmental; Environmental Policy; Power & Energy; Renewable Energy; Supply Chain Management

Keywords: energy portfolio; Ghana; energy balance; energy sector; electricity demand; renewable energy resources; developing countries; Africa; sustainable development; Ghana Energy Commission

1. Introduction

Energy demand and its associated service to meet social and economic development and improve human health and welfare is increasing due to the requirement to meet basic human needs and productivity (Edenhofer et al., 2011). The energy development is closely linked with the economic development of a country. Supplying the energy needs of the citizens of a country will directly affect the economic growth of the country. Access to energy plays a critical role to achieve the Millennium Development Goals (MDGs). Certainly, there is a close connection between the energy inadequacies and poverty indicators like, illiteracy, life expectancy, infant mortality, total fertility rate, and rapid urbanization in developing countries like Ghana; this is because people in the rural areas migrate to search for better living conditions and social amenities in the urban areas (Lipton & Ravallion, 1993). Access to energy has been linked to improving human development. Yet, 1.3 billion people, which is equivalent to 10% of the world's population, lack access to electricity. From this percentage, 22% are those living in developing countries with almost 97% of this percentage without access to electricity living in sub-Saharan Africa and development Asia (International Energy Agency, 2015).

Globally, per capita income has been established to have a positive correlation with per capita energy use. Therefore, economic growth can be identified as the most pertinent driver behind increasing energy consumption within the last decades. Nonetheless, no agreement on the trend of the causal relationship between energy-use and increased macroeconomic output has been established (Edenhofer et al., 2011; Narayan & Smyth, 2008). For example, there is evidence from a multitude of studies using individual countries as a reference which has failed to reach a consensus as to the direction of causation. A study using United States of America was able to find a bidirectional causality between energy consumption or its usage and Gross Domestic Product (Abosedra & Baghestani, 1989; Lee, 2006; Stern, 2000). Another study found that Gross Domestic Product causes energy consumption (Asafu-Adjaye, 2000; Meng & Niu, 2015) in the contrary, other studies found out that Gross Domestic Product and energy consumption were independent (Cheng, 1995; Eden & Hwang, 1984; Oh & Lee, 2004).

The major global energy challenges are in three folds: securing energy supply to meet growing demand, providing everybody with access to energy services, and curbing energy's contribution to climate change (Asif & Muneer, 2007; Helm, 2002; IPCC, 2011). Africa as a continent is rich in energy resources but poor in its supply. The Sub-Saharan Africa accounts for 13% of the world's population, but only 4% of its populace have access to energy. Yet, there has been a rapid economic growth and energy by 45% since 2000 (International Energy Agency, 2014).

The main backbone of industrialized countries is the extensive nature of technological advancement through innovation and scientific research toward nation building. However, developing countries like Ghana have suffered many drawbacks in the field of health (Asumadu-Sarkodie & Owusu, 2015), water management (Asumadu-Sarkodie, Owusu, & Jayaweera, 2015; Asumadu-Sarkodie, Owusu, & Rufangura, 2015; Asumadu-Sarkodie, Rufangura, Jayaweera, & Owusu, 2015), energy management (Asumadu-Sarkodie & Owusu, 2016c, 2016d), etc., due to limited and sporadic scientific research in these areas that provide local and private investors with the required information to make decisive choices toward its investment. Therefore, a multidisciplinary approach that unveils the core issues in Ghana in the scientific space would boost nation building. Ghana is one of the most successful countries in Sub-Saharan Africa in improving access to electricity, having showed long-term and strong political commitment since the establishment of its National Electrification Scheme in 1989 (International Energy Agency, 2014). Ghana has a huge potential to grow into a modernized economy and to reduce the high incidence rate of poverty to an acceptable level. The ultimate goal of the Growth and Poverty Reduction Strategy (GPRS II) by the Government of Ghana's development agenda is to grow the economy to a middle-income status of US$1,000 per capita and to reduce the incidence rate of poverty among Ghanaians. To meet this development target, it requires an annual economic growth rate of about 10% from the current level of 6.7% through increasing accessibility and supply of energy (Ministry of Energy, 2009).

An effective energy policy creates measures that can support the innovation of new, alternative energy technologies required to meet a growing demand and meet the challenges of creating a sustainable future in the energy sector. The complexity of creating a sustainable future that meets the demands of a populace is not easy to create without an extensive energy policy framework. In order to create an extensive and effective energy policy framework, there is the need for a viable and working energy sector (Omer, 2008). Successful energy sectors follow some framing objectives which include but not limited: (a) stabilizing supply of energy which is an important infrastructural precondition for modern society to function; (b) reducing environmental impacts from energy production and usage; (c) introducing more efficient energy usage per unit of outcome; and (d) continual development of alternative energy sources in order to reduce dependency on degradable and limited natural resources (Jørgensen, 2005) which will serve as a bequest for future generations.

The energy sector is defined by the Intergovernmental Panel on Climate Change (IPCC) as "comprising all energy extraction, conversion, storage, transmission and distribution processes with the exception of those that use final energy in the end-use sectors (industry, transport, building, agriculture, and forestry)" (Climate everyone's Business, 2014). Since climate change has become a global issue, a growing number of developing countries have become interested in energy policies with emerging renewable energy sectors (S. Asumadu-Sarkodie & P. Owusu, 2016; S. Asumadu-Sarkodie & P. A. Owusu, 2016a, 2016b; Asumadu-Sarkodie & Owusu, 2016c, 2016d; Choi, Park, & Lee, 2011). The energy sector is the largest contributor to global greenhouse gas emissions. In 2010, 35% of direct greenhouse gas emissions came from energy production. In recent years, the long-term trend of gradual de-carbonization of energy has reversed. From 2000 to 2010, the growth in energy sector emissions outpaced the growth in overall emissions by about 1% per year (Climate everyone's Business, 2014). Developing a functional and proactive energy sector has been established to have a positive relationship with economic growth (AC00804425, 1993; Dowrick, 1994) puts forward qualitative changes such as interactions, competition, and the entry of new actors like renewable energy to describe a country's economic growth.

This study reviews Ghana's energy sector national energy statistics and policy framework. The review of Ghana's energy sector policy is to answer the question of, what has been done so far? And what is the way forward? The future research in Ghana cannot progress without consulting the past and institutional capabilities. Both qualitative and quantitative research methodologies were employed in the study. As part of the research techniques, a secondary data on the national energy statistics were adopted from the archives of the Ministry of Energy and Petroleum, the Volta River Authority (VRA), the Ghana Grid Company (GRIDCo), Ghana National Petroleum Corporation (GNPC), the National Petroleum Authority (NPA), Tema Oil Refinery (TOR), the Public Utility Regulatory Commission (PURC), the Electricity Company of Ghana (ECG), the Northern Electricity Distribution Company (NEDCo), the West African Gas Pipeline Company (WAPCo), the West African Gas Pipeline Authority (WAGPA), the Bank of Ghana (BoG), and the Ghana Statistical Service (GSS). In this case, concepts, experiences, techniques, information gathered from the actors in the field of the study, peer-reviewed journals and other literature relevant to this study were analyzed and reviewed. The study brings to bear a number of intervention strategies by Ghana's Energy sector which provides new, high-quality and cost-competitive energy services to poor people and communities, thus alleviating poverty. Understanding Ghana's energy sector strategic planning and energy policies would serve as a useful tool for both local and foreign investments, help in national decision-making for the efficient development and utilization of energy resources.

2. Energy indicators and energy balance

The 1992 Earth Summit recognized the significant role that indicators can play in helping countries to make informed decisions concerning sustainable development. 'This recognition is articulated in Chapter 40 of Agenda 21 which calls on countries at the national level, as well as international, governmental and non-governmental organizations to develop and identify indicators of sustainable development that can provide a solid basis for decision-making at all levels (United Nations, 3–14 June 1992; Vera & Langlois, 2007). Moreover, Agenda 21 specifically calls for the harmonization of efforts to develop sustainable development indicators at the national, regional and global levels, including the incorporation of a suitable set of these indicators in common, regularly updated and widely accessible reports and databases' (Summit, 1997; Vera & Langlois, 2007).

Energy indicators are not just energy statistics; rather, they go beyond basic statistics to provide a deeper understanding of causal relationships in the energy–environment, economics nexus, and to highlight linkages that may not be evident from basic statistics. Collectively, indicators can give a clear picture of the whole energy system, including interlinkages and trade-offs among various dimensions of sustainable development, as well as the longer term implications of current decisions and behavior (Vera & Langlois, 2007). In Table 1, the energy indicators of Ghana's energy sector are given.

In Table 1, the Population of Ghana increases from almost 22 million in 2006 to about 27 million in 2013. With at constant 2006 prices, the Gross Domestic Product increases from about 19 million Ghana Cedis in 2006 to about 32 million Ghana Cedis in 2013. With a correlation coefficient r (8)=0.98, $p < 0.05$, increasing Population has a positive correlation with the increasing Gross Domestic Product. In other words, there is a strong relationship between the two indicators. Total Energy Generated increased from 8,430 GWh in 2006 to 12,870 GWh in 2013; Total Final Energy Consumed, that is, the energy which is not being used for transformation into other forms of energy, increased from 5,177 ktoe to 6,886 ktoe; Total Electricity Consumed which includes commercial losses increased from 7,322 GWh in 2006 to 10,583 GWh in 2013; There were no significant changes in Total Biomass Consumed in 2006 (2,671 ktoe) and 2012 (2,676 ktoe) yet, there was decline in its consumption in 2007 till 2011 before it began to rise again; Total Petroleum Products Consumed increased from 1,873 ktoe in 2006 to 3,422 ktoe in 2013. Total Final Energy Consumed per capita increased from 0.24 TOE/capita in 2006 to 0.26 TOE/capita; Total Electricity Generated per capita increased from 387 kWh/capita in 2006 to 486 kWh/capita; Total Electricity Consumed per capita increased from 338 kWh/capita in 2006 to 399 kWh/capita in 2013. Yet, there was a decline in the Energy Intensity of the Economy from 0.28 TOE/GHS 1,000 of GDP to 0.21 TOE/GHS 1,000 of GDP and the Total Electricity Consumed per GDP also declined from 394 kWh/GHS 1,000 of GDP in 2006 to 327 kWh/GHS 1,000 of GDP. In Table 2, the Energy Balance as at 2013 is given.

Energy Balance shows in a consistent accounting framework, the production, transformation, and final consumption of all forms of energy for a given country in a given period of time, with quantities expressed in terms of a single accounting unit for purposes of comparison and aggregation. The Energy balance presents an overview of the energy produced and consumed in a system, matching input and output for a specific period of time, usually one year (Bhattacharyya, 2011; Mauritius, 2014). The energy balance (Table 2) shows the supply and final uses (demand) of energy and the different types of fuel (crude oil, natural gas, petroleum products, wood, charcoal, hydro, solar, and electricity). The energy supply is presented as the Total Primary Energy Supply. The energy demand is presented as the Total Final Consumption. The difference between the supply and the demand is mainly due to fuel transformed into electricity. Statistical Difference in Table 2 shows the difference between calculated and observed inland consumption (International Energy Agency, 2010). It

Table 1. Energy indicators (2006–2013)								
Energy indicator	Unit	2006	2007	2008	2009	2010	2011	2012	2013
Total final energy Consumed	KTOE	5,176.9	5,274.1	5,209.8	5,731.7	5,670.2	6,192.1	6,687.9	6,886.0
Total electricity generated	GWh	8,430.0	6,978.0	8,324.0	8,958.0	10,167.0	11,200.0	12,023.8	12,870.0
Total electricity consumed	GWh	7,361.9	6,440.5	7,219.4	7,452.4	8,317.4	9,186.6	9,258.0	10,583.2
Total petroleum products consumed	KTOE	1,872.6	2,126.6	2,071.3	2,597.7	2,491.1	2,826.6	3,317.5	3,422.3
Total biomass consumed	KTOE	2,671.3	2,593.7	2,517.8	2,493.3	2,463.9	2,575.6	2,588.8	2,676.0
Population	million	21.8	22.3	22.9	23.4	24.7	25.3	25.9	26.5
GDP (Constant 2006 prices)	million Ghana Cedis	18,705.1	19,913.4	21,592.2	22,454.0	24,252.0	27,891.0	30,099.0	32,322.0
Energy intensity of the economy	TOE/GHS 1,000 of GDP	0.28	0.26	0.24	0.26	0.23	0.22	0.22	0.21
Total final energy consumed/capita	TOE/capita	0.24	0.24	0.23	0.24	0.23	0.24	0.26	0.26
Total electricity generated/capita	kWh/capita	386.7	312.9	363.5	382.8	411.6	442.7	464.2	485.7
Total electricity consumed/capita	kWh/capita	337.7	288.8	315.3	318.5	336.7	363.1	357.5	399.4
Total petroleum consumed/capita	TOE/capita	0.09	0.10	0.09	0.11	0.10	0.11	0.13	0.13
Total biomass consumed/capita	TOE/capita	0.12	0.12	0.11	0.11	0.10	0.10	0.10	0.10
Total electricity consumed/GDP	kWh/GHS 1,000 of GDP	393.6	323.4	334.4	331.9	343.0	329.4	307.6	327.4

Source: GDP and population data from Ghana Statistical Service.

includes the sum of the unexplained differences for individual fuels as they appear in the energy statistics.

3. Primary energy supply and final energy consumption
In Figure 1, Ghana's primary energy supply is given. Ghana's primary energy supply is calculated as the sum of imported fuels and locally available fuel, less re-exports to bunkers after adjusting for stock changes. Oil supply grew from 1,812 ktoe in 2000 to 4,011 ktoe in 2013; Natural gas supply started in 2009 with 5 ktoe and increased to 292 ktoe; hydro supply increased from 609 ktoe to 700 ktoe and wood supply declined from 3,888 ktoe to 3,553. This decline is due to a Government biomass policy which promotes the use of alternative fuels such as LPG as a substitute for wood fuel and charcoal by addressing the institutional and market constraints that hamper increasing access of LPG in Ghana (Ministry of Energy Ghana, 2009).

In Figure 2, Ghana's final energy consumption is given. Ghana's final energy consumption is the energy which is not being used for transformation into other forms of energy and it is calculated as the total amount of energy required by end-users as a final product. Final electricity consumption increased from 597 ktoe in 2000 to 910 ktoe in 2013; petroleum consumption increased from grew from 1,533 ktoe in 2000 to 3,303 ktoe in 2013; biomass consumption declined from 3,432 ktoe in 2000 to 2,588 ktoe in 2013 due to a switch in end-user preferences to Liquid Petroleum Gas.

Table 2. Energy balance (KTOE)									
Supply and consumption	Crude oil	Natural gas	Petroleum Products	Wood	Charcoal	Hydro	Solar	Electricity	Total
Indigenous production	5,371.8	–	–	3,553.9	–	708.0	0	–	9,633.7
Imports	1,328.3	291.6	3,070.4	–	–	–	–	2.3	4,692.7
Exports	−5,210.9	–	−216.5	–	−0.7	–	–	−10.5	−5,438.6
Stock changes	−160.9	–	−171.0	–	–	–	–	–	−331.9
Total primary energy supply	1,328.4	291.6	2,682.9	3,553.9	−0.7	708.0	0	−8.2	8,555.9
Electricity plants	−881.1	−290.0	−5.2	–	–	−708.0	0	1,106.6	−777.7
Petroleum refinery	−446.5	–	437.8	–	–	–	–	–	−8.7
Charcoal kilns	–	–	–	−1,989.5	1,112.2	–	–	–	−877.2
Own use	−41.2	–	–	–	–	–	–	−6.9	−48.1
Losses	−27.3	–	–	–	–	–	–	−132.7	−160.0
Final energy consumption	–	–	3,300.1	1,564.4	1,111.6	–	–	910.2	6,886.2
Residential sector	–	–	151.3	1,311.8	899.4	–	–	433.2	2,795.7
Commerce and services sector	–	–	22.9	31.0	86.1	–	–	152.9	292.9
Industry	–	–	380.1	221.6	126.1	–	–	322.2	1,050.0
Agriculture and fisheries sector	–	–	101.0	–	–	–	–	1.8	102.8
Transport	–	–	2,644.8	–	–	–	–	–	2,644.8
Statistical difference	−67.8	1.7	−126.2	–	–	–	–	48.6	−143.7

Source: Ministry of Energy and Petroleum, Ghana.

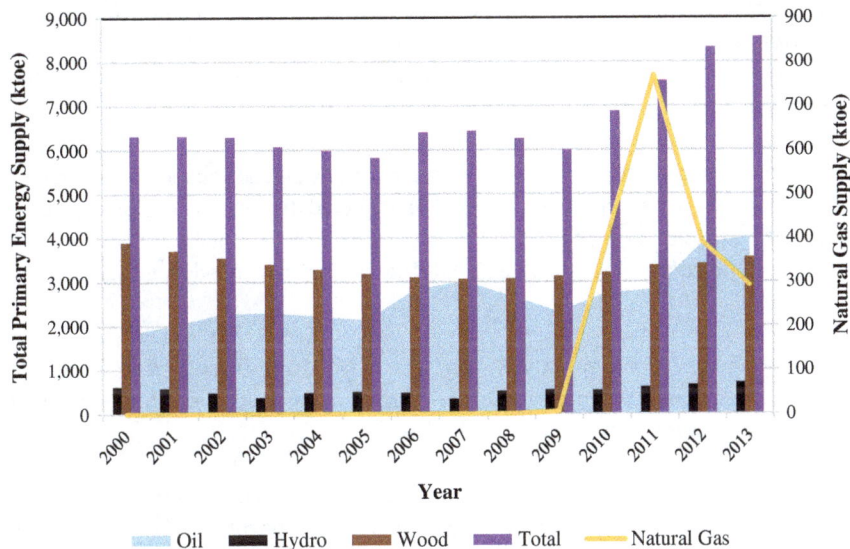

Figure 1. Total primary energy supply.

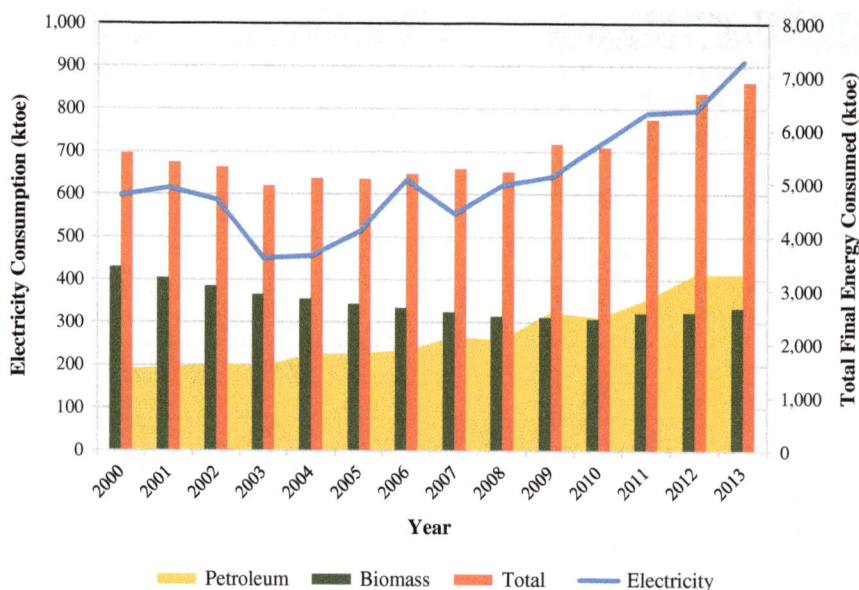

Figure 2. Final energy consumed.

4. Electricity

Electrification is considered primarily in the context of providing electrical energy to urban households and industries. In 2000, primary electricity generation was about 7,224 GWh, which dropped to 5,901 GWh by 2003. The electricity energy balance of Ghana from 2000 to 2003 is presented in Table 3 (Ghana Energy Commission, 2012; SWERA, 2012). It includes the officially recorded total national power outages and private diesel generation employed to meet shortfalls.

There is a constant increase in power supply from 0.9 to 1.8 percent per annum with the exception of 2003. The drop in power supply was generally due to the suspension of the operations of the Volta Aluminium Company (VALCO) aluminum smelter. Nevertheless, electricity demand was projected to increase from 6.58 TWh in 1997 to about 20 - 21 TWh by 2020. Volta River Authority's (VRA) forecast shows an increase of 20 TWh by 2020 considering average economic growth from 1990 to 2000 (Volta Aluminium, 2015).

As the electricity generation capacity of the Ghana increased, many decisions concerning different power generation options had to be considered to meet the projected demand. A decision regarding the use of centralized power solutions had to be made as the present power sector is dominated by hydroelectric power which constitutes about 63% of the nation's installed capacity (Energy Commission Ghana, 2013).

For the low-growth scenario, Volta River Authority's projection was 16 TWh at an average demand growth of between 6 and 8% per annum. In Electricity Company of Ghana, the average demand growth was about 7% per annum (Electricity Company of Ghana, 2012).

The demand for petroleum was projected to rise from 1.6 million tonnes in 2000 to over 3 million tonnes in 2012 if the Ghana Poverty Reduction Strategy targets were to be met (Webapps01.un.org, 2012).

Demand for grid electricity grew from about 6,900 GWh in 2000 to a projected figure of about double by 2012 (Ghana Energy Commission, 2012; SWERA). Natural gas was expected to replace light crude oil by 2007 as a fuel for electricity generation should the West African gas pipeline project to commence as planned but was not the case. It was projected to overtake hydropower as the dominant primary fuel for power generation by 2010. However, overdependence on natural gas

Table 3. Electricity supplies balance (SWERA)					
electricity energy balance		**2000**	**2001**	**2002**	**2003**
Primary electricity generation	Gigawatt-hour (GWh)	7,224	7,860	7,297	5,901
	Percentage thermal	8.49	15.91	30.97	33.83
	Percentage hydro	91.50	84.07	69.01	65.84
	Percentage solar	0.01	0.02	0.02	0.02
Carbon dioxide emissions due to electricity generation	Thousand tones	443	894	1,742	1,405
Total imports	Gigawatt-hour	864	462	1,146	940
Total exports including wheeling	Gigawatt-hour	675	846	833	833
Net import (+)/export (−)	Gigawatt-hour	188	−384	313	79
Total primary supply	Gigawatt-hour	7,411	7,476	7,610	5,980
	Percentage thermal	8.27	16.00	29.30	33.39
	Percentage net import	2.54	0	4.11	1.32
	Percentage hydro	89.19	84.38	66.18	64.97
	Percentage solar	0.01	0.02	0.02	0.02
Transmission losses		229	259	368	333
Distribution (technical losses)		552	561	549	615
Total technical losses		*781*	*820*	*917*	*948*
	Percentage losses over primary production	10.8	10.4	12.6	16.07
	Percentage losses over primary supply	10.5	11.0	12.0	15.85
Final supply expected	Gigawatt-hour	*6891*	*7284*	*7,061*	*5,524*
Officially recorded total nationwide power outages in hours		NA	704	501	485
Actual electricity reaching consumers (Gigawatt-hour)		*6890.5*	*7166.5*	*6,867.6*	*5,243.1*
Private diesel generation to meet shortfall (Gigawatt-hour)			117.5	193.4	281
Percentage share of total electricity consumption					
	Residential	32.26	30.85	34.17	46.85
	Agriculture and Fisheries	0.04	0.04	0.06	0.10
	Commercial and Services	6.46	9.71	11.02	15.00
	Industry	24.89	24.19	25.53	33.53
	VALCO	36.35	35.21	29.22	4.52

from the West African gas pipeline for electricity generation could have put the nation's energy security at risk.

Without Volta Aluminium Company (VALCO), the residential sector of the economy consumes about 54% of the country's generated electricity supply. Residential electricity consumption increased from 688.03 GWh in 1990 at an average growth rate of 11% to 2373.8 GWh in 2000 (Ghana. Valcotema.com, 2015). A comprehensive demand output and corresponding generation requirement for 2008, 2012, and 2020 are presented in Table 4. The two scenarios considered are the business-as-usual scenario and the Ghana Poverty Reduction Strategy (GPRS) high economic growth scenario. In both scenarios, the power consumed by Volta Aluminium Company (VALCO) is considered because it is the biggest power consumer in Ghana.

In Table 5, Ghana's installed electricity generation capacity is given. From 1999 through 2002, approximately 75% of the overall supply of electricity in Ghana was hydro-generated. Since 2003, it has consistently been reduced to about 65%. The per capita electricity consumption was 358 kWh in 2000 indicating an increase of 15.9% as compared to 309 kWh in 1999. The per capita electricity consumption in Ghana is lower than the weighted average in the sub-Saharan region (Ministry of Energy Ghana, 2009). Currently, Ghana's electricity generation consist of hydropower, thermal, embedded generation, and renewables. Although a larger proportion of the nation's known hydropower potential, including the Akosombo and Kpong, have already been developed, there are some undeveloped sites. Currently, Ghana's installed capacity from hydropower is 1, 580 MW (Akosombo-1,020 MW, Bui-400 MW and Kpong-160 MW); thermal generation constitutes 1,494 MW; embedded generation (Genser Power) constitutes 5 MW and renewables (Volta River Authority installed Solar) constitute 2.5 MW. In total, Ghana's current installed capacity as of December, 2013, is 3,081 MW. The target is to achieve installed power generation capacity of 4,000 MW and also universal access to affordable electricity by 2015. This would be achieved through the Private Public Partnership (PPP) in the development of new power plants as well as ensuring cost-recovery for the production, transmission, and distribution of electricity (Ministry of Energy Ghana, 2009, Figure 3).

The Volta River Authority (VRA) generates all the electricity consumed in the country and also for the export market. Over the last five (5) years, VRA imported approximately 1,000 GWh of electricity, which represents about 14% of overall supply. Crude oil imports account for approximately 10% of total commodity trade (i.e., import plus exports), and consume between 15and 40% of the nation's export earnings.

In 2000, an amount of US$ 528 million, about 27% of the country's total export earning was spent on the importation of about 1.1 million tonnes of crude oil and 0.8 million tonnes of petroleum products (Beg.utexas.edu, 2005). The nature of world crude oil prices and the negative impact on the nation's balance of payment made the Ministry of Energy to constitute the National Petroleum Tender Board (NPTB) to coordinate the procurement of crude oil and petroleum products (Ministry of Energy Ghana, 2009).

Ghana imports all her crude oil requirements, which amount to 60,000 bpsd. Out of this, 30000 bpsd imports are from Nigeria on government-to-government contract with an additional 15,000 bpsd purchased by bid-offers, through the National Petroleum Tender Board. 15,000 bpsd of light

Table 4. Electricity demand to meet economic growth scenario				
	Operation	Electricity demand and generation requirement in Gigawatt-hour (GWh)		
		2008	2012	2020
Business—as—usual economic growth				
Demand	Without VALCO	6,960	8,530	12,780
	With VALCO	9,550	11,450	15,500
Generation required	Without VALCO	8,400 (±3%)	10,600 (±2%)	15,500(±2%)
	With VALCO	10,900 (±3%)	13,100 (±2%)	18,000 (±2%)
GPRS high economic growth				
Demand	Without VALCO	15,640	19,065	27,380
	With VALCO	17,540	21,065	29,180
Generation required	Without VALCO	17,500 (±3%)	20,150 (±2%)	30,400 (±2%)
	With VALCO	20,000 (±3%)	22,650 (±2%)	32,900 (±2%)

Table 5. Installed electricity generation capacity (end of December, 2013)			
Plant	Fuel type	Installed capacity (MW)	
		Name Plate	Dependable
Hydro generation			
Akosombo	Water	1,020	900
Bui	Water	400	342
Kpong	Water	160	140
Sub-Total		*1,580*	*1,382*
Thermal Generation			
Takoradi Power Company (TAPCO)	LCO/Natural Gas	378	300
Takoradi International Company (TICO)	LCO/Natural Gas	252	200
Sunon Asogli Power (Ghana) Limited (SAPP) -	Natural Gas	220	180
IPP			
Cenit Energy Ltd (CEL)	LCO/Natural Gas	126	110
Tema Thermal 1 Power Plant (TT1PP)	LCO/Natural Gas	126	110
Tema Thermal 2 Power Plant (TT2PP)	Natural Gas	49.5	45
Takoradi T3	LCO	132	120
Mines Reserve Plant (MRP)	Diesel/Gas	85	80
Effasu Power Barge	Natural Gas	125	100
Sub-Total		*1,494*	*1,245*
Embedded generation			
Genser Power—IPP	LPG	5	2.1
Sub-Total		*5*	*2.1*
Renewables			
VRA Solar	Sunshine	2.5	1.9
Sub-total		*2.5*	*1.9*
Total		3,081.0	2,631.0

Source: Ghana Grid Company.

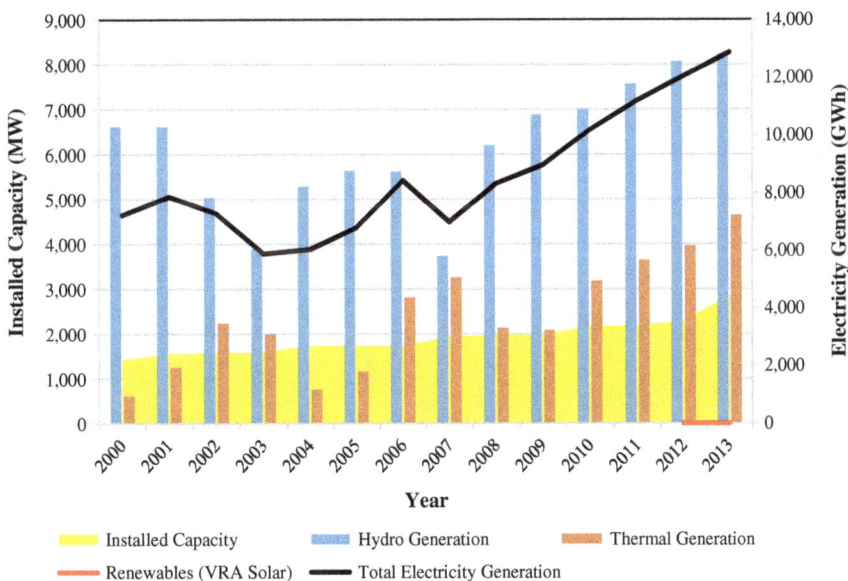

Figure 3. Trends electricity generation.

Table 6. Ghana's fossil fuel imports (1997–1999)				
Year	1997	1998	1999	2000
Oil imports (US$ m)	234	221	323	NA
% of imports (fob)	11	10	12	20

crude oil is also purchased by the Volta River Authority (VRA) to feed its thermal plants at Aboadze for electricity generation (Npa.gov.gh). Importation of crude oil and petroleum products and debt servicing have been a drain on the nation's hard currency earnings chopping off about 40% of export earnings in the early 1980s and dropping to an average of 13% by 2000. The pattern of importation of fossil fuels into Ghana since 1997–2000 is presented in Table 6.

In Figure 4, electricity import, export, and net import of Ghana's energy sector is given. The main imports include capital equipment, crude oil and petroleum products, food, consumer, and intermediate goods. Import of electricity decreased from 864 GWh in 2000 to 27 GWh in 2013. This decline is due to Ghana's energy efficiency and conservation policy that sought to discontinue, through legislation on standardization and labeling, the local production, importation, and use of inefficient electricity consuming equipment and appliances (Ministry of Energy Ghana, 2009). Electricity export in Ghana fluctuates depending on electricity generated in the country. The minimum electricity export occurred in 2013 selling 122 GWh and the maximum export made in 2010 selling 1,036 GWh. Volta River Authority exports about 4% of the electricity it generates to the neighboring countries. Communaute du Benin requested an additional supply of 387 GWh recently which Volta River Authority is supplying. Volta River Authority also supplies electricity to some border towns in Burkina Faso. In 2011, the total energy transmitted by the Ghana Grid Company (GRIDCo) outside Ghana to La Compagnie Ivoirienne d'Electricité ("CIE"), La Communaute Electrique Du Benin ("CEB"), La Societe Nationale D'Electricite Du Burkina ("SONABEL"), and the Youga Mine in Burkina Faso came to 774.991GWh (Hogan Lovells, 2013). The highest net import is made in 2002 for 534 GWh, with the minimum net import made in 2001 for 160 GWh. This is because Ghana had a net export of electricity only in 2001 when there was unexpected abundant water in the Akosombo and Kpong hydroelectric reservoirs. The negative net import represents the net export made. Maximum net export was made in 2010 for 930 GWh, whiles the minimum net export was made in 2013 for 95 GWh. The negative values mean that the value of imports exceeded the value of export earnings.

In Figure 5, trends in peak load in Ghana are given. Peak load is an energy demand management term that describes a period in which electrical power is expected to be provided for a sustained period at a significantly higher than the average supply level. Ghana's load at peak consists of the maximum demand of the Electricity Company of Ghana, Northern Electricity Distribution Company, direct customers of the Volta River Authority and the Mines. System peak consists of Ghana load at

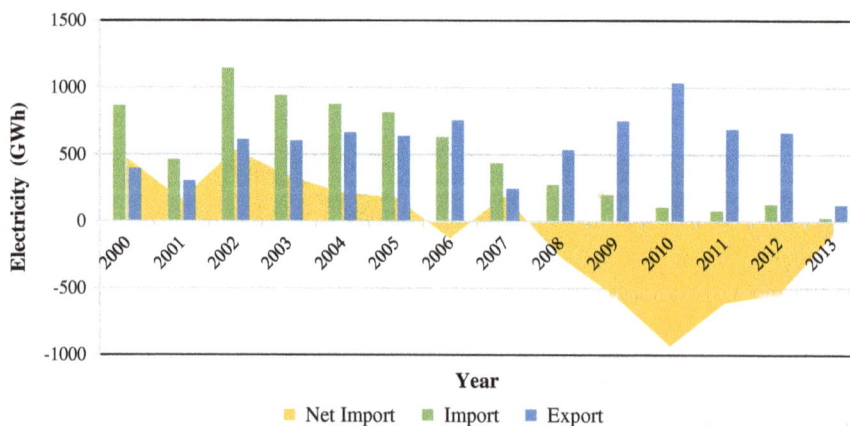

Figure 4. Electricity net import, import, and export.

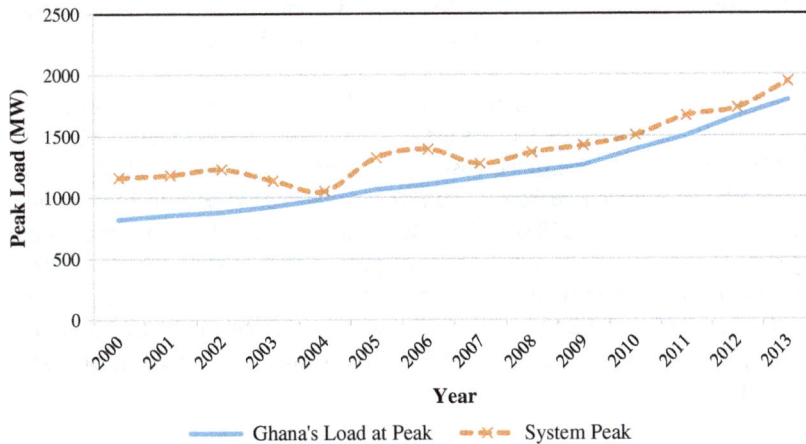

Figure 5. Trends in peak load.

peak, Volta Aluminium Company and export load. Ghana's maximum peak load and the total system peak of the grid transmission system occurred in 2013 about 1,800 MW and 1,900 MW, respectively. For the Volta Aluminium Company to be operating between 3–4 potlines, Ghana's peak load and the total system peak would increase to 1,980 MW and 2,500 MW, respectively (Energy Commission Ghana, 2013).

In Figure 6, Akosombo Dam Month End Elevation is given. The total electricity made available for gross transmission in 2013 was 12, 870 GWh, as against 12,164 GWh in 2012 and 11,200 GWh in 2011. The 2012 generation comprised 8,071 GWh (67%) hydropower and 3,639 GWh (33%) of thermal power. Although hydropower generation share decreased by about 0.5 percentage points over 2011, energy produced increased by about 510 GWh due to significant water inflows into the Akosombo reservoir in 2012 (Energy Commission Ghana, 2013). The maximum water inflows into the Akosombo reservoir occurred in October, 2010, was almost 280 feet in height. Higher than expected average annual precipitation is expected this year – 2013, according to the Ghana Meteorological Agency (GMA). Higher inflows into the hydropower reservoir would improve the overall power generation to offset inadequate or delayed gas supply.

In Figure 7, the trends in transmission losses are given. Total power transmission losses in 2013 was 4.8% of gross transmission, which is 0.5 percentage point improvement over 2012; 2012 was 4.3% of gross transmission, 0.4 percentage point improvement over 2011 but the minimum

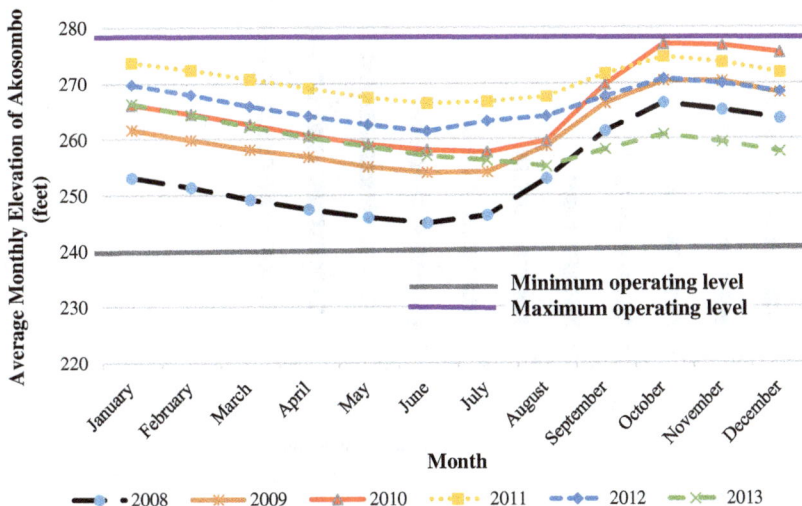

Figure 6. Akosombo dam month end elevation.

transmission losses occurred in 2000 which were 2.8%% of gross transmission yet, the maximum transmission losses occurred in 2003 which were 5.9% of gross transmission, respectively.

In Table 6 and 7, electricity purchases and sales by the Electricity Company of Ghana (ECG) and Northern Electricity Distribution Company (NEDco) are given. The total purchases made by ECG grew from 3,989 GWh in 2000 to 8,479 GWh due to increasing population and increasing electricity demand. In the same vein, total sales increased from 2,910 GWh in 2000 to 6,476 GWh in 2013 coupled with distribution losses increasing from 1,079 GWh in 2000 to 2,003 GWh in 2013. However, there was a 0.4 percentage loss improvement over 2012. The total purchases made by NEDco grew from 330 GWh in 2000 to 937 GWh due to increasing population and increasing electricity demand. In the same vein, total sales increased from 232 GWh in 2000 to 737 GWh in 2013 coupled with distribution losses increasing from 1,079 GWh in 2000 to 2,003 GWh in 2013. However, the percentage losses increased by 1.3% over 2012.

In Figure 8, the distribution losses by the Electricity Company of Ghana (ECG) and the Northern Electricity Distribution Company (NEDco) are given. Distribution losses from Electricity Company of Ghana (ECG) between the years 2000 and 2010 were better-off than the Northern Electricity Distribution Company (NEDco). However, there were improvements in distribution losses from the Northern Electricity Distribution Company (NEDco) between the years 2010 and 2013, whereas distribution losses started increasing from Electricity Company of Ghana (ECG) in the same years.

In Figure 9, the electricity consumption by customer class is given. The total electricity consumption increased from 6,367 GWh in 2000 to 9,355 GWh in 2013. The major energy consuming industries in Ghana are: the Volta Aluminium Company (VALCO), Electricity Company of Ghana, and the Northern Electricity Department. Industrial electricity consumption amounts to 4,224 GWh (45%) in 2013 compared to 4,153 GWh (48%) in 2012 with a decline of 3%; residential electricity consumption amounts to 3,228 GWh (35%) in 2013 compared to 2,805 GWh (35%) in 2012 with no change in percentage; non-residential electricity consumption amounts to 1,525 GWh (16%) in 2013 compared to 1,153 GWh (13%) in 2012 with a decline of 3% and street lighting electricity consumption was 377 GWh (4%) in 2013 compared to 315 GWh (3%) with an increase of 1%, respectively.

5. Petroleum

The discovery of the Jubilee field in Ghana in 2007 has fed expectations of more to come in this relatively under-explored basin stretching from Mauritania to the Niger Delta. The area under license has doubled in the last 5 years, with technical discoveries being made in Liberia, Sierra Leone, and Côte

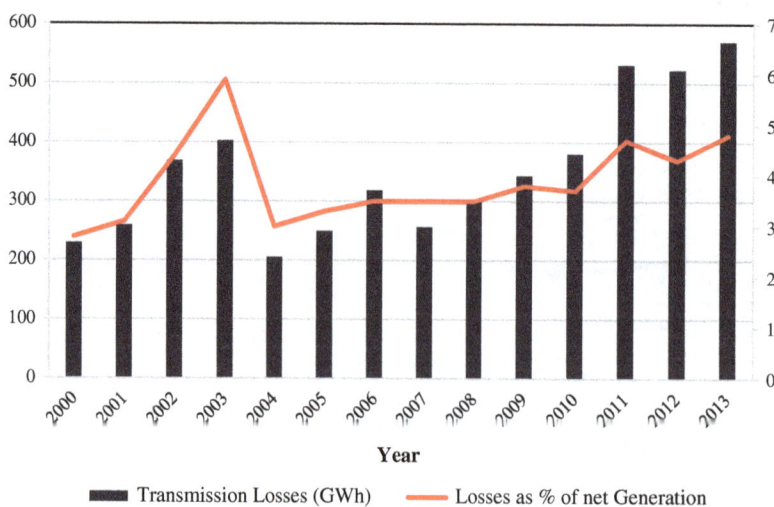

Figure 7. Trends in transmission losses.

Table 7. Electricity purchases and sales by ECG														
Year	2000	2001	2002	2003	2004	2005	2006	2007	2008	2009	2010	2011	2012	2013
Total purchases (GWh)	3,989	4,175	4,326	4,496	4,818	5,045	5,253	5,146	5,799	6,052	6,771	7,259	7,944	8,479
Total sales (GWh)	2,910	3,080	3,200	3,343	3,542	3,761	3,978	3,906	4,335	4,442	4,952	5,339	6,041	6,476
Distribution losses (GWh)	1,079	1,095	1,126	1,153	1,276	1,285	1,275	1,240	1,464	1,610	1,819	1,920	1,903	2,003
Percentage losses	27.0	26.2	26.0	25.6	26.5	25.5	24.3	24.1	25.2	26.6	26.9	26.4	24.0	23.6

Source: GRIDCo, VRA and ECG.

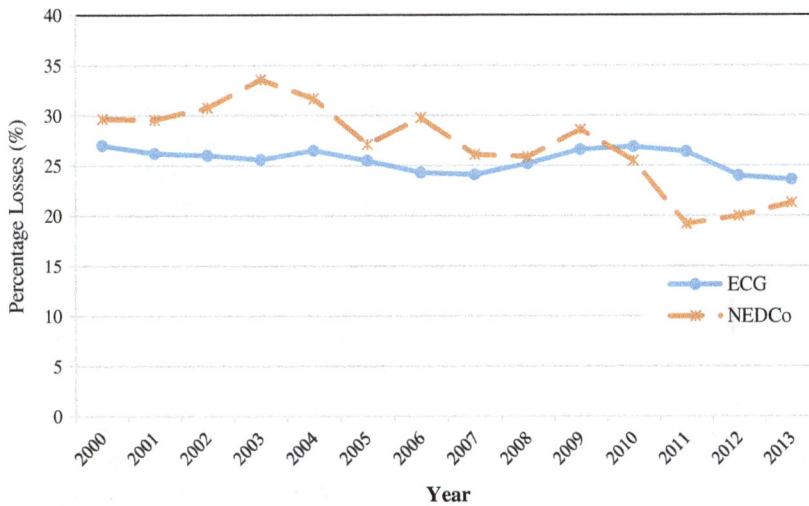

Figure 8. Trends in distribution losses.

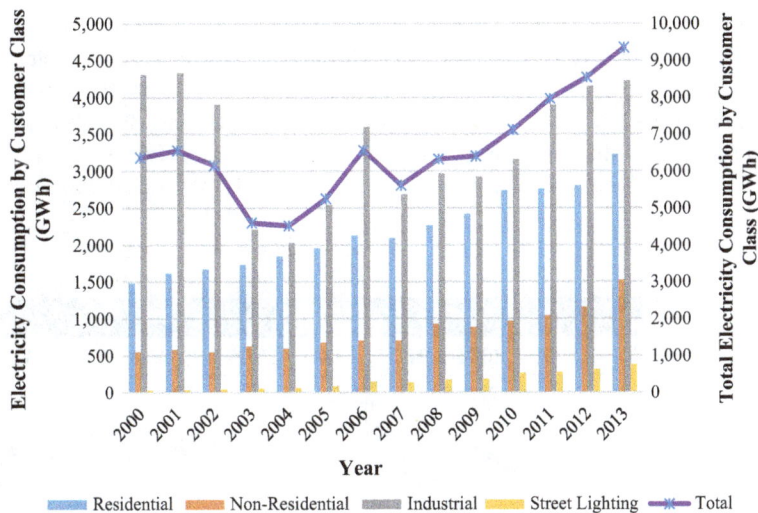

Figure 9. Electricity consumption by customer class.

Table 8. Electricity purchases and sales by NEDCo														
Year	2000	2001	2002	2003	2004	2005	2006	2007	2008	2009	2010	2011	2012	2013
Total purchases (GWh)	330	355	383	426	473	501	507	494	529	566	635	719	822	937
Total sales (GWh)	232	250	265	283	323	365	356	365	392	404	473	581	658	737
Distribution losses (GWh)	98	105	118	143	150	136	151	129	137	162	162	138	164	200
Percentage losses	29.7	29.6	30.8	33.6	31.7	27.1	29.8	26.1	25.9	28.6	25.5	19.2	20.0	21.3

Source: GRIDCo, VRA, and NEDCo.

D'Ivoire, but a further appraisal is required to ascertain their commerciality (International Energy Agency, 2014).

In Table 8, the crude oil production is given. Crude oil is produced from both Saltpond and Jubilee Field. The maximum crude oil produced was 30.5 kilotons in 2008. Presently, the crude oil produced is 15 kilotons from Saltpond field. Crude oil production in the Jubilee field began in 2010 at 181.1 kilotons. As at 2013, the crude oil produced from Jubilee oil was 5251.5 kilotons, which brings the total crude oil production to 5,266.5 kilotons, respectively.

In Table 9, the crude oil export is given. The quantity of crude oil exported grew from 62,474 bbls in 2002 to 36,048,290 bbls in 2013 at a value of US$ 3,885 million. In Table 10, the crude oil import is given. The maximum total import of crude oil made was in 2007 at about 2,000 kilotons. Crude oil imports are for two reasons: for refinery and electricity generation. Majority of crude oil import is used in the refinery. The maximum crude oil import for the refinery was in 2004 at 1,406.2 kilotons, whiles the maximum crude oil import for electricity generation was in 2013 at 927.8 kilotons (Figure 10). In Figure 10, the crude oil export is given.

In Figure 11, natural gas import through the West African Gas Pipeline is given. The maximum natural gas import of almost 31,000,000 MMBtu was made in 2011. In 2012, the total natural gas required to run all the dual-fueled thermal plants in optimum mode was almost 180 million standard cubic feet per day (mmscfd). Nonetheless, only an average of 65 mmscfd was available, in consonance with the forecast of the energy commission for that year. West Africa Gas Pipeline (WAGP) gas flow was truncated in August 2012, due to an accident on the undersea-pipeline in the Togolese waters that very month (International Energy Agency, 2014).

In 2013, the average annual volume of natural gas expected from the West Africa Gas Pipeline (WAGP) is likely to reduce further to about 35–40 mmscfd (35,000–40,000 MMBtu), due to technical and demand challenges being encountered in Nigeria. Nevertheless, domestic gas from the Jubilee field is likely to ramp the annual average up to 45–50 mmscfd by the end of the year. 2013 also saw a commencement of development of other fields neighboring Jubilee, namely Sankofa, TEN, Sankofa East which are expected to bring along more associated gas by 2017–2018 depending upon the

Table 9. Crude oil production (kilotons)												
Year	2002	2003	2004	2005	2006	2007	2008	2009	2010	2011	2012	2013
From saltpond field	8.9	10.3	22.9	11.8	22.9	27.1	30.5	24.8	13.0	10.0	15.1	15.0
From jubilee field	NE[a]	NE	NE	NE	NE	NE	NE	NE	181.1	3,394.0	4,118.7	5,251.5
Total	8.9	10.3	22.9	11.8	22.9	27.1	30.5	24.8	195.0	3,404.8	4,133.8	5,266.5

Note: [a]NE means non-existence.

Source: Ghana National Petroleum Corporation.

Table 10. Crude oil export												
Year	2002	2003	2004	2005	2006	2007	2008	2009	2010	2011	2012	2013
Quantity (bbls)	62,474	71,996	160,115	82,447	160,457	189,378	213,730	173,444	97,642	24,731,475	26,430,934	36,048,290
Value (million US$)	N.A	N.A	N.A	N.A	N.A	N.A	N.A	N.A	N.A	2,779	2,976	3,885

Note: NA means not-available.

Source: Adapted from Bank of Ghana.

timing and rate of the development of the fields. These new fields are projected to yield an average, ranging from 100 to 500 mmscfd from the year 2020 (Ministry of Energy Ghana, 2009).

In Tables 10–13, petroleum products production, petroleum products, import, petroleum products, export and petroleum products supplied to the economy are given. The main petroleum products produced are: Liquid Petroleum Gas (LPG), gasoline, kerosene, Aviation Turbine Kerosene, gas oil, and fuel oil. The maximum amount of LPG was produced in 2005 at 75.3 kilotons; the maximum amount of gasoline was produced in 2004 at 553.1 kilotons; the maximum amount of kerosene was produced in 2008 at 168.6 kilotons; the maximum amount of ATK was produced in 2005 at 119 kilotons; the maximum amount of gas oil was produced in 2004 at 568.4 kilotons; and the maximum amount of fuel oils was produced in 2000 at 261.9 kilotons, respectively (Table 10).

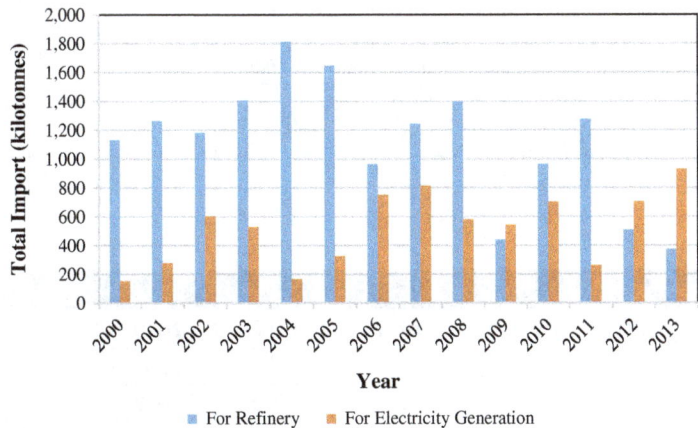

Figure 10. Crude oil import.

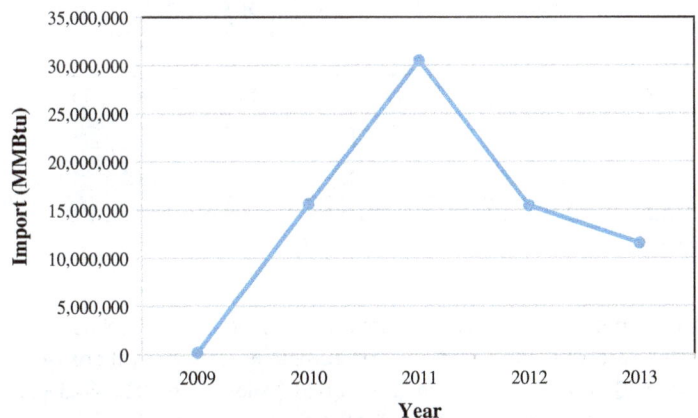

Figure 11. Natural gas import through the West African gas pipeline.

Table 11. Petroleum products production (kilotons)

Year	2000	2001	2002	2003	2004	2005	2006	2007	2008	2009	2010	2011	2012	2013
LPG	9.7	7.0	24.4	52.6	65.5	75.3	35.8	67.3	54.6	14.0	31.6	44.6	26.8	25.6
Gaso-lines	238.6	286.3	346.2	433.8	553.1	567.1	294.4	493.0	391.2	135.0	337.7	344.3	157.7	167.3
Kero-sene	51.8	98.1	61.1	109.6	111.1	87.7	65.1	122.0	168.6	48.7	71.0	52.6	21.1	14.6
ATK	108.3	64.0	81.6	85.6	106.9	119.0	46.2	65.8	21.3	1.3	116.7	116.1	47.6	59.8
Gas Oil	358.1	353.5	446.5	506.6	568.4	486.3	294.2	398.2	360.5	102.8	292.6	309.8	121.5	113.3
Fuel Oils	261.9	261.1	195.7	163.5	199.1	205.4	155.5	48.7	225.4	25.3	96.8	90.6	79.2	43.5

Source: Tema Oil Refinery.

Table 12. Petroleum products import (kilotons)

Year	2000	2001	2002	2003	2004	2005	2006	2007	2008	2009	2010	2011	2012	2013
LPG	35.4	35.6	32.0	16.7	11.0	7.1	67.8	47.2	67.8	150.6	148.0	177.8	241.6	203.9
Premium gasoline	387.0	389.4	370.8	232.1	255.4	167.5	360.5	274.9	254.5	563.4	570.1	712.8	811.5	1,017.4
Kero-sene	30.4	21.5	48.8	34.6	0.0	0.0	99.9	66.7	136.4	77.7	0.0	0.0	0.0	0.0
Gasoil	363.2	354.3	298.0	285.7	313.1	403.7	780.0	806.9	579.0	969.5	871.7	1,200.6	1,309.4	1,638.7
Fuel oil	0.3	0.1	0.1	0.0	0.0	0.0	0.0	0.0	0.0	0.0	0.0	0.0	0.0	44.3
DPK	0.0	0.0	0.0	0.0	0.0	0.0	0.0	0.0	0.0	0.0	0.0	17.5	115.0	0.0
ATK	0.0	0.0	0.0	0.0	0.0	0.0	0.0	55.9	0.0	0.0	0.0	0.0	0.0	65.6

Source: National Petroleum Authority (2015).

Table 13. Petroleum products export (kilotons)

Year	2000	2001	2002	2003	2004	2005	2006	2007	2008	2009	2010	2011	2012	2013
LPG	6.2	1.2	4.5	11.2	6.0	12.5	10.4	9.6	5.0	1.1	0.0	0.0	0.0	0.0
Gas oil	0.6	1.0	1.9	12.0	42.4	37.7	66.1	52.7	88.4	381.9	290.9	356.5	80.8	51.8
Re-sidual fuel oil	190.7	215.7	151.7	89.4	168.9	162.8	45.9	26.2	148.4	30.2	40.6	43.5	44.5	3.7
Heavy gaso-line	97.1	126.7	129.2	103.0	146.5	161.9	99.8	133.7	73.0	20.5	93.6	141.1	54.3	36.0
Pre-mium gaso-line	0.0	0.0	0.0	1.1	4.4	42.0	13.5	30.1	38.8	51.6	119.4	116.9	0.0	0.0
ATK	0.0	0.0	0.0	0.8	0.0	0.1	0.4	2.5	0.3	0.0	0.0	18.0	0.0	0.0

Source: Tema Oil Refinery and National Petroleum Authority.

The maximum petroleum product import for Liquid Petroleum Gas (LPG), premium gasoline, kerosene, gas oil, fuel oil, Dual Purpose Kerosene (DPK), and Aviation Turbine Kerosene (ATK) are 241.6 kilotons in 2012, 1,017.4 kilotons in 2013, 136.4 kilotons in 2008, 1,638.7 kilotons in 2013, 44.3 kilotons in 2013, 115 kilotons in 2012, and 65.6 kilotons in 2013 (Table 11).

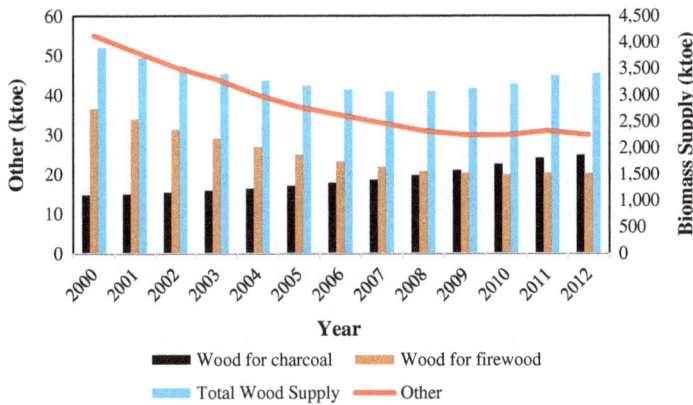

Figure 12. Biomass supply.

The maximum petroleum product export for Liquid Petroleum Gas (LPG), gas oil, residual fuel oil, heavy gasoline, premium gasoline, and Aviation Turbine Kerosene (ATK) are 12.5 kilotons in 2005, 381.9 kilotons in 2009, 215.7 kilotons in 2001, 161.9 kilotons in 2005, 119.4 kilotons in 2010, and 18 kilotons in 2011 (Table 12).

The maximum petroleum product supplied to the economy for Liquid Petroleum Gas (LPG), premix gasoline, premix, kerosene, Aviation Turbine Kerosene (ATK), gas oil, and Residual Fuel Oil (RFO) are; 268.5 kilotons in 2012, 1,080.6 kilotons in 2013, 58.9 kilotons in 2012, 89.3 kilotons in 2009, 141.3 kilotons in 2012, 1,722.6 kilotons in 2013, and 57.1 kilotons in 2000, respectively (Table 13).

6. Biomass

Biomass is Ghana's dominant energy resource in terms of endowment and consumption. Biomass resources cover about 20.8 million hectares of the land mass of Ghana (23.8 million hectares) and is the source of supply of about 60% of total energy used in the country (Energy Commission Ghana, 2013). The enormous arable and degraded land mass of Ghana has the potential for the cultivation of crops and plants which can be converted into a wide range of solid and liquid biofuels Ministry of Energy Ghana, 2009; Reegle, 2015; Sustainable Development Action Plan (SDAP), 2015). Since the mid-1990s, the composition of energy consumption in Ghana has been approximately 71% biomass, 20% crude oil, 8% hydropower, and less than 1% solar energy. In 2000, the composition changed significantly to 60% biomass, 28% oil products, and 11% electricity (Ministry of Energy Ghana, 2009). In Figure 12, the biomass supply is given. The main supply of biomass is in the form of charcoal, firewood, and others (saw dust, saw mill residue, etc.). The maximum biomass supply in the form of charcoal, firewood, and others is as follows: 1,989 ktoe in 2013, 2,742 ktoe in 2000, and 55 ktoe in 2000 which brings the total biomass supply from 3,891 ktoe in 2000 to 3,554 ktoe, respectively (Figure 12). This decline is due to the replacement of biomass with biogas and LPG for cooking and heating purposes.

In Table 14, the biomass consumption is given. Biomass consumption in the form of firewood declines from 2,747 ktoe in 2000 to 1,535 ktoe in 2013; other biomass consumption also declined from 55 ktoe in 2000 to 30 ktoe in 2013. Nevertheless, biomass consumption in the form of charcoal increased from 636 ktoe in 2000 to 1,112 ktoe in 2013. This sharp increase is due to increased energy demand for cooking and heating purposes. Moreover, the cost of charcoal is far cheaper than any of the alternative sources of energy for cooking and heating purposes.

In Table 15, charcoal export is given. The quantity of charcoal export declines from 3.0 kilotons in 2000 to 0.8 kilotons in 2013 representing a growth rate of -61.4%. This decline in export is due to overconsumption by local consumers.

7. Energy prices

High prices of crude oil and petroleum products in the world market also create some difficulties for oil-importing developing countries, including Ghana because when domestic energy prices in developing countries fall below opportunity costs, price increases are recommended to conserve fiscal revenue and to ensure efficient use of resources (Hope, 1995). In Table 16, the average crude oil prices are given. The maximum average price of crude oil in the first quarter was US$ 121.02/ barrel in 2012, the maximum average price of crude oil in the second quarter was US$ 122.84/ barrel in 2008, the maximum average price of crude oil in the third quarter was US$ 116.92/ barrel in 2008, and the maximum average price of crude oil in the fourth quarter was US$ 110.08/ barrel in 2012. In Figure 13, the trends in crude oil prices are given from January, 2005–December, 2013. It is evident that April, 2008, had the highest crude oil price getting closer to almost US$ 140/barrel (Figure 13).

In Table 17, detailed retail prices of major petroleum product are given. A summary of it is displayed in Figure 14. In Figure 14, a summary of retail prices of major petroleum product is given. The major petroleum products are: premium gasoline, gas oil, kerosene, LPG, and refined fuel oils. Apart from a decline in 2009 due to a drastic oil reduction in the world market, there is a constant rise of retail prices in Ghana with increasing years (Figure 14).

In Table 18, the average electricity end-user tariff is given. There is a rapid increase in average end-user tariff from 0.017 GHS/kWh in 2000 to 0.307 GHS/kWh in 2013.

Average prices of charcoal in the Ghana rose from GH¢9 per mini bag and GH¢15 per maxi bag in 2011 to GH¢11 per mini bag and GH¢18 per maxi bag in 2012. Regions with the high price of charcoal in 2012 were Western and Central. Regions with the low price were Northern, Brong-Ahafo, and Ashanti. Northern region saw a drop in the average mini bag price of charcoal. There was a drop in

Table 14. Petroleum products supplied to the economy (kilotons)

Year	2000	2001	2002	2003	2004	2005	2006	2007	2008	2009	2010	2011	2012	2013
LPG	45.0	42.5	50.0	56.7	65.7	70.5	88.0	93.3	117.6	220.6	178.4	214.4	268.5	251.8
Premium gasoline	524.4	535.1	570.2	479.8	575.6	537.8	511.9	544.2	545.0	701.4	737.8	807.0	992.7	1,080.6
Premix	30.6	27.0	26.8	28.9	27.5	31.4	33.7	41.0	50.7	55.1	32.4	45.6	58.9	53.4
Kerosene	67.6	70.5	74.8	68.8	73.2	74.3	76.5	63.3	34.6	89.3	49.3	62.4	45.6	27.8
ATK	96.9	76.4	90.5	89.8	107.4	119.3	114.7	122.8	119.2	124.7	108.4	135.3	141.3	131.9
Gas oil	665.8	685.4	717.8	755.3	848.9	880.4	934.0	1,147.0	1,092.1	1,280.0	1,271.9	1,431.2	1,665.0	1,722.6
RFO	57.1	52.0	51.9	45.7	45.2	47.8	56.8	51.3	47.9	40.3	30.9	37.5	33.5	39.3

Source: National Petroleum Authority.

Table 15. Biomass consumption (ktoe)

Year	2000	2001	2002	2003	2004	2005	2006	2007	2008	2009	2010	2011	2012	2013
Firewood	2,742	2,539	2,350	2,176	2,017	1,873	1,742	1,644	1,566	1,520	1,490	1,535	1,520	1,535
Charcoal	636	649	684	705	782	835	894	917	921	943	944	1,010	1,039	1,112
Other	55	51	47	44	40	37	35	33	31	30	30	31	30	30
Total biomass	3,432	3,238	3,082	2,925	2,839	2,745	2,671	2,594	2,518	2,493	2,464	2,576	2,589	2,676

Table 16. Charcoal export (kilotons)													
Year	2000	2001	2002	2003	2004	2005	2006	2007	2008	2009	2010	2011	2012	2013
Quan-tity	3.0	2.8	3.5	4.6	4.6	5.7	2.9	3.6	2.9	4.3	1.4	0.8	2.0	0.8
Growth rate (%)	–	−6.7	25.0	31.4	0.0	23.9	−49.1	24.1	−19.4	48.3	−67.4	−42.9	150.0	−61.4

the average prices for maxi bag in Greater Accra. Average charcoal price for a mini bag doubled in the Upper East and Upper West Regions. Eastern and Western Regions also experienced significant charcoal price increment. Energy Commission of Ghana estimated that an average charcoal price in 2013 could range between 20 and 25% over the 2012 average price nationwide due to a general

Table 17. Average crude oil prices (US$/barrel)									
Year	2005	2006	2007	2008	2009	2010	2011	2012	2013
First quarter	48.00	62.67	58.63	96.47	45.56	77.19	105.18	121.02	112.64
Second quarter	52.89	70.43	68.67	122.84	59.71	79.44	117.19	110.38	103.31
Third quarter	61.83	70.53	74.67	116.92	69.01	76.94	112.15	109.67	109.61
Fourth quarter	57.75	60.89	88.68	57.31	75.54	87.32	109.04	110.08	109.27

Source: Adapted from Bank of Ghana.

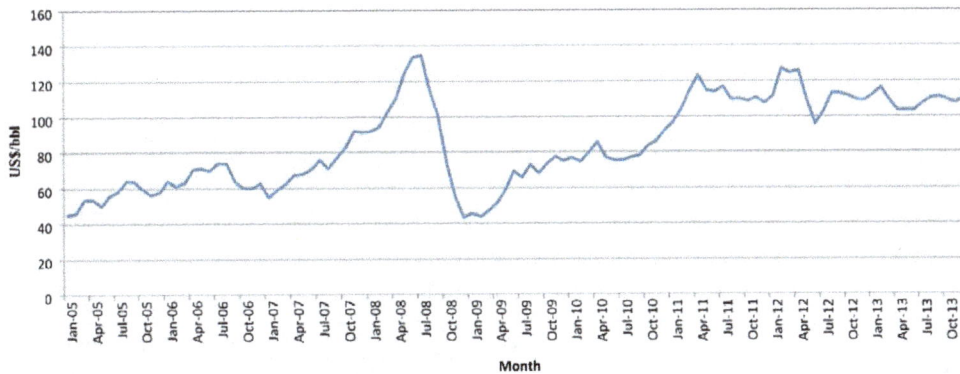

Figure 13 Trend in crude oil prices since January, 2005–December, 2013 (Energy Commission Ghana, 2013).

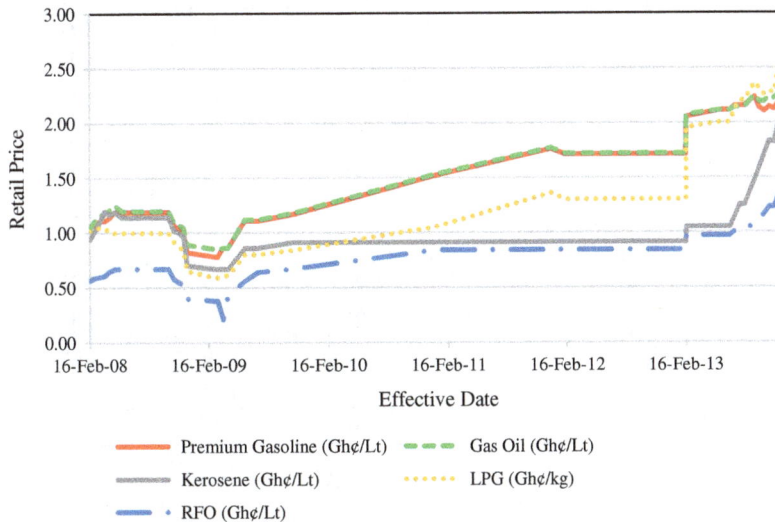

Figure 14. Retail prices of major petroleum product.

Table 18. Retail prices of major petroleum products (Energy Commission Ghana, 2013)					
Effective date	Exchange rate (Gh¢/US$)	Premium gasoline (Gh¢/Lt)	Gas oil (Gh¢/Lt)	Kerosene (Gh¢/Lt)	LPG (Gh¢/kg)	RFO (Gh¢/Lt)
16-Feb-08	0.98	1.04	1.04	0.94	1.02	0.57
01-Mar-08	0.98	1.09	1.11	1.01	1.04	0.59
16-Mar-08	0.98	1.11	1.16	1.09	1.05	0.60
01-Apr-08	0.98	1.11	1.18	1.20	1.05	0.61
16-Apr-08	0.98	1.14	1.21	1.17	1.01	0.65
03-May-08	0.98	1.19	1.25	1.19	1.00	0.67
26-May-08	0.98	1.19	1.20	1.14	1.00	0.67
16-Oct-08	1.14	1.19	1.20	1.14	1.00	0.67
01-Nov-08	1.15	1.07	1.10	1.02	0.92	0.58
16-Nov-08	1.16	1.03	1.08	1.00	0.88	0.55
01-Dec-08	1.17	0.99	1.04	0.97	0.84	0.53
12-Dec-08	1.20	0.82	0.89	0.70	0.65	0.40
09-Mar-09	1.33	0.78	0.85	0.67	0.59	0.38
16-Mar-09	1.36	0.78	0.85	0.67	0.59	0.38
01-Apr-09	1.38	0.86	0.86	0.67	0.61	0.21
16-Apr-09	1.40	0.86	0.86	0.67	0.61	0.43
06-Jun-09	1.44	1.11	1.12	0.86	0.80	0.56
16-Jul-09	1.49	1.11	1.12	0.86	0.80	0.64
31-Oct-09	1.45	1.17	1.18	0.91	0.84	0.67
04-Jan-11	1.46	1.52	1.53	0.91	1.05	0.84
29-Dec-11	1.55	1.76	1.77	0.91	1.36	0.84
11-Feb-12	1.66	1.71	1.72	0.91	1.30	0.84
01-Mar-12	1.68	1.71	1.72	0.91	1.30	0.84
16-Mar-12	1.68	1.71	1.72	0.91	1.30	0.84
01-Apr-12	1.68	1.71	1.72	0.91	1.30	0.84
16-Apr-12	1.70	1.71	1.72	0.91	1.30	0.84
01-May-12	1.71	1.71	1.72	0.91	1.30	0.84
16-May-12	1.71	1.71	1.72	0.91	1.30	0.84
01-Jun-12	1.73	1.71	1.72	0.91	1.30	0.84
16-Jun-12	1.83	1.71	1.72	0.91	1.30	0.84
01-Jul-12	1.87	1.71	1.72	0.91	1.30	0.84
16-Jul-12	1.88	1.71	1.72	0.91	1.30	0.84
01-Aug-12	1.88	1.71	1.72	0.91	1.30	0.84
16-Aug-12	1.89	1.71	1.72	0.91	1.30	0.84
01-Sep-12	1.89	1.71	1.72	0.91	1.30	0.84
16-Sep-12	1.89	1.71	1.72	0.91	1.30	0.84
01-Oct-12	1.89	1.71	1.72	0.91	1.30	0.84
16-Oct-12	1.89	1.71	1.72	0.91	1.30	0.84
01-Nov-12	1.89	1.71	1.72	0.91	1.30	0.84
16-Nov-12	1.88	1.71	1.72	0.91	1.30	0.84
01-Dec-12	1.88	1.71	1.72	0.91	1.30	0.84

Table 18. (Continued)

Effective date	Exchange rate (Gh¢/US$)	Premium gasoline (Gh¢/Lt)	Gas oil (Gh¢/Lt)	Kerosene (Gh¢/Lt)	LPG (Gh¢/kg)	RFO (Gh¢/Lt)
16-Dec-12	1.88	1.71	1.72	0.91	1.30	0.84
01-Jan-13	1.89	1.71	1.72	0.91	1.30	0.84
16-Jan-13	1.89	1.71	1.72	0.91	1.30	0.84
01-Feb-13	1.89	1.71	1.72	0.91	1.30	0.84
16-Feb-13	1.89	1.71	1.72	0.91	1.30	0.84
17-Feb-13	1.89	2.05	2.07	1.05	1.95	0.97
01-Jun-13	1.92	2.11	2.11	1.05	2.00	0.97
16-Jun-13	1.94	2.11	2.11	1.05	2.00	0.97
01-Jul-13	1.99	2.11	2.11	1.05	2.00	0.97
17-Jul-13	1.99	2.15	2.14	1.15	2.12	1.01
01-Aug-13	1.99	2.15	2.14	1.25	2.20	1.05
16-Aug-13	2.00	2.15	2.14	1.25	2.20	1.05
16-Sep-13	2.00	2.23	2.24	1.48	2.36	1.05
01-Oct-13	2.00	2.13	2.19	1.59	2.31	1.11
16-Oct-13	2.00	2.10	2.19	1.71	2.25	1.17
01-Nov-13	2.01	2.14	2.23	1.83	2.28	1.23
16-Nov-13	2.04	2.12	2.22	1.82	2.27	1.23
01-Dec-13	2.07	2.19	2.26	2.02	2.52	1.35
16-Dec-13	2.09	2.19	2.26	2.02	2.52	1.35

increase in Liquid Petroleum Gas price, an alternative but cleaner cooking fuel. Increases could rise between 30 and 35% on an average in the southern sector due to Liquid Petroleum Gas (LPG) supply shortages and a further expected rise in transportation and labor costs (Energy Commission Ghana, 2013).

8. The way forward

The provision of energy services contributes directly or indirectly to poverty alleviation. In order to ensure access to affordable, reliable, sustainable, and modern energy for all, Ghana has begun expanding her economy with the growing Ghanaian population as a way to meet the Sustainable Development Goal 1, which seeks to end poverty and improve well-being (Griggs et al., 2013). Increasing the country's prosperity and sustainable modern way of life requires improved universal, affordable access to clean energy that minimizes local pollution and health impacts and mitigates global warming (Griggs et al., 2013). In order to stimulate economic growth and reduce poverty, the Government of Ghana has, since 2001, initiated the Ghana Poverty Reduction Strategy (GPRS) as a strategic framework to tackle both economic growth and poverty reduction. The Ghana Poverty

Table 19. Average electricity end-user tariff

Year	2000	2001	2002	2003	2004	2005	2006	2007	2008	2009	2010	2011	2012	2013
Average end user tariff (Gh¢/kWh)	0.017	0.034	0.065	0.071	0.074	0.073	0.078	0.097	0.148	0.148	0.211	0.245	0.232	0.307
Exchange rate (Gh¢/US$)	0.70	0.73	0.84	0.88	0.90	0.91	0.92	0.97	1.20	1.43	1.45	1.55	1.88	1.97
Average end user tariff (US$/kWh)	0.024	0.047	0.077	0.080	0.082	0.080	0.084	0.100	0.123	0.104	0.145	0.158	0.124	0.156

Source: Bank of Ghana.

Reduction Strategy was to transform Ghana's low-income status to middle-income status with a target of about US$ 1,000 per capita by 2012 (Ministry of Energy Ghana, 2009). Such a transformation requires energy-intensive economic activities, such as job creation, improving quality of education, and reaching out to communities where the national electricity grid is inaccessible, through decentralized sustainable energy systems which are already in place (Ministry of Energy).

There are a number of intervention strategies by Ghana's Energy sector which provides new, high-quality, and cost-competitive energy services to poor people and communities, thus alleviating poverty. There are provisions of services to support small, local, income generating activities, thereby enhancing employment opportunities, such as the provision of jobs in industries that result from the macroeconomic growth enabled by access to energy services, infrastructure, and technical capacity (Ministry of Energy Ghana, 2009).

The Ghana energy sector has initiated the universal access to electricity since 1988 known as the National Electrification Scheme (NES) at a time when access to electricity by the whole population was 33% (International Energy Agency, 2014). The initiative deems to support the economic recovery program which had been initiated by the Government to increase the overall socioeconomic development of the nation. The extension of electricity to rural areas was a way to open up the country for socioeconomic development thereby slowing down rural–urban migration. Currently, all regional and district capitals have been connected to the national grid system since its inception (Energy Commission Ghana, 2013).

As a follow-up to the National Electrification Scheme, a Self-Help Electrification Program (SHEP) was initiated with the sole aim of assisting rural communities to get access to electricity. Since its inception, Ghana's population access to electricity has increased from 33% to 76% (Ghana Energy Commission, 2015).

The Energy Commission of Ghana initiated a National Off-grid Rural Electrification Program (OEP) which targets remote communities for the provision of electricity through renewable energy technologies. The aim of the initiative was to achieve a substantial level of penetration of solar electrification as a platform for the promotion of solar photovoltaic (PV) systems for basic lighting in rural off-grid communities. The initiative established solar battery charging service centers for the promotion of solar photovoltaics (PVs), thereby endorsing ownership of solar home systems. The initiative led to the establishment of a solar photovoltaic (PV) market in Ghana while improving the socioeconomic conditions in rural communities as a result of the extension of electricity coverage in rural areas. Currently, there are more than solar battery charging stations and over 5,000 solar home systems in Ghana. The National Off-grid Rural Electrification Program was scheduled for implementation in six phases throughout the country for a period of 15 years. A total of 19,000 communities was targeted for electrification and 2,000 satellite solar battery charging centers are planned for installation to serve communities within a 5 km radius. Table 19 shows the phases of the off-grid solar electrification program (Table 20).

Ghana's Energy sector continues to improve access to electricity by the introduction of the Eight Hundred and Thirty-Second ACT of the Parliament of the Republic of Ghana entitled: Renewable Energy Act, 2011, which has assented to provide for the development, management, utilization,

Table 20. Off-grid solar electrification program (Ghana Energy Commission, 2015)					
Phase 1	Phase 2	Phase 3	Phase 4	Phase 5	Phase 6
2005–2005	2006–2008	2009–2011	2012–2014	2015–2017	2018–2020
1-year	3-years	3-years	3-years	3-years	3-years
14 towns	2000 towns	2000 towns	4000 towns	800 towns	2986 towns
0.1% penetration rate	10% penetration rate	10% penetration rate	22% penetration rate	42% penetration rate	16% penetration rate

sustainability, and adequate supply of renewable energy for the generation of heat and power by the year 2020. Ghana's Renewable Energy Development Program aims to: (i) assess the availability of renewable energy resources; (ii) to examine the technical feasibility and cost-effectiveness of promising renewable energy technologies; (iii) to ensure the efficient production and use of the country's renewable energy resources; and (iv) develop the relevant information base that will facilitate the establishment of a planning framework for the rational development and use of the country's renewable energy resources (Ghana Energy Commission, 2011).

9. Conclusion

In this study, a review of Ghana's energy sector national energy statistics and policy framework was done to create an awareness of the strategic planning and energy policies of Ghana's energy sector to serve as a useful tool for both local and foreign investors, help in national decision-making for the efficient development and utilization of energy resources. The review of Ghana's energy sector policy answered the question, what has been done so far? And what is the way forward? The future research in Ghana cannot progress without consulting the past and institutional capabilities. Both qualitative and quantitative research methodologies were employed in the study. As part of the research techniques, a secondary data on the national energy statistics were adopted from the archives of the Ministry of Energy and Petroleum, the Volta River Authority (VRA), the Ghana Grid Company (GRIDCo), Ghana National Petroleum Corporation (GNPC), the National Petroleum Authority (NPA), Tema Oil Refinery (TOR), the Public Utility Regulatory Commission (PURC), the Electricity Company of Ghana (ECG), the Northern Electricity Distribution Company (NEDCo), the West African Gas Pipeline Company (WAPCo), the West African Gas Pipeline Authority (WAGPA), the Bank of Ghana (BoG), and the Ghana Statistical Service (GSS). In this case, concepts, experiences, techniques, information gathered from the actors in the field of the study, peer-reviewed journals and other literature relevant to this study were analyzed and reviewed. The study brought to bear a number of intervention strategies by Ghana's Energy sector which provides new, high-quality, and cost-competitive energy services to poor people and communities, thus alleviating poverty. The summary of findings is as follows:

- With a correlation coefficient r (8)=0.98, $p < 0.05$, Ghana's increasing Population has a positive correlation with the increasing Gross Domestic Product.
- Total Energy Generated increased from 8,430 GWh in 2006 to 12,870 GWh in 2013.
- Total Final Energy Consumed, that is, the energy which is not being used for transformation into other forms of energy, increased from 5,177 ktoe to 6,886 ktoe.
- The total electricity consumption increased from 6,367 GWh in 2000 to 9,355 GWh in 2013.
- There was a decline in the Energy Intensity of the Economy from 0.28 TOE/GHS 1,000 of GDP to 0.21 TOE/GHS 1,000 of GDP.
- Ghana's primary energy supply is made up of oil, hydro, wood, and natural gas.
- Final electricity consumption increased from 597 ktoe in 2000 to 910 ktoe in 2013.
- Biomass consumption declined from 3,432 ktoe in 2000 to 2,588 ktoe in 2013 due to a switch in end-user preferences to Liquid Petroleum Gas.
- Import of electricity decreased from 864 GWh in 2000 to 27 GWh in 2013
- Ghana's installed capacity from hydropower is 1,580 MW (Akosombo-1,020 MW, Bui-400 MW and Kpong-160 MW); thermal generation constitutes 1,494 MW; embedded generation (Genser Power) constitutes 5 MW; and renewables (Volta River Authority installed Solar) constitute 2.5 MW. In total, Ghana's current installed capacity as of December, 2013, is 3,081 MW.

It was evident in the study that the current shortage in power supply prevalent in Ghana originates from inadequate and unreliable fuel supply for the operation of the thermal power plant, transmission, and distribution losses. There is the need for expansion of power generation capacity to meet the growing power demand. In the light of this, a research on how Ghana's energy portfolio can be upgraded to meet the increasing energy demand is encouraged.

Acknowledgment

The authors are grateful to the Ghana Energy Commission for making the National Energy Policy document available and accessible. The Editor (Shashi Dubey) and the anonymous reviewers are commended for their useful comments and suggestions. Any errors are the sole responsibility of the authors.

Funding

The authors received no direct funding for this research.

Author details

Samuel Asumadu-Sarkodie[1]
E-mail: samuel.sarkodie@metu.edu.tr
Phebe Asantewaa Owusu[1]
E-mail: phebe.owusu@metu.edu.tr
[1] Sustainable Environment and Energy Systems, Middle East Technical University, Northern Cyprus Campus, Guzelyurt, Turkey.

References

Abosedra, S., & Baghestani, H. (1989). New evidence on the causal relationship between United States energy consumption and gross national product. *Journal of Energy Development, 14*, 285–292.

AC00804425, A. (1993). *The East Asian miracle: Economic growth and public policy.* Oxford University Press.

Asafu-Adjaye, J. (2000). The relationship between energy consumption, energy prices and economic growth: Time series evidence from Asian developing countries. *Energy Economics, 22*, 615–625. doi:10.1016/S0140-9883(00)00050-5

Asif, M., & Muneer, T. (2007). Energy supply, its demand and security issues for developed and emerging economies. *Renewable and Sustainable Energy Reviews, 11*, 1388–1413.
http://dx.doi.org/10.1016/j.rser.2005.12.004

Asumadu-Sarkodie, S., & Owusu, P. (2016). Feasibility of biomass heating system in Middle East Technical University, Northern Cyprus Campus. *Cogent Engineering.* doi:10.1080/23311916.2015.1134304

Asumadu-Sarkodie, S., & Owusu, P. A. (2015). Media impact on students' body image. *International Journal for Research in Applied Science and Engineering Technology, 3*, 460–469.

Asumadu-Sarkodie, S., & Owusu, P. A. (2016a). Multivariate co-integration analysis of the Kaya factors in Ghana. *Environmental Science and Pollution Research.* doi:10.1007/s11356-016-6245-9

Asumadu-Sarkodie, S., & Owusu, P. A. (2016b). The relationship between carbon dioxide and agriculture in Ghana, a comparison of VECM and ARDL model. *Environmental Science and Pollution Research.* doi:10.1007/s11356-016-6252-x

Asumadu-Sarkodie, S., & Owusu, P. A. (2016c). The potential and economic viability of solar photovoltaic in Ghana. *Energy Sources, Part A: Recovery, Utilization, and Environmental Effects.* doi:10.1080/15567036.2015.1122682

Asumadu-Sarkodie, S., & Owusu, P. A. (2016d). The potential and economic viability of wind farm in Ghana. *Energy Sources, Part A: Recovery, Utilization, and Environmental*

Effects. doi:10.1080/15567036.2015.1122680

Asumadu-Sarkodie, S., Owusu, P. A., & Jayaweera, H. M. (2015). Flood risk management in Ghana: A case study in Accra. *Advances in Applied Science Research, 6*, 196–201.

Asumadu-Sarkodie, S., Owusu, P. A., & Rufangura, P. (2015). Impact analysis of flood in Accra, Ghana. *Advances in Applied Science Research, 6*, 53–78.

Asumadu-Sarkodie, S., Rufangura, P., Jayaweera, H. M., & Owusu, P. A. (2015). Situational analysis of flood and drought in Rwanda. *International Journal of Scientific and Engineering Research, 6*, 960–970. doi:10.14299/ijser.2015.08.013
http://dx.doi.org/10.14299/000000

Beg.utexas.edu. (2005). *Guide to electric power in Ghana.* Retrieved from http://www.beg.utexas.edu/energyecon/IDA/USAID/RC/Guide_to_Electric%20Power_in_Ghana.pdf

Bhattacharyya, S. C. (2011). Energy data and energy balance. In *Energy economics* (pp. 9–39). Springer.

Cheng, B. S. (1995). An investigation of cointegration and causality between energy consumption and economic growth. *Journal of Energy and Development, 21*(1).

Choi, H., Park, S., & Lee, J.-d. (2011). Government-driven knowledge networks as precursors to emerging sectors: A case of the hydrogen energy sector in Korea. *Industrial and Corporate Change,* dtr002.

Climate Everyone's Business. (2014). Climate change: Implications for the energy sector. Retrieved from https://www.worldenergy.org/wp-content/uploads/2014/06/Climate-Change-Implications-for-the-Energy-Sector-Summary-from-IPCC-AR5-2014-Full-report.pdf

Dowrick, S. (1994). The East Asian miracle: Economic growth and public policy. *Economic Record, 70*, 469.

Eden, S., & Hwang, B.-K. (1984). The relationship between energy and GNP: Further results. *Energy Economics, 6*, 186–190.

Edenhofer, O., Pichs-Madruga, R., Sokona, Y., Seyboth, K., Kadner, S., Zwickel, T., ... von Stechow, C. (2011). *Renewable energy sources and climate change mitigation.* Cambridge: Cambridge University Press.
http://dx.doi.org/10.1017/CBO9781139151153

Electricity Company of Ghana. (2012). *Electricity Company of Ghana (ECG) ECOWREX.* Retrieved from http://www.ecowrex.org/stakeholder/electricity-company-ghana-ecg

Ghana Energy Commission (2011). *Ghana renewable energy bill.* Retrieved 17 October, 2014, from http://www.ecowrex.org/system/files/documents/2011_renewable-energy-bill_government-of-ghana.pdf

Ghana Energy Commission. (2012). *The SWERA Ghana Project.* Retrieved from http://www.energycom.gov.gh/downloads/Technical%20Reports/SWERA%20-%20National%20Report.pdf

Ghana Energy Commission. (2015). Ghana energy commission report. Retrieved 29 May, 2015, from http://www.energycom.gov.gh/

Energy Commission Ghana. (2013). *Energy commission supply and demand 2013.* Retrieved from http://www.energycom.gov.gh/files/Energy%20Commission%20-%202013%20Energy%20Outlook%20for%20Ghana.pdf

Volta Aluminium (Ghana.Valcotema.com). (2015). Retrieved 8 June, 2015, from http://www.valcotema.com/

Griggs, D., Stafford-Smith, M., Gaffney, O., Rockström, J., Öhman, M. C., Shyamsundar, P., ... Noble, I. (2013). Policy: Sustainable development goals for people and planet. *Nature, 495*, 305–307. Retrieved from http://www.nature.com/nature/journal/v495/n7441/abs/495305a.html#supplementary-information
http://dx.doi.org/10.1038/495305a

Helm, D. (2002). Energy policy: Security of supply, sustainability and competition. *Energy Policy, 30*, 173–184.
http://dx.doi.org/10.1016/S0301-4215(01)00141-0

Hogan Lovells. (2013). *The power market in Ghana.*

Retrieved from http://www.hoganlovells.com/files/
Publication/1502043f-6c0f-4cc3-81ef-65365d230b24/
Presentation/PublicationAttachment/a071edc5-abc7-
46bf-9a13-6ae1f369d559/The%20Power%20Market%20
in%20Ghana.pdf

Hope, E. (1995). *Energy price increases in developing countries:
Case studies of Colombia, Ghana, Indonesia, Malaysia,
Turkey, and Zimbabwe.* World Bank Policy Research
Working Paper (1442).

International Energy Agency. (2010). *Natural gas annual
questionnaire.* Retrieved from http://wds.iea.org/wds/pdf/
Gas_documentation.pdf

International Energy Agency. (2014). World energy outlook
special report. In I. E. Agency (Ed.), *Africa energy outlook.*

International Energy Agency. (2015). *Energy access database.*
Retrieved from http://www.worldenergyoutlook.org/
resources/energydevelopment/energyaccessdatabase/

IPCC (2011). *Special report on renewable energy sources and
climate change mitigation.* New York, NY: Cambridge
University Press.

Jørgensen, U. (2005). Energy sector in transition—technologies
and regulatory policies in flux. *Technological Forecasting
and Social Change, 72,* 719–731.
http://dx.doi.org/10.1016/j.techfore.2004.12.004

Lee, C.-C. (2006). The causality relationship between energy
consumption and GDP in G-11 countries revisited. *Energy
Policy, 34,* 1086–1093.
http://dx.doi.org/10.1016/j.enpol.2005.04.023

Lipton, M., & Ravallion, M. (1993). *Poverty and policy* (Vol. 1130):
World Bank Publications.

Mauritius, S. (2014). *Energy and water statistics-2013.*
Retrieved from http://statsmauritius.govmu.org/English/
Publications/Documents/Regular%20Reports/energy%20
and%20water/Energy2013.pdf

Meng, M., & Niu, D. (2015). The relationship between energy
consumption and economic growth in China: An
application of the partial least squares method. *Energy
Sources, Part B: Economics, Planning, and Policy, 10,* 75–81.
http://dx.doi.org/10.1080/15567249.2011.604067

Ministry of Energy Ghana. (2009). *National energy policy
(Revised).* Retrieved from http://www.energymin.gov.gh/

Narayan, P. K., & Smyth, R. (2008). Energy consumption and
real GDP in G7 countries: New evidence from panel

cointegration with structural breaks. *Energy Economics,
30,* 2331–2341. doi:10.1016/j.eneco.2007.10.006

National Petroleum Authority. (Npa.gov.gh). (2015). Retrieved
8 June, 2015, from http://www.npa.gov.gh/npa_new/
index.p

Oh, W., & Lee, K. (2004). Causal relationship between energy
consumption and GDP revisited: The case of Korea
1970–1999. *Energy Economics, 26,* 51–59. doi:10.1016/
S0140-9883(03)00030-6

Omer, A. M. (2008). Energy, environment and sustainable
development. *Renewable and Sustainable Energy Reviews,
12,* 2265–2300. doi:10.1016/j.rser.2007.05.001

Reegle. (2015). *Clean energy information gateway.*
Retrieved 8 June, 2015, from http://www.reegle.info/
policy-and-regulatory-overviews/GH

Stern, D. I. (2000). A multivariate cointegration analysis of
the role of energy in the US macroeconomy. *Energy
Economics, 22,* 267–283.
http://dx.doi.org/10.1016/S0140-9883(99)00028-6

Summit, E. (1997). *5: Programme for the further
implementation of Agenda 21.*

Sustainable Development Action Plan. (2015). *Securing the
future for the next generation of Ghanaians.* National
Programme on Sustainable Consumption and Production.
Retrieved from http://www.unep.org/roa/Portals/137/
Docs/pdf/Thematic/RE/Ghana%20SDAP%20Final%20
Report%20Volume%202.pdf

SWERA. (2012). SWERA/Data Open Energy Information.
Retrieved 8 June, 2015, from http://en.openei.org/wiki/
SWERA/Data

United Nations. (1992, June 3–14). *Agenda 21: Programme
of action for sustainable development.* United Nations
Conference on Environment and Development, Chapters
3, 4, 24, 25, and 26.

Vera, I., & Langlois, L. (2007). Energy indicators for sustainable
development. *Energy, 32,* 875–882.
http://dx.doi.org/10.1016/j.energy.2006.08.006

Volta River Authority Ghana. (2015). *VRA completes first solar
plant.* Retrieved 29 May, 2015, from http://www.vra.com

Webapps01.un.org. (2012). *Ghana poverty reduction strategy.*
Retrieved 8 June, 2015, from http://webapps01.un.org/

nvp/indpolicy.action?id=136

3

A review of Ghana's water resource management and the future prospect

Phebe Asantewaa Owusu[1], Samuel Asumadu-Sarkodie[1]* and Polycarp Ameyo[2]

*Corresponding author: Samuel Asumadu-Sarkodie, Sustainable Environment and Energy Systems, Middle East Technical University, Northern Cyprus Campus, Guzelyurt, 99738 TRNC, Turkey
E-mail: samuel.sarkodie@metu.edu.tr
Reviewing editor: Shashi Dubey, Hindustan College of Engineering, India

Abstract: Water covers about 70% of the earth's surface and it exists naturally in the earth in all the three physical states of matter and it is always moving around because the water flows with the current. Out of the earth's percentage of water covering the surface, only about 2.5% is fresh water and due to the fact that most are stored in deep groundwater, a small amount is readily available for human use. Water scarcity is becoming a major concern for people around the world and the need to protect the existing ones and find ways or means to provide safe water for individuals around the globe in adequate quantities with keeping the needs of future generations in mind. Water is life, and it is linked with lots of services either directly or indirectly, such as; human health and welfare and social and economic development of a community or country. The need to delve into Ghana's water resources management is essential. The study reviewed existing literature on the various members of the Water Resource Commission (WRC) in Ghana; the various basins in the country; the existing measures that the WRC authorities have in place to deal with water resources management issues; the challenges that hinder the progress of their achievements and some suggestions that if considered can improve the current water resources management situations in Ghana.

ABOUT THE AUTHORS

Phebe Asantewaa Owusu studies Masters in Sustainable Environment and Energy Systems at Middle East Technical University, Northern Cyprus Campus where she's also a Graduate Assistant in the Chemistry Department. Her research interest includes but not limited to: proteomics, econometrics, energy economics, climate change and sustainable development.

Samuel Asumadu-Sarkodie is a multidisciplinary researcher who currently studies Masters in Sustainable Environment and Energy Systems at Middle East Technical University, Northern Cyprus Campus where he's also a Graduate Assistant in the Chemistry Department. His research interest includes but not limited to: renewable energy, econometrics, energy economics, climate change and sustainable development.

Polycarp Ameyo was an undergraduate student in the Civil Engineering Department at Middle East Technical University, Northern Cyprus Campus who completed in 2016.

PUBLIC INTEREST STATEMENT

Ghana as a country has a standing history with small scale mining of gold, it existed as far back as the eighth century as a household economic activity. In spite of these influences, the industry has a number of negative effects on the environment and most significantly water bodies. River bodies that were the main source of water for drinking, household chores and other activities have been destroyed by these activities of illegal small scale mining of gold. The paper reviewed existing literature on the basins in Ghana, the existing measures put in place by Ghana Water Commission to deal with water resources management issues and the challenges that hinder the progress of their achievement. Recommendations are made to ensure the efficient use of water across all the sectors that ensures sustainable water withdrawals, the demand and supply of freshwater in order to address water scarcity in the future prospects.

Subjects: Civil, Environmental and Geotechnical Engineering; Pollution; Water Engineering; Water Science

Keywords: water resources management; Ghana; basins; illegal mining; water weeds; renewable energy; climate change; groundwater; environmental sustainability engineering

1. Introduction

Water resource management refers to the skilful and efficient planning of the scarce fresh water resources available for the consumption of the entire human population. These water resources are composed of all surface water and underground water that can be treated for human use. In water resource engineering the focus is mainly turned towards the conception planning design tendering construction and operation of the water resource systems such as dams, water treatment systems, water supply systems waste water collection and removal systems among others (Yanmaz, 2013).

Proper management of water resources is necessary for quality control operation and treatment of water resource systems. According to Pahl-Wostl (2007), having an uncontrolled urbanization in developing and threshold countries will lead to excessive pressure on the available water resources. Water has to be controlled in order to prevent loss of lives, excessive damage to private and public properties. As such the branches of urban hydrology and hydraulics in civil engineering deals with the design and proper planning of every kind of control facilities which includes storm drainage systems, measures of flood mitigation like dams and diversion weirs, design of main road culvert systems among others (Yanmaz, 2013).

The bio-assimilation strategy which involves the use of physical and biological processes that are employed to reduce, convert and store pollutants on land before they are released to the aquatic system is considered to be the most ecological, sound sustainable and cost effective approach that can be used to restore water quality conditions in the low lying streams (Osborne & Kovacic, 1993). This helps a great deal in making sure that river basins and streams are protected from pollution and can be effectively used for various purpose.

Water is all around us yet not every individual or country has access to potable water that meets their basic needs (health, sanitation and recreational activities) and development. The earth's surface is covered with about 70% of water and it is the only substance that exist naturally on earth in all the three physical states of matter (solid, liquid and gas) and it is always on the move. With all the water on earth only about 2.5% is fresh water and because most of the water is stored in deep groundwater, only a small amount is easily available for human use. Therefore, ensuring that adequate supply of water is available for human use is essential (Oki & Kanae, 2006). The aim of the study is to review existing literature on the various members of the Water Resource Commission (WRC) in Ghana; the various basins in the country; the existing measures that the WRC authorities have in place to deal with water resources management issues; the challenges that hinder the progress of their achievements and some suggestions that if considered can improve the current water resources management situations in Ghana. In essence, the study will inform the public on the importance of improving water quality by reducing natural and artificial pollution and inform Government on the need to protect the water ecosystems towards achieving equitable access to safe and affordable drinking water.

The increasing number of population around the globe, especially in developing countries poses a threat to the water resource and whether it would be sustained for future generations hence the need to manage the resource. As water moves with time and space steadily with the hydrologic cycle, the term 'water management' extends over a wide variety of activities and discipline. Water management can be divided into three classifications: managing the resource, managing water services, and managing the trade-offs needed to balance supply and demand (United Nations, 2014). Water resource management is of great importance to countries and the world at large of which Ghana is no exception.

The management of water resources in Ghana is regulated by the WRC of Ghana. The WRC of Ghana was established by an Act of Parliament (Act 522 of 1996) with the obligation to regulate and mange Ghana's water resources and co-ordinate government policies in relation to them. Act 522 of 1996 specifies that ownership and control of all waters are vested in the President on behalf of the people, and plainly defines the WRC as the overall body responsible for the management of water resources in Ghana (WRC Ghana, 2015a).

The aim of the study is to review Ghana's water resource management in order to create awareness on water conservation and the efficient use of water across all the sectors that ensures sustainable water withdrawals, the demand and supply of freshwater in order to address water scarcity in the future prospects.

1.1. Members of the WRC
The WRC, Ghana is made up of 15 members, including the Chairperson, the Executive Secretary, a Chief and two other persons, of whom one shall be a woman (WRC Ghana, 2015a). The rest of the representative institutions are depicted in Figure 1.

1.2. Basins in Ghana
According to the WRC of Ghana, Ghana has five basins namely (Figure 2): the Densu River basin, Ankobra basin, Pra basin, Tano basin and White Volta basin (WRC Ghana, 2015c).

1.2.1. Densu River basin
The Densu basin is located in the south-eastern part of Ghana and lies within longitude 10° 30′W–10° 45′W and latitude 50° 45′N–60° 15′N. It shares its catchment boundary with the Odaw and Volta basins to the east and north, the Birim in the north-west and Ayensu and Okrudu in the west. The basin

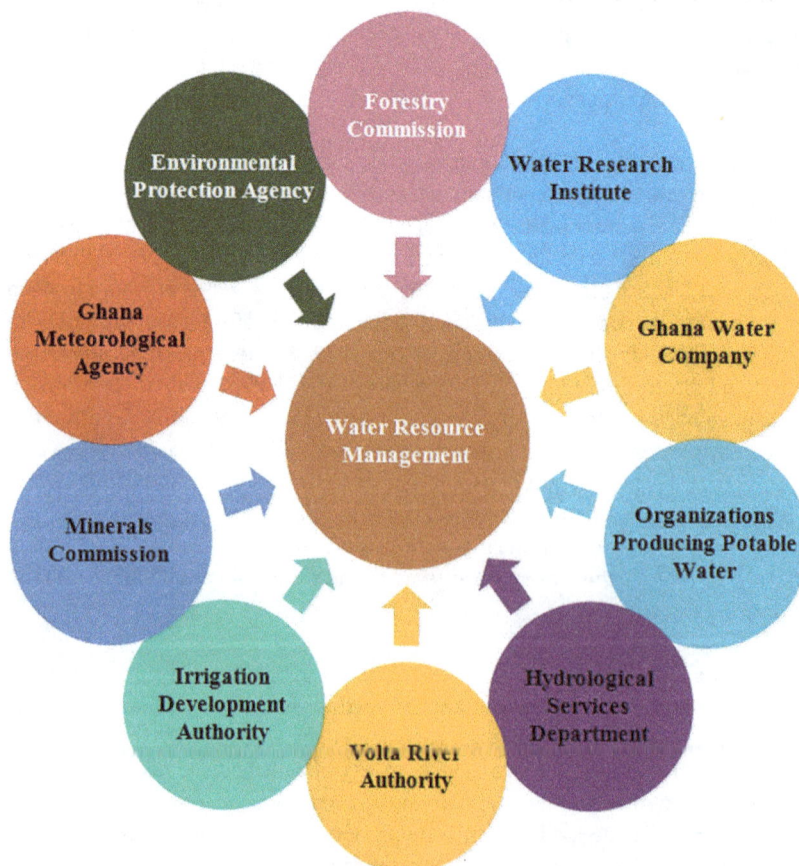

Figure 1. Representative institutions of Ghana's water resource management.

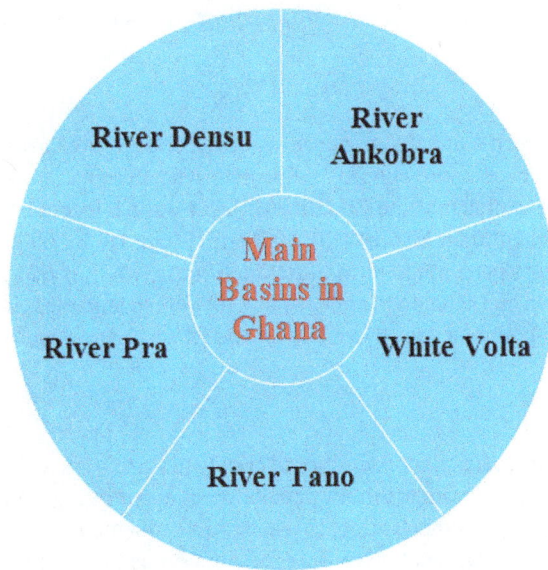

Figure 2. The main basins in Ghana.

has an area of about 2,490 km^2 and spans eleven local government assemblies in the three regions, namely; Greater Accra, eastern and central regions respectively. There are about 200 settlements in the basin and the total population is over 600,000 equivalent to 240 individuals per kilometer square. The soils are mainly well drained, friable, leaky, loamy savanna mostly reddish brown.

1.2.2. Ankobra basin

The Ankobra basin is bounded to the east, west and south by the Pra basin, Tano basin and the Gulf of Guinea respectively. It is located within latitudes 40° 52′N and 60° 27′N and longitudes 10° 42′W and 20° 33′W (WRC Ghana, 2015d). According to WRC Ghana (2015b), the basin has an area of about 8,403 km^2 covering eleven districts in the area of Wassa Amenfi, Wassa West and Nzema East Districts. The Ankobra basin falls under the south-western equatorial and the wet semi-equatorial climatic regions. The south-western equatorial is the wettest climatic region in Ghana with an annual mean rainfall above 1,900 mm and vegetation consists of rainforest as well as semi deciduous forest.

1.2.3. Pra basin

The Pra basin is located between latitudes 50° N and 70° 30′N and longitudes 20° 30′W, and 00° 30′W in the south central of Ghana. The drainage network comprises the main Pra and its major tributaries of Birim, Anum and Offin rivers and their tributaries with a drainage area of about 22,106 km^2, with an average elevation of about 300 m and generally less than 600 m above sea level. It features Lake Bosomtwe, which is a natural lake that stands out as prominent protected area (WRC Ghana, 2015e).

1.2.4. White Volta basin

The White Volta river system extents through Togo, Burkina Faso and Ghana. It spans three administrative regions in Ghana, i.e. all the upper east region, 70% of the upper west region, and 50% of the northern region. The White Volta River basin in Ghana is located between latitudes 7° 30′N and longitudes 0° 0′W. The drainage area within Ghana is about 44% of the total area of the White Volta river system.

The White Volta River and its main tributaries in the northern part, the Red Volta (Nazinon) and the Kulpawn/Sissili rivers, take their sources in the central and north-eastern portions of Burkina Faso. It contributes on an annual basis, on average some 20% of the inflow to the Volta Lake and hence is an essential element of the hydropower generated at Akosombo Dam and Kpong power stations in the lower Volta (WRC Ghana, 2015c).

1.2.5. Tano basin

The landscape of the Tano basin, ranges between 0 and 700 m above sea level. The climate of the basin falls partly under the wet, semi-equatorial and partly under south-western equatorial climate zones of Ghana, thus characterized by double rainfall maxima (USAID, 2011). The Basin is located in the south western part of Ghana and lies amid latitudes 50′N and 70° 40′N and longitudes 20° 00′W and 30° 15′N. Arable lands occupy the highest percentage of the total land mass, which is 50%. There is commercial farming of cocoa, plantain and other food crops. About 10% of the land mass is used for human settlement and the remaining 40% of it is covered by forests which are principally protected (WRC Ghana, 2015c).

2. Water resource management

Water is essential to all forms of life and needed in sufficient quantity and quality to sustain life. With the increasing human population around the globe, with a current population of about over seven billion people and the growth of technology require human society to devote more and more attention to protection of adequate supplies of water. Degradation of water resources has long been a concern of human society (Karr, 1991). Water can be managed in various number of ways provided there is a joint effort to achieve the goal of protecting the fresh water resources available.

2.1. Water resource management in Ghana

Ghana is privileged to have adequate water supply for its indigenous if managed properly. According to USAID (2011), the occurrence of groundwater in Ghana is related with three main geological formations as showed in Figure 3.

One percent is obtained from the Mesozoic and Cenozoic sediments, 54% in the basement complex and 45% over the Volta in the formation of the country. The aquifer depths of the country are normally between 10 and 60 m, and yields seldom surpass 6 m³/h. In the Mesozoic and Cenozoic formations occurring in the extreme south-eastern and western part of the country, the aquifer depths vary amid 6–120 m. There are as well limestone aquifers some of which are 120–300 m in depth, with an average yield as high as 180 m³/h. Groundwater resource in Ghana is generally good except for some cases of localized pollution and areas with high levels of iron, fluoride and other minerals (USAID, 2011). Some coastal aquifers especially have their groundwater being salinized (WRC Ghana, 2015f).

2.1.1. Activities undertaken to manage Ghana's groundwater resource

There are a lot of activities the regulators of water resources in Ghana put in place to ensure a sustainable water resource for the country and future generations. These are as follows (WRC Ghana, 2011);

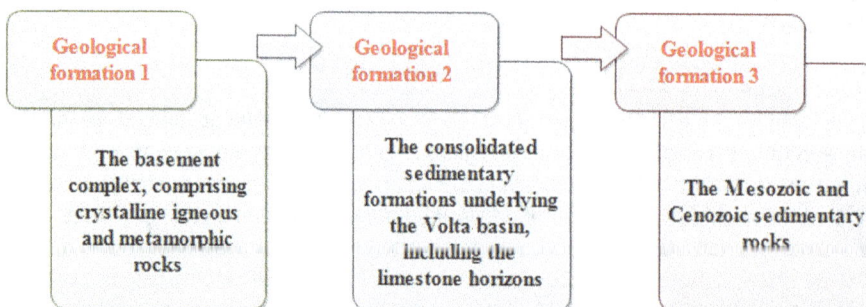

Figure 3. Geological formation of the occurrence of groundwater in Ghana.

- Awareness creation, education and training of stakeholders through radio and television programs, newspaper articles, community Durbar's, etc.
- Carry out field operations to control pollution.
- Check illegal mining activities which of late has become a challenge to regulators because lots of water bodies are being destroyed.
- Stop encroachments around water bodies.
- Dredge sections of the river to prevent flooding where necessary.
- Running of ecological and water quality monitoring.
- Initiating establishment of regulated buffer areas to protect the quality of open water bodies (rivers and lakes).
- Adopting programs targeting the management of water resources, such as Integrated water resources management (IWRM) and ensuring its completion.
- Strengthen policies and regulations for the protection and long-term sustainability of Ghana's groundwater resources.

2.1.2. Usage of water resources in Ghana

In Ghana, the major consumptive uses of the water resources are water consumption, irrigation and livestock watering. Domestic and industrial urban water supplies are based almost entirely on surface water, either lock up behind small dams or diverted by levees in rivers. Water supplies in rural areas, however, are obtained almost entirely from groundwater sources. The various groundwater development programs have resulted in the drilling of more than 10,000 boreholes nationwide (UNFCC, 2011).

The main non-consumptive uses are hydropower generation, inland fisheries and water navigation. In Akosombo, is the Akosombo Dam, located 100 km from the mouth of the Volta, the first Volta hydroelectric dam which was constructed in 1964, which created one of the largest man-made lakes in the world, covering an average area of about 8,300 km^2 (Mensah, 2010). Another hydroelectric dam was constructed in 1981 with an area covering about 40 km^2 and it is located 20 km downstream of Akosombo. Rainwater harvesting has also become common and has a great potential to increase water availability in certain localized areas (WRC Ghana, 2015f).

2.2. Integrated water resources management

The basic ideas of IWRM emerged decades ago. According to Giordano and Shah (2014), they are a call to consider water holistically, to manage it across sectors, and to ensure a broader participation in decision making. It is to bring to an end the fragmentary approaches to water resource management and other development decisions made for the benefit of a single group or section. These concepts an exceptional point of considering the improvement in water management and governance, and they have been formalized as what has now become in capitals, IWRM, with specific prescriptive principles whose is operated frequently sustained by donor funding and international advocacy.

The roots of IWRM can be traced back to the establishment of the Tennessee Valley Authority in 1993 (Stålnacke & Gooch, 2010). The next major landmark in the development of the concept was in 1977 at the International Water Conference in Mal del Plata. The emphasis of the conference was on coordination between different water sectors, primarily at the national level. The issues were given further attention at the 1992 Dublin conference, preparing for the Earth Summit in Rio de Janeiro (Stålnacke & Gooch, 2010). The Dublin statement included the following principles showed in Figure 4.

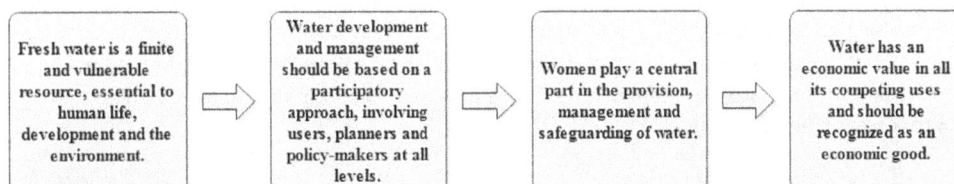

| Fresh water is a finite and vulnerable resource, essential to human life, development and the environment. | Water development and management should be based on a participatory approach, involving users, planners and policy-makers at all levels. | Women play a central part in the provision, management and safeguarding of water. | Water has an economic value in all its competing uses and should be recognized as an economic good. |

Figure 4. Principles of IWRM.

2.2.1. Implementation of IWRM in Ghana

Ghana is primarily rural; about 56.6% of its population live in rural areas. These rural people are the most deprived in terms of access to safe drinking water and other socio-economic infrastructure. According to Anokye and Gupta (2012), elements of the IWRM include a holistic, integrated approach and the main principles of public participation, the role of gender and the notion of recognizing the economic value of water.

Ghana as a country is keen on protecting its water resources, with traditional authorities in the primitive years used to give regulations and taboos in order to protect these water bodies. Even before the WRC was given the sole mandate in 1996 of managing water resources in Ghana, a policy was launched in 1994—The National Community Water and Sanitation Program to ensure that water resources were protected. The WRC of Ghana adopting the IWRM in its approach to manage its water resources was a good decision made.

As the IWRM stresses the significance of involving all stakeholders; authorities, organizations, the public and private sectors, non-state actors and civil society in the management of water resources, Ghana is doing same with the WRC decentralizing its roles by the generation of boards in the various major basins in the country, involving municipal authorities and district authorities as well as equipping them with appropriate tools to aid their work, to explain to the community stakeholders their roles as well as members of the community and particularly the involvement of women at all stages.

It also creates awareness of the importance of water among policy-makers and the public (Lonergan & Brooks, 1994), also involving users of water in the planning and implementation of water projects, as decisions arrived at might cater more for the needs of the public. It also in cooperates members of the community in the management, operation and maintenance of facilities (Anokye & Gupta, 2012; WRC Ghana, 2015c).

3. Challenges in the management of water resources in Ghana

During this search, there were a number of challenges identified to be hampering the progress of water resource management and sustainability in Ghana by the WRC and its affiliated bodies. This paper would like to focus on these; illegal mining activities, pollution, improper agricultural activities, climate change and proliferation of water weeds or otherwise known as seaweeds.

3.1. Illegal mining activities

Ghana as a country has a standing history with small scale mining of gold, it existed as far back as the eighth century as a household economic activity. It was legalized when the Small Scale Mining Law (PNDCL 218) 1989 was passed and public policies formulated to ease the implementation of the law (Kessey & Arko, 2013). Since then, the industry has become a major contributor to the total quantity of gold Ghana produced and an employer to a lot of the rural labor force. The household name for small scale illegal gold mining is "Galamsey".

In spite of these influences, the industry has a number of negative effects on the environment and most significantly water bodies. River bodies that were the main source of water for drinking, household chores and other activities have all been destroyed by these activities of illegal small scale mining of gold. Subsequently, the small scale mining industry of gold is getting more damaging as a second largest pollution after agriculture in Africa (Kessey & Arko, 2013) and in Ghana causing more deaths each day as well as destroying the water resource available making the work of the WRC more challenging dealing with. All the chemicals, especially mercury that is from the mercury-gold amalgamation process is dumped back into the river bodies as well as the washing of their gold dust and oils from the generators used for drilling. Figure 5 shows a pictorial view of the extent of damage done to water bodies in Ghana.

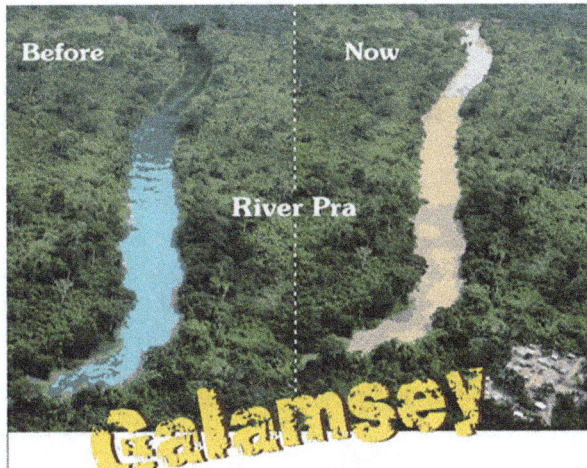

Figure 5. Illegal mining activities destroys river body (MyJoy, 2015).

Aside illegal mining activities of gold, sand winning in and around water bodies is also destroying water bodies in Ghana. In the sea it enhances sea erosion and disrupts aquatic life which are essential in the management of water resources. Figure 6 shows a pictorial view of illegal sand winning along the coastal shores of Ghana.

3.2. Waste and chemical pollution

Water is in limited amount all over the world and without water, the rest of our earthly habitat could not have sustained us; almost everywhere, increasing amounts of organic waste and industrial pollutants threaten water resources quality and availability (Karr & Chu, 2000). At Ghana indiscriminate disposal of organic waste from homes and chemical waste from industries been dumped into water pollutes the water bodies a lot.

The Korle Lagoon in Accra, Ghana, has become one of the most polluted water bodies in Ghana. It is the main outlet through which all major drainage channels in the city empty their waste into the sea (Asumadu-Sarkodie, Owusu, & Jayaweera, 2015; Asumadu-Sarkodie, Owusu, & Rufangura, 2015). Large quantities of untreated industrial waste emptied into open drains has led to severe pollution in the lagoon and disrupted its natural ecology. This has made it challenging for the authorities to manage the situation, but efforts are still being made. The net effect of these activities is that, at the slightest rainfall observed, the lagoon overflows causing flooding in its environs (Owusu Boadi & Kuitunen, 2002).

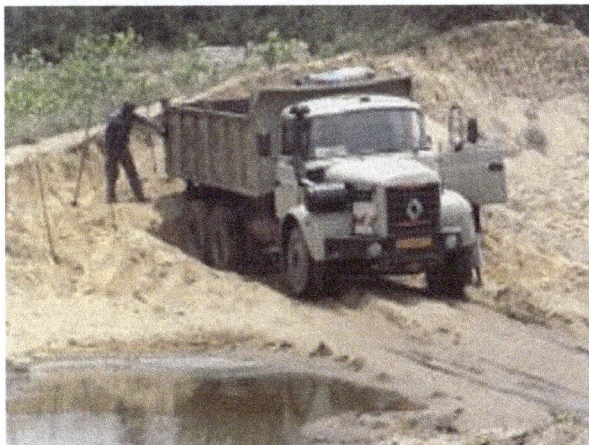

Figure 6. Illegal sand winning activity (GNA, 2015).

3.3. Improper agricultural activities

In Ghana, most of the occupation of the rural communities are crop production or livestock raring and most of Ghana's water resources are located in these rural areas and the improper activities such as improper application of fertilizers, livestock's going out to graze and drink by themselves in water bodies, chemical such as DDT used for fishing, high intensity lights for fishing and the cultivation crops along river banks ends up polluting these bodies.

Research in the Densu River basin by Karikari and Ansa-Asare (2006), concluded in their results that the poor microbial activity observed might be due to contamination caused by human and livestock. With the humans, it was a common practice for people living along the catchment to discharge their domestic and agricultural waste into the river and the livestock's defecating and urinating in the river bodies. The high turbidity also observed by the group in the basin was attributed to poor farming activities such as intensive agriculture, resulting in large quantities of top soil ending up in the river after a heavy rainfall.

3.4. Climate change

Climate change a known situation in the 21st century has taken the centre stage in the development agenda of both developed and developing countries. Its main cause is credited to the increasing emissions of Carbon dioxide and other greenhouse gases (Nitrous Oxide, Nitric acid, Methane, Chlorofluorocarbons, etc.) and subsequently increasing temperatures in the atmosphere (Asumadu-Sarkodie & Owusu, 2016b, 2016c, 2016f; Asumadu-Sarkodie, Rufangura, Jayaweera, & Owusu, 2015; Kankam-Yeboah, Amisigo, & Obuobi, 2011).

There is a known fact that the climate in Ghana has changed significantly with impacts being felt everywhere in the country, especially on water resource management; with research showing that there is increased evaporation, decreased and highly variable rainfall pattern, and frequent pronounced flood and drought situations (Asumadu-Sarkodie, Owusu, & Jayaweera, 2015; Asumadu-Sarkodie, Owusu, & Rufangura, 2015). The impacts of the rising temperature are during the dry season (December–March). Also, there have been intensive rainfall events causing flood situations such as the recent event one, on the 3rd June 2015 in Accra, Ghana's capital, together with an explosion at a Goil fuel oil filling station that resulted in a death toll of over 152 lives and the loss of properties worth millions of Cedi's and the pollution of water bodies around the catchment area, affecting the freshwater quality (Asumadu-Sarkodie, Owusu, & Rufangura, 2015).

A research conducted by the Council for Scientific and Industrial Research-Water Research Institute (CSIR-WRI) showed that aside climate change considerations, Ghana is predicted to become a water stress country by 2025. In the aspect of increases in the frequency and severity of extreme weather events, it is most likely that the climate change scenario will worsen future water scarcity in many places in the country (Asumadu-Sarkodie, Owusu, & Rufangura, 2015; Kankam-Yeboah et al., 2011).

3.5. Proliferation of water weeds

Aquatic weeds are sophisticated plants that grow in water or in wet soils, which are mostly not desired. They grow in profusion and usually occur along the shores of water bodies like dams, lakes, along rivers and river mouths. The menace of water weeds is reaching worrying extents in many parts of the world, particularly in tropical water bodies where they have led to ecological, economic losses and the challenge of managing water resources efficiently and Ghana is no exception (Aloo, Ojwang, Omondi, Njiru, & Oyugi, 2013).

In Ghana the most affected area is the Tano/Abby/Ehy Lagoon complex as well as along 25 km of the river from the convergence of the Tano river with the Lagoon complex, which poses a threat to the aquatic life, the quality of the water and the management of the water body for the authorities (WRC Ghana, 2015c).

4. The future prospect
Section 4 outlines possible solutions towards water resource management in Ghana.

4.1. Incorporating customary water management
Prior to the colonial era traditional leaders used to manage water resources and they did a good job to maintain and manage these resources. This was achieved by the leaders by ploughing, contouring, clay pots in storing water, wells being dug, homestead ponds, rainwater harvesting, taboos/totems, etc. Some examples of taboos/totems are: in some communities' water bodies are not to be visited or, so to speak disturbed on a particular day within a week, because water bodies are termed to be "spirits"; also livestock were not to be found around or graze along streams and river bodies; No cutting of trees along the river banks as well as the pollution of river bodies; Fishing activities were not allowed in any river bodies, etc.

These methods and practices epitomized the practices of ethnic groups. It helped to minimize deforestation, soil erosion and also was to allow the ecosystem replenish itself (Atampugre, Botchway, Esia-Donkoh, & Kendie, 2015). If these are incorporated into the governments' management it would go a long way in improving and sustaining water resources in Ghana.

4.2. Increasing awareness of the dangers of illegal mining and education
In dealing with the problem of illegal mining destroying a lot of water bodies in Ghana, the governments with their task force going to around these sites, arresting the culprits but still after the raid people would still go back. Even though the pits they dug for the mining ended up killing some of their members they would still go back, this is evident that all the measures in place to stop these illegal miners so far it has not been assisting in the combating of this act.

An increased awareness of the dangers of illegal mining to these miners and also to their environment in relation to the future of generations yet to be born. Behaviour change cannot happen overnight, but with proper advocacy and well planned awareness creation as well as behaviour change communications to these miners in the affected areas can go a long way to help improve the situation.

Education on the license being given for small scale mining should be intensified and after giving the miners the license, they should be educated on how cover their pits to avoid dangers posed to their lives as well as others and also how it should be away from water bodies and they not destroy these bodies.

4.3. Renewable energy resource for mitigating climate change
Climate change has become a global phenomenon and of interest to all parties involved around the world. Even if a particular country is not emitting most of these greenhouse gases, they have been affected since the atmosphere is a unit. In Ghana, if there is a contribution to this, it would be the felling of trees for timber and burning some to be used as a source of fuel (charcoal) for cooking (Asumadu-Sarkodie & Owusu, 2016a; Owusu & Asumadu-Sarkodie, 2016). Also, liquid fuels used in cars with bad engines, ending up producing fumes from the exhaust pipes of these vehicles which are mainly Carbon monoxide gases.

To address the above, the use of the charcoal must be minimized and replaced with Liquefied Petroleum Gas which the government of Ghana is trying to achieve by giving out some cylinders to some rural areas since they use a lot of wood in cooking and also with stricter measures by the Forestry Commission of Ghana to the indiscriminate felling of trees.

All round the world today there is an advocate for the use of renewable sources of energy to mitigate climate change and is proven to help. Ghana can also go into new emerging, renewable sources of energy such as Solar and Wind and also increase investments in hydropower, which is already present in Ghana but facing challenges (Asumadu-Sarkodie & Owusu, 2016d, 2016e).

4.4. Stakeholder specific responsibilities

In Ghana most of the water bodies and resources are mostly in the rural areas. In this, assigning specific duties to stakeholders of small communities as to how they can help protect and restore their water resources and its related ecosystems by these roles, as well as equipping them with the adequate materials needed to do so. In addition to this, improving sanitation in communities would also help protect water resources, which should also be added to their responsibilities.

4.5. Investment in wastewater treatment

In Ghana a lot of untreated waste-water is likely to end-up in storm water drains or gutters, lakes, streams, etc. Which intern pollutes some good drinking water bodies in case of flood situations or disruptions in drainage pathways. Total waste-water production annually in Ghana is yet to be evaluated, because there is little or no data on industrial and commercial production except that from domestic sources and out of this only about 8% of it (Waste-water – Domestic) undergo some degree of treatment in Ghana (Gyampo, 2012).

If a lot of research goes into the treatment of waste-water in Ghana and invested in it, as well as the assurance given to the population that it is safe and can be used. In attaining this it would go a long way to minimize shortages of water and improve groundwater recharge since a lot are used for irrigation purposes, which in turn will improve the management of water situations in Ghana.

5. Conclusion

Water covers about 70% of the earth's surface and it exists naturally in the earth in all the three physical states of matter and it is always moving around because the water flows with the current. Out of the earth's percentage of water covering the surface, only about 2.5% is fresh water and due to the fact that most are stored in deep groundwater, a small amount is readily available for human use. Hence, the goal of ensuring sufficient supply of water is available for everyone in our world at present and considering the future generations.

The paper aimed at considering the whole picture of water resource management in Ghana. A qualitative research was employed by reviewing papers and some credible web pages in the area of study. The study looked at the available water resources in Ghana (Basins-Ankobra, Densu, Pra, Tano and Volta), Water resource management in Ghana, Incorporating IWRM in Ghana, Challenges that the Ghana Water Commission is facing in the management of water and some recommendations that can help improve the management of water resources in Ghana.

IWRM adoption by Ghana is a good decision, but other measures or recommendations are needed to be considered to better the measures already put in place by the WRC. The Challenges identified during the study, that tends to hinder the work of the WRC of Ghana are as follows; illegal mining activities, waste and chemical pollution, improper agricultural activities, climate change and proliferation of water weeds.

Moreover, adopting scientific planning and strategies that reduces the factors that affect water loss such as rainfall harvesting, drip irrigation, landscape water use, soil improvements (soil moisture monitoring), furrow diking, metering and sub-metering are efficient measures of promoting water conservation in Ghana.

The aforementioned challenges can be addressed by taking these recommendations into consideration; incorporating customary water management, renewable energy resource for mitigating climate change, investment in waste-water treatment, stakeholder specific responsibilities and increasing awareness of the dangers of illegal mining and education.

In putting the above suggestions into account and implemented, the achievements of the WRC would improve a lot more and water resources protected and kept in adequate quantities for the populace and future generations to come.

Funding
The authors received no direct funding for this research.

Author details
Phebe Asantewaa Owusu[1]
E-mail: phebe.owusu@metu.edu.tr
ORCID ID: http://orcid.org/0000-0001-7364-1640
Samuel Asumadu-Sarkodie[1]
E-mail: samuel.sarkodie@metu.edu.tr
ORCID ID: http://orcid.org/0000-0001-5035-5983
Polycarp Ameyo[2]
E-mail: polycarpameyo@gmail.com
ORCID ID: http://orcid.org/0000-0002-6141-1220
[1] Sustainable Environment and Energy Systems, Middle East Technical University, Northern Cyprus Campus, Guzelyurt, 99738 TRNC, Turkey.
[2] Department of Civil Engineering, Middle East Technical University, Northern Cyprus Campus, Morphou, 99738 TRNC, Turkey.

References
Aloo, P., Ojwang, W., Omondi, R., Njiru, J. M., & Oyugi, D. (2013). A review of the impacts of invasive aquatic weeds on the bio-diversity of some tropical water bodies with special reference to Lake Victoria (Kenya). *Biodiversity Journal, 4*, 471–482.

Anokye, N. A., & Gupta, J. (2012). Reconciling IWRM and water delivery in Ghana – The potential and the challenges. *Physics and Chemistry of the Earth, Parts A/B/C, 47–48*, 33–45. doi:10.1016/j.pce.2011.06.010

Asumadu-Sarkodie, S., & Owusu, P. (2016a). A review of Ghana's energy sector national energy statistics and policy framework. *Cogent Engineering, 3*, 1155274. doi:10.1080/23311916.2016.1155274

Asumadu-Sarkodie, S., & Owusu, P. A. (2016b). Feasibility of biomass heating system in Middle East Technical University, Northern Cyprus Campus. *Cogent Engineering, 3*, 1134304. doi:10.1080/23311916.2015.1134304

Asumadu-Sarkodie, S., & Owusu, P. A. (2016c). Multivariate co-integration analysis of the Kaya factors in Ghana. *Environmental Science and Pollution Research.* doi:10.1007/s11356-016-6245-9

Asumadu-Sarkodie, S., & Owusu, P. A. (2016d). The potential and economic viability of solar photovoltaic in Ghana.*Energy Sources, Part A: Recovery, Utilization, and Environmental Effects.* doi:10.1080/15567036.2015.1122682

Asumadu-Sarkodie, S., & Owusu, P. A. (2016e). The potential and economic viability of wind farms in Ghana. *Energy Sources, Part A: Recovery, Utilization, and Environmental Effects.* doi:10.1080/15567036.2015.1122680

Asumadu-Sarkodie, S., & Owusu, P. A. (2016f). The relationship between carbon dioxide and agriculture in Ghana, a comparison of VECM and ARDL model. *Environmental Science and Pollution Research.* doi:10.1007/s11356-016-6252-x

Asumadu-Sarkodie, S., Owusu, P. A., & Jayaweera, H. M. (2015). Flood risk management in Ghana: A case study in Accra. *Advances in Applied Science Research, 6*, 196–201.

Asumadu-Sarkodie, S., Owusu, P. A., & Rufangura, P. (2015). Impact analysis of flood in Accra, Ghana. *Advances in Applied Science Research, 6*, 53–78.

Asumadu-Sarkodie, S., Rufangura, P., Jayaweera, H. M., & Owusu, P. A. (2015). Situational analysis of flood and drought in Rwanda. *International Journal of Scientific and Engineering Research, 6*, 960–970. doi:10.14299/ijser.2015.08.013

Atampugre, G., Botchway, D. -V. N., Esia-Donkoh, K., & Kendie, S. (2015). Ecological modernization and water resource management: A critique of institutional transitions in Ghana. *GeoJournal*, 1–12.

Giordano, M., & Shah, T. (2014). From IWRM back to integrated water resources management. *International Journal of Water Resources Development, 30*, 364–376. doi:10.1080/07900627.2013.851521

GNA. (2015). MP worried over illegal sand winning. *Social.* Retrieved November 24, 2015, from http://www.ghananewsagency.org/social/mp-worried-over-illegal-sand-winning--76550

Gyampo, M. A. (2012). Wastewater production, treatment, and use in Ghana. Paper presented at the Third Regional Workshop of the Project 'Safe Use of Wastewater in Agriculture.

Kankam-Yeboah, K., Amisigo, B., & Obuobi, E. (2011). *Climate change impacts on water resources in Ghana.*

Karikari, A., & Ansa-Asare, O. (2006). Physico-chemical and microbial water quality assessment of Densu River of Ghana. *West African Journal of Applied Ecology, 10*(1).

Karr, J. R. (1991). Biological integrity: A long-neglected aspect of water resource management. *Ecological Applications, 1*, 66–84.
http://dx.doi.org/10.2307/1941848

Karr, J. R., & Chu, E. W. (2000). *Introduction: Sustaining living rivers.* Springer.

Kessey, K. D., & Arko, B. (2013). Small scale gold mining and environmental degradation, in Ghana: Issues of mining policy implementation and challenges. *Journal of Studies in Social Sciences, 5.*

Lonergan, S. C., & Brooks, D. B. (1994). *Watershed: The role of fresh water in the Israeli-Palestinian conflict.* IDRC.

Mensah, F. (2010). *Integrated water resource management in Ghana: Past, present, and future.* Paper presented at the Proceedings from the World Environmental and Water Resources Congress.

MyJoy. (2015). *Illegal sand winning in Ghana.* Retrieved November 15, 2015, from www.myjoyonline.com

Oki, T., & Kanae, S. (2006). Global hydrological cycles and world water resources. *Science, 313*, 1068–1072.
http://dx.doi.org/10.1126/science.1128845

Osborne, L. L., & Kovacic, D. A. (1993). Riparian vegetated buffer strips in water-quality restoration and stream management. *Freshwater Biology, 29*, 243–258.
http://dx.doi.org/10.1111/fwb.1993.29.issue-2

Owusu Boadi, K. O., & Kuitunen, M. (2002). Urban waste pollution in the Korle Lagoon, Accra, Ghana. *The Environmentalist, 22*, 301–309. doi:10.1023/A:1020706728569

Owusu, P. A., & Asumadu-Sarkodie, S. (2016). A review of renewable energy sources, sustainability issues and climate change mitigation. *Cogent Engineering.* doi:10.1080/23311916.2016.1167990

Pahl-Wostl, C. (2007). Transitions towards adaptive management of water facing climate and global change. *Water Resources Management, 21*, 49–62.

Stålnacke, P., & Gooch, G. D. (2010). Integrated water resources management. *Irrigation and Drainage Systems, 24*, 155–159. http://dx.doi.org/10.1007/s10795-010-9106-6

UNFCC. (2011). *Ghana's second national communication.* Retrieved November 15, 2015, from http://unfccc.int/resource/docs/natc/ghanc2.pdf

United Nations. (2014, October 07). *Water resources management.* Retrieved November 15, 2015, from http://www.unwater.org/topics/water-resources-management/en/

USAID. (2011). *Biodiversity and tropical forests environmental threats and opportunities assessment*. Retrieved from http://www.encapafrica.org/documents/biofor/ETOA_Ghana_FINAL.pdf

WRC Ghana. (2011). *Groundwater management strategy*. Retrieved from http://www.wrc-gh.org/documents/reports/

WRC Ghana. (2015a). *About us*. Retrieved November 15, 2015, from http://www.wrc-gh.org/about-us/

WRC Ghana. (2015b). *Ankobra basin*. Retrieved November 15, 2015, from http://wrc-gh.org/en/basins/18/ankobra

WRC Ghana. (2015c). *Basins*. Retrieved November 15, 2015, from http://www.wrc-gh.org/basins/

WRC Ghana. (2015d). *Densu basin*. Retrieved November 15, 2015, from http://wrc-gh.org/en/basins/15/densu

WRC Ghana. (2015e). *Pra basin*. Retrieved November 15, 2015, from http://wrc-gh.org/en/basins/17/pra-basin

WRC Ghana. (2015f). *Water resource management and governance*. Retrieved November 23, 2015, from http://www.wrc-gh.org/water-resources-management-and-governance/ground-water-management/

Yanmaz, A. M. (2013). *Applied water resources engineering*. Metu Press.

4

A review of renewable energy sources, sustainability issues and climate change mitigation

Phebe Asantewaa Owusu[1] and Samuel Asumadu-Sarkodie[1*]

*Corresponding author: Samuel Asumadu-Sarkodie, Sustainable Environment and Energy System, Middle East Technical University, Northern Cyprus Campus, Kalkanli, Guzelyurt 99738, TRNC, Turkey

E-mail: samuel.sarkodie@metu.edu.tr

Reviewing editor: Shashi Dubey, Hindustan College of Engineering, India

Abstract: The world is fast becoming a global village due to the increasing daily requirement of energy by all population across the world while the earth in its form cannot change. The need for energy and its related services to satisfy human social and economic development, welfare and health is increasing. Returning to renewables to help mitigate climate change is an excellent approach which needs to be sustainable in order to meet energy demand of future generations. The study reviewed the opportunities associated with renewable energy sources which includes: Energy Security, Energy Access, Social and Economic development, Climate Change Mitigation, and reduction of environmental and health impacts. Despite these opportunities, there are challenges that hinder the sustainability of renewable energy sources towards climate change mitigation. These challenges include Market failures, lack of information, access to raw materials for future renewable resource deployment, and our daily carbon footprint. The study suggested some measures and policy recommendations which when considered would help achieve the goal of renewable energy thus to reduce emissions, mitigate climate change and provide a clean environment as well as clean energy for all and future generations.

Subjects: Bio Energy; Clean Tech; Clean Technologies; Environmental; Renewable Energy

Keywords: renewable energy sources; climate change mitigation; sustainability issues; clean energy; carbon footprint; environmental sustainability engineering

ABOUT THE AUTHORS

Phebe Asantewaa Owusu studies Masters in Sustainable Environment and Energy Systems at Middle East Technical University, Northern Cyprus Campus where she is also a graduate assistant in the Chemistry Department.

Samuel Asumadu-Sarkodie is a multidisciplinary researcher who currently studies Masters in Sustainable Environment and Energy Systems at Middle East Technical University, Northern Cyprus Campus where he is also a graduate assistant in the Chemistry Department. His research interest includes but is not limited to: renewable energy, econometrics, energy economics, climate change and sustainable development.

PUBLIC INTEREST STATEMENT

Energy is a requirement in our everyday life as a way of improving human development leading to economic growth and productivity. The return-to-renewables will help mitigate climate change is an excellent way but needs to be sustainable in order to ensure a sustainable future and bequeath future generations to meet their energy needs. Knowledge regarding the interrelations between sustainable development and renewable energy in particular is still limited. The aim of the paper is to ascertain if renewable energy sources are sustainable and examine how a shift from fossil fuel-based energy sources to renewable energy sources would help reduce climate change and its impact. A qualitative research was employed by reviewing peer-reviewed papers in the area of study. This study brought to light the opportunities associated with renewable energy sources; energy security, energy access, social and economic development and climate change mitigation and reduction of environmental and health impacts.

1. Introduction

The world is fast becoming a global village due to the increasing daily requirement of energy by all population across the world while the earth in its form cannot change. The need for energy and its related services to satisfy human social and economic development, welfare and health is increasing. All societies call for the services of energy to meet basic human needs such as: health, lighting, cooking, space comfort, mobility and communication and serve as generative processes (Edenhofer et al., 2011). Securing energy supply and curbing energy contribution to climate change are the two-over-riding challenges of energy sector on the road to a sustainable future (Abbasi & Abbasi, 2010; Kaygusuz, 2012). It is overwhelming to know in today's world that 1.4 billion people lack access to electricity, while 85% of them live in rural areas. As a result of this, the number of rural communities relying on the traditional use of biomass is projected to rise from 2.7 billion today to 2.8 billion in 2030 (Kaygusuz, 2012).

Historically, the first recorded commercial mining of coal occurred in 1,750, near Richmond, Virginia. Momentarily, coal became the most preferred fuel for steam engines due to its more energy carrying capacity than corresponding quantities of biomass-based fuels (firewood and charcoal). It is noteworthy that coal was comparatively cheaper and a much cleaner fuel as well in the past centuries (Abbasi, Premalatha, & Abbasi, 2011). The dominance of fossil fuel-based power generation (Coal, Oil and Gas) and an exponential increase in population for the past decades have led to a growing demand for energy resulting in global challenges associated with a rapid growth in carbon dioxide (CO_2) emissions (Asumadu-Sarkodie & Owusu, 2016a). A significant climate change has become one of the greatest challenges of the twenty-first century. Its grave impacts may still be avoided if efforts are made to transform current energy systems. Renewable energy sources hold the key potential to displace greenhouse gas emissions from fossil fuel-based power generating and thereby mitigating climate change (Edenhofer et al., 2011).

Sustainable development has become the centre of recent national policies, strategies and development plans of many countries. The United Nations General Assembly proposed a set of global Sustainable Development Goals (SDGs) which included 17 goals and 169 targets at the UN in New York by the Open Working Group. In addition, a preliminary set of 330 indicators was introduced in March 2015 (Lu, Nakicenovic, Visbeck, & Stevance, 2015). The SDGs place greater value and demands on the scientific community than did the Millennium Development Goals. In addressing climate change, renewable energy, food, health and water provision requires a coordinated global monitoring and modelling of many factors which are socially, economically and environmentally oriented (Hák, Janoušková, & Moldan, 2016; Owusu, Asumadu-Sarkodie, & Ameyo, 2016).

Research into alternate sources of energy dated back in the late 90s when the world started receiving shock from oil produces in terms of price hiking (Abbasi et al., 2011). It is evidential in literature that replacing fossil fuel-based energy sources with renewable energy sources, which includes: bioenergy, direct solar energy, geothermal energy, hydropower, wind and ocean energy (tide and wave), would gradually help the world achieve the idea of sustainability. Governments, intergovernmental agencies, interested parties and individuals in the world today look forward to achieving a sustainable future due to the opportunities created in recent decades to replace petroleum-derived materials from fossil fuel-based energy sources with alternatives in renewable energy sources. The recent launch of a set of global SDGs is helping to make sure that climate change for twenty-first century and its impacts are combated, and a sustainable future is ensured and made as a bequest for future generations (Edenhofer et al., 2011; Lu et al., 2015).

Against this backdrop, the study seeks to examine the potentials and trends of sustainable development with renewable energy sources and climate change mitigation, the extent to which it can help and the potential challenges it poses and how a shift from fossil to renewable energy sources is a sure way of mitigating climate change. To achieve this objective, concepts, techniques and peer-reviewed journals are analysed and reviewed judiciously.

The remainder of the paper is sectioned into five: Section 2 discusses renewable energy sources and sustainability and climate change, Section 3 elaborates on the various renewable energy sources and technologies, Section 4 elaborates on the renewable energy sources and sustainable development, Section 5 elaborates on challenges affecting renewable energy sources and policy recommendations and Section 6 concludes the study.

2. Renewable energy sources and sustainability

Renewable energy sources replenish themselves naturally without being depleted in the earth; they include bioenergy, hydropower, geothermal energy, solar energy, wind energy and ocean (tide and wave) energy. The main renewable energy forms and their uses are presented in Table 1.

Tester (2005) defines sustainable energy as, "a dynamic harmony between the equitable availability of energy-intensive goods and services to all people and preservation of the earth for future generations".

The world's growing energy need, alongside increasing population led to the continual use of fossil fuel-based energy sources (Coal, Oil and Gas) which became problematic by creating several challenges such as: depletion of fossil fuel reserves, greenhouse gas emissions and other environmental concerns, geopolitical and military conflicts, and the continual fuel price fluctuations. These problems will create unsustainable situations which will eventually result in potentially irreversible threat to human societies (UNFCC, 2015). Notwithstanding, renewable energy sources are the most outstanding alternative and the only solution to the growing challenges (Tiwari & Mishra, 2011). In 2012, renewable energy sources supplied 22% of the total world energy generation (U.S. Energy Information Administration, 2012) which was not possible a decade ago.

Reliable energy supply is essential in all economies for heating, lighting, industrial equipment, transport, etc. (International Energy Agency, 2014). Renewable energy supplies reduce the emission of greenhouse gases significantly if replaced with fossil fuels. Since renewable energy supplies are obtained naturally from ongoing flows of energy in our surroundings, it should be sustainable. For renewable energy to be sustainable, it must be limitless and provide non-harmful delivery of environmental goods and services. For instance, a sustainable biofuel should not increase the net CO_2 emissions, should not unfavourably affect food security, nor threaten biodiversity (Twidell & Weir, 2015). Is that really what is happening today? I guess not.

In spite of the outstanding advantages of renewable energy sources, certain shortcoming exists such as: the discontinuity of generation due to seasonal variations as most renewable energy resources are climate-dependent, that is why its exploitation requires complex design, planning and control optimization methods. Fortunately, the continuous technological advances in computer hardware and software are permitting scientific researchers to handle these optimization difficulties using computational resources applicable to the renewable and sustainable energy field (Baños et al., 2011).

Table 1. Renewable energy sources and their use (Panwar et al., 2011)	
Energy sources	**Energy conversion and usage options**
Hydropower	Power generation
Morden biomass	Heat and power generation, pyrolysis, gasification, digestion
Geothermal	Urban heating, power generation, hydrothermal, hot dry rock
Solar	Solar home systems, solar dryers, solar cookers
Direct solar	Photovoltaic, thermal power generation, water heaters
Wind	Power generation, wind generators, windmills, water pump
Wave and tide	Numerous design, barrage, tidal stream

2.1. Renewable energy and climate change

Presently, the term "climate change" is of great interest to the world at large, scientific as well as political discussions. Climate has been changing since the beginning of creation, but what is alarming is the speed of change in recent years and it may be one of the threats facing the earth. The growth rate of carbon dioxide has increased over the past 36 years (1979–2014) (Asumadu-Sarkodie & Owusu, 2016c, 2016f), "averaging about 1.4 ppm per year before 1995 and 2.0 ppm per year thereafter" (Earth System Research Laboratory, 2015). The United Nations Framework Convention on Climate Change defines climate change as being attributed directly or indirectly to human activities that alters the composition of the global atmosphere and which in turn exhibits variability in natural climate observed over comparable time periods (Fräss-Ehrfeld, 2009).

For more than a decade, the objective of keeping global warming below 2 °C has been a key focus of international climate debate (Asumadu-Sarkodie, Rufangura, Jayaweera, & Owusu, 2015; Rogelj, McCollum, Reisinger, Meinshausen, & Riahi, 2013). Since 1850, the global use of fossil fuels has increased to dominate energy supply, leading to a rapid growth in carbon dioxide emissions. Data by the end of 2010 confirmed that consumption of fossil fuels accounted for the majority of global anthropogenic greenhouse gas (GHG) emissions, where concentrations had increased to over 390 ppm (39%) above preindustrial levels (Edenhofer et al., 2011).

Renewable technologies are considered as clean sources of energy and optimal use of these resources decreases environmental impacts, produces minimum secondary waste and are sustainable based on the current and future economic and social needs. Renewable energy technologies provide an exceptional opportunity for mitigation of greenhouse gas emission and reducing global warming through substituting conventional energy sources (fossil fuel based) (Panwar, Kaushik, & Kothari, 2011).

3. Renewable energy sources and technology

Renewable energy sources are energy sources from natural and persistent flow of energy happening in our immediate environment. They include: bioenergy, direct solar energy, geothermal energy, hydropower, wind and ocean energy (tide and wave).

3.1. Hydropower

Hydropower is an essential energy source harnessed from water moving from higher to lower elevation levels, primarily to turn turbines and generate electricity. Hydropower projects include Dam project with reservoirs, run-of-river and in-stream projects and cover a range in project scale. Hydropower technologies are technically mature and its projects exploit a resource that vary temporarily. The operation of hydropower reservoirs often reflects their multiple uses, for example flood and drought control (Asumadu-Sarkodie, Owusu, & Jayaweera, 2015; Asumadu-Sarkodie, Owusu, & Rufangura, 2015), irrigation, drinking water and navigation (Edenhofer et al., 2011). The primary energy is provided by gravity and the height the water falls down on to the turbine. The potential energy of the stored water is the mass of the water, the gravity factor ($g = 9.81$ ms^{-2}) and the head defined as the difference between the dam level and the tail water level. The reservoir level to some extent changes downwards when water is released and accordingly influences electricity production. Turbines are constructed for an optional flow of water (Førsund, 2015). Hydropower discharges practically no particulate pollution, can upgrade quickly, and it is capable of storing energy for many hours (Hamann, 2015).

3.1.1. Hydropower source potential

Hydropower generation technical annual potential is 14,576 TWh, with an estimated total capacity potential of 3,721 GW; but, currently the global installed capacity of hydropower is much less than it's potential. According to the World Energy Council Report, about 50% of hydropower installed capacity is among four countries namely China, Brazil, Canada and USA (World Energy Council, 2013). The resource potential of hydropower could be altered due to climate change. Globally, the alterations caused by climate change in the existing hydropower production system are estimated to be less than 0.1%, even though additional research is needed to lower the uncertainties of these projection (Edenhofer et al., 2011).

3.1.2. Hydropower environmental and social impact

Hydropower generation does not produce greenhouse gases and thus mostly termed as a green source of energy. Nonetheless, it has its advantages and disadvantages. It improves the socio-economic development of a country; but, also considering the social impact, it displaces a lot of people from their homes to create it, though they are compensated but are not enough. The exploitation of the sites for hydropower such as, reservoirs that are often artificially created leading to flooding of the former natural environment. In addition, water is drained from lakes and watercourses and transported through channels over large distances and to pipelines and finally to the turbines that are often visible, but they may also go through mountains by created tunnels inside them (Førsund, 2015). Hydroelectric structures affect river body's ecology, largely by inducing a change into its hydrologic characteristics and by disturbing the ecological continuity of sediment transport and fish migration through the building of dams, dikes and weirs (Edenhofer et al., 2011). In countries where substantial plants or tree covers are flooded during the construction of a dam, there may be formation of methane gas when plants start rotting in the water, either released directly or when water is processed in turbines (Førsund, 2015).

3.2. Bioenergy

Bioenergy is a renewable energy source derived from biological sources. Bioenergy is an important source of energy, which can be used for transport using biodiesel, electricity generation, cooking and heating. Electricity from bioenergy attracts a large range of different sources, including forest by-products such as wood residues; agricultural residues such as sugar cane waste; and animal husbandry residue such as cow dung. One advantage of biomass energy-based electricity is that fuel is often a by-product, residue or waste product from the above sources. Significantly, it does not create a competition between land for food and land for fuel (Urban & Mitchell, 2011). Presently, global production of biofuels is comparatively low, but continuously increasing (Ajanovic, 2011). The annual biodiesel consumption in the United States was 15 billion litres in 2006. It has been growing at a rate of 30–50% per year to achieve an annual target of 30 billion litres at the end of year 2012 (Ayoub & Abdullah, 2012).

3.2.1. Bioenergy source potential

Biomass has a large potential, which meets the goal of reducing greenhouse gases and could insure fuel supply in the future. A lot of research is being done in this area trying to quantify global biomass technology. According to Hoogwijk, Faaij, Eickhout, de Vries, and Turkenburg (2005) the theoretical potential of bioenergy at the total terrestrial surface is about 3,500 EJ/year. The greater part of this potential is located in South America and Caribbean (47–221 EJ/year), sub-Saharan Africa (31–317 EJ/year) and the Commonwealth of Independent States (C.I.S) and Baltic states (45–199 EJ/year). The yield of biomass and its potential varies from country to country, from medium yields in temperature to high level in sub tropic and tropic countries. With biomass, a lot of research is focusing on an environmentally acceptable and sustainable source to mitigate climate change (Demirbas, Balat, & Balat, 2009).

3.2.2. Bioenergy environmental and social impact

The use of biological components (plant and animal source) to produce energy has always been a cause of worry especially to the general public and as to whether its food produce are to be used to provide fuel since there are cases of food aid needed around the world in deprived countries. About 99.7% of human food is obtained from the terrestrial environment, while about 0.3% comes from the aquatic domain. Most of the suitable land for biomass production is already in use (Ajanovic, 2011). Current studies have underlined both positive and negative environmental and socio-economic effects of bioenergy. Like orthodox agriculture and forestry systems, bioenergy can worsen soil and vegetation degradation related with the overexploitation of forest, too exhaustive crop and forest residue removal, and water overuse (Koh & Ghazoul, 2008; Robertson et al., 2008). Diversion of crops or land into bioenergy production can induce food commodity prices and food security (Headey & Fan, 2008). Proper operational management, can bring about some positive effects which includes enhanced biodiversity (Baum, Leinweber, Weih, Lamersdorf, & Dimitriou, 2009; Schulz, Brauner, & Gruß, 2009), soil carbon increases and improved soil productivity (Baum, Weih, Busch, Kroiher, & Bolte, 2009; Edenhofer et al., 2011; Tilman, Hill, & Lehman, 2006).

3.3. Direct solar energy
The word "direct" solar energy refers to the energy base for those renewable energy source technologies that draw on the Sun's energy directly. Some renewable technologies, such as wind and ocean thermal, use solar energy after it has been absorbed on the earth and converted to the other forms. Solar energy technology is obtained from solar irradiance to generate electricity using photovoltaic (PV) (Asumadu-Sarkodie & Owusu, 2016d) and concentrating solar power (CSP), to produce thermal energy, to meet direct lighting needs and, potentially, to produce fuels that might be used for transport and other purposes (Edenhofer et al., 2011). According to the World Energy Council (2013), "the total energy from solar radiation falling on the earth was more than 7,500 times the World's total annual primary energy consumption of 450 EJ" (Urban & Mitchell, 2011).

3.4. Geothermal energy
Geothermal energy is obtained naturally from the earth's interior as heat energy source. The origin of the heat is linked with the internal structure of the planet and the physical processes occurring there. Although heat is present in the earth's crust in huge quantities, not to mention the deepest parts, it is unevenly distributed, rarely concentrated, and often at depths too great to be exploited mechanically.

Geothermal gradient averages about 30 °C/km. There are areas of the earth's interior which are accessible by drilling, and where the gradient is well above the average gradient (Barbier, 2002). Heat is mined from geothermal reservoirs using wells and other means. Reservoirs that are naturally adequately hot and permeable are called hydrothermal reservoirs, while reservoirs that are satisfactorily hot but are improved with hydraulic stimulation are called enhanced geothermal systems (ESG). Once drawn to the surface, fluids of various temperatures can be used to generate electricity and other purposes that require the use of heat energy (Edenhofer et al., 2011).

3.5. Wind energy
The emergence of wind as an important source of the World's energy has taken a commanding lead among renewable sources. Wind exists everywhere in the world, in some places with considerable energy density (Manwell, McGowan, & Rogers, 2010). Wind energy harnesses kinetic energy from moving air. The primary application of the importance to climate change mitigation is to produce electricity from large turbines located onshore (land) or offshore (in sea or fresh water) (Asumadu-Sarkodie & Owusu, 2016e). Onshore wind energy technologies are already being manufactured and deployed on large scale (Edenhofer et al., 2011). Wind turbines convert the energy of wind into electricity.

3.6. Ocean energy (tide and wave)
Surface waves are created when wind passes over water (Ocean). The faster the wind speed, the longer the wind is sustained, the greater distance the wind travels, the greater the wave height, and the greater the wave energy produced (Jacobson & Delucchi, 2011). The ocean stores enough energy to meet the total worldwide demand for power many times over in the form of waves, tide, currents and heat. The year 2008 saw the beginning of the first generation of commercial Ocean energy devices, with the first units being installed in the UK-SeaGen and Portugal-Pelamis. There are presently four ways of obtaining energy from sea areas, namely from Wind, Tides, Waves and Thermal differences between deep and shallow Sea water (Esteban & Leary, 2012).

4. Renewable energy and sustainable development
Renewable energy has a direct relationship with sustainable development through its impact on human development and economic productivity (Asumadu-Sarkodie & Owusu, 2016b). Renewable energy sources

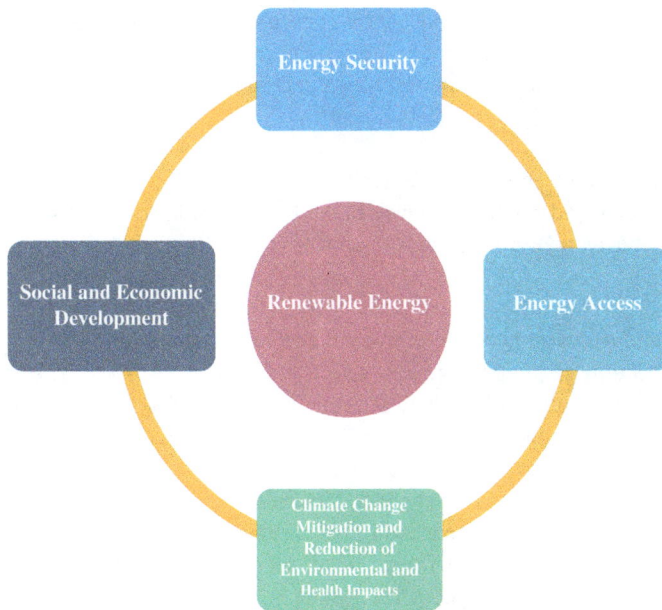

Figure 1. Opportunities of renewable energy sources.

provide opportunities in energy security, social and economic development, energy access, climate change mitigation and reduction of environmental and health impacts (Asumadu-Sarkodie & Owusu, 2016g). Figure 1 shows the opportunities of renewable energy sources towards sustainable development.

4.1. Energy security
The notion of energy security is generally used, however there is no consensus on its precise interpretation. Yet, the concern in energy security is based on the idea that there is a continuous supply of energy which is critical for the running of an economy (Kruyt, van Vuuren, de Vries, & Groenenberg, 2009). Given the interdependence of economic growth and energy consumption, access to a stable energy supply is of importance to the political world and a technical and monetary challenge for both developed and developing countries, because prolonged interferences would generate serious economic and basic functionality difficulties for most societies (Edenhofer et al., 2011; Larsen et al., 2009). Renewable energy sources are evenly distributed around the globe as compared to fossils and in general less traded on the market. Renewable energy reduces energy imports and contribute diversification of the portfolio of supply options and reduce an economy's vulnerability to price volatility and represent opportunities to enhance energy security across the globe. The introduction of renewable energy can also make contribution to increasing the reliability of energy services, to be specific in areas that often suffer from insufficient grid access. A diverse portfolio of energy sources together with good management and system design can help to enhance security (Edenhofer et al., 2011).

4.2. Social and economic development
Generally, the energy sector has been perceived as a key to economic development with a strong correlation between economic growth and expansion of energy consumption. Globally, per capita incomes are positively correlated with per capita energy use and economic growth can be identified as the most essential factor behind increasing energy consumption in the last decades. It in turn creates employment; renewable energy study in 2008, proved that employment from renewable energy technologies was about 2.3 million jobs worldwide, which also has improved health, education, gender equality and environmental safety (Edenhofer et al., 2011).

4.3. Energy access
The sustainable development goal seven (affordable and clean energy) seeks to ensure that energy is clean, affordable, available and accessible to all and this can be achieved with renewable energy

source since they are generally distributed across the globe. Access concerns need to be understood in a local context and in most countries there is an obvious difference between electrification in the urban and rural areas, this is especially true in sub-Saharan Africa and South Asian region (Brew-Hammond, 2010).

Distributed grids based on the renewable energy are generally more competitive in rural areas with significant distances to the national grid and the low levels of rural electrification offer substantial openings for renewable energy-based mini-grid systems to provide them with electricity access (Edenhofer et al., 2011).

4.4. Climate change mitigation and reduction of environmental and health impacts

Renewable energy sources used in energy generation helps to reduce greenhouse gases which mitigates climate change, reduce environmental and health complications associated with pollutants from fossil fuel sources of energy. The change in total GHG emissions in European Environmental Agency (EEA) countries for 1990–2012 and their GHG emissions per capita are depicted in Figures 2 and 3. Figure 2 shows that greenhouse gas emissions declined by 14% in 33 EEA countries between the years 1990–2012. Nevertheless, there was variation in individual member countries, while there was a decrease in GHG emissions in 22 EEA countries, there was an increase in 11 EEA countries. GHG emissions per capita declined by 22% between the years 1990–2012 in the EEA countries as depicted in Figure 3 (EEA, 2016).

Figure 4 shows United States carbon dioxide gas emissions from 1990–2013. Figure 2 shows an example of carbon dioxide emission levels being reduced from 1990–2013 in United States, a shift from mainly fossil fuel-based energy sources to renewable energy sources (United States Environmental Protection Agency, 2014).

5. Challenges affecting renewable energy sources

Renewable energy sources could become the major energy supply option in low-carbon energy economies. Disruptive alterations in all energy systems are necessary for tapping widely available renewable Energy sources. Organizing the energy transition from non-sustainable to renewable energy is often described as the major challenge of the first half of the twenty-first century (Verbruggen et al., 2010). Figure 5 shows the interconnection of factors affecting renewable energy supplies and sustainability. It is evident from Figure 5 that a major barrier towards the use of renewable energy source depends on a country's policy and policy instrument which in turn affect the cost and technological innovations. In addition, technological innovations affect the cost of renewable energy

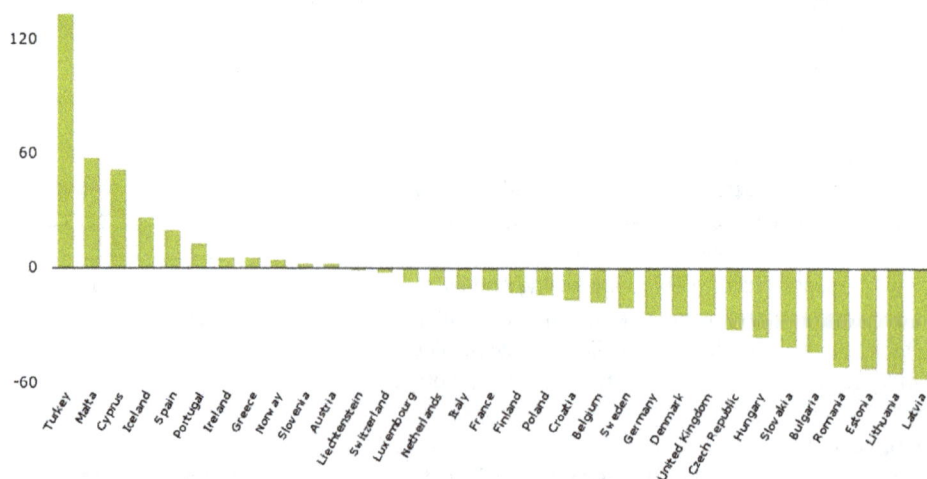

Figure 2. Change in total GHG emissions in EEA-33 countries (1990–2012) (EEA, 2016).

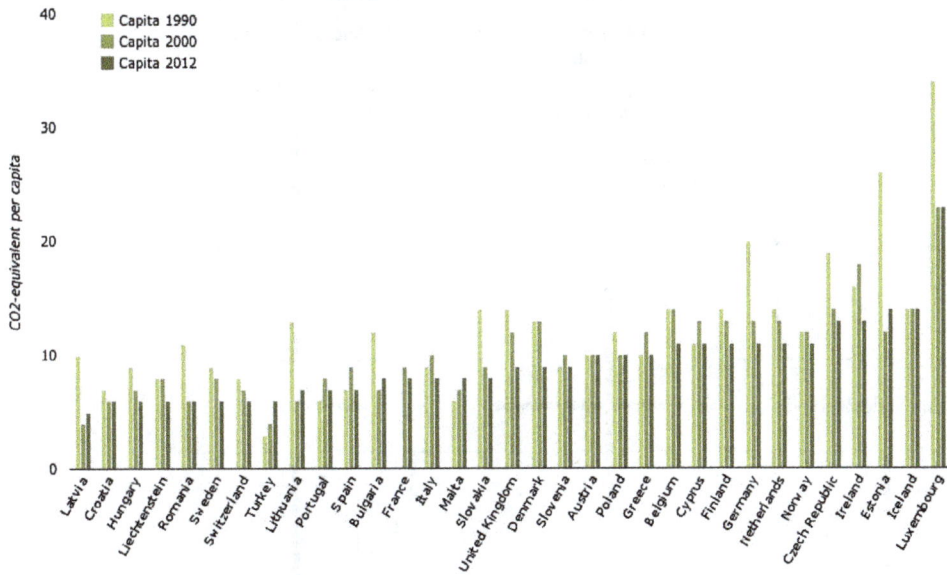

Figure 3. GHG emissions per capita in EEA-33 countries (EEA, 2016).

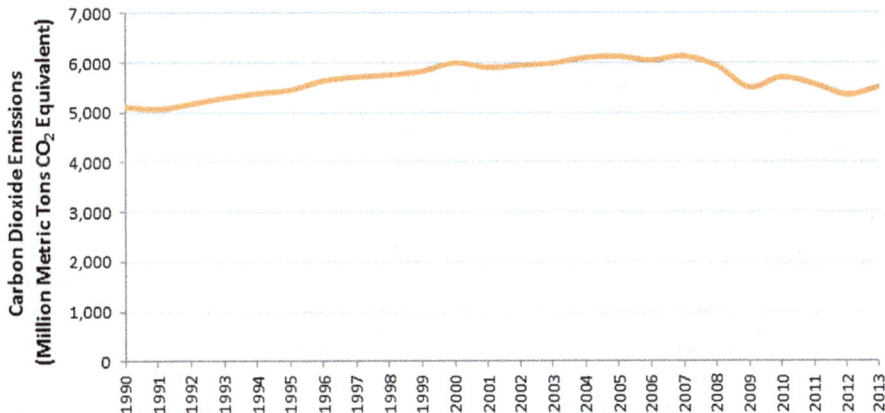

Figure 4. United States carbon dioxide gas emissions, 1990–2013 (United States Environmental Protection Agency, 2014).

technologies which in turn leads to market failures and low patronization of the renewable energy technology. In the light of this, an effective renewable energy policy should take the interconnection of factors affecting renewable energy supplies and sustainability into consideration.

The following are policy recommendations emanating from the study that can help mitigate climate change and its impacts:

• All sectors and regions have the potential to contribute by investing in Renewable energy technologies and policies to help reduce it.

• Reducing our carbon footprint through the changes in lifestyle and behaviour patterns can contribute a great deal to the mitigation of climate change.

• Research into innovations and technologies that can reduce land use and also reduce accidents from renewable energy sources and the risk of resource competition, for example in Bioenergy where food for consumption competing with energy production.

• Enhancing international cooperation and support for developing countries towards the expansion of infrastructure and upgrading technology for modern supply and sustainable energy services as a way of mitigating climate change and its impacts.

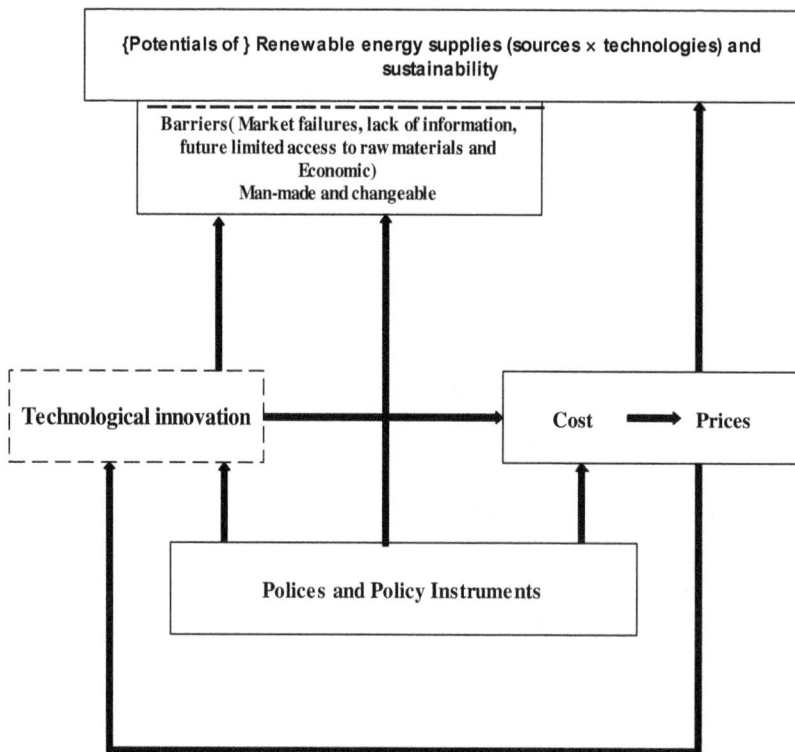

Figure 5. Interconnection of factors affecting renewable energy supplies and sustainability, adapted from Edenhofer et al. (2011); Verbruggen et al. (2010).

6. Conclusion

Energy is a requirement in our everyday life as a way of improving human development leading to economic growth and productivity. The return-to-renewables will help mitigate climate change is an excellent way but needs to be sustainable in order to ensure a sustainable future for generations to meet their energy needs. Knowledge regarding the interrelations between sustainable development and renewable energy in particular is still limited. The aim of the paper was to ascertain if renewable energy sources were sustainable and how a shift from fossil fuel-based energy sources to renewable energy sources would help reduce climate change and its impact. A qualitative research was employed by reviewing papers in the scope of the study. Even though, the complete lifecycle of renewable energy sources have no net emissions which will help limit future global greenhouse gas emissions. Nevertheless, the cost, price, political environment and market conditions have become barriers preventing developing, least developed and developed countries to fully utilize its potentials. In this way, a creation of global opportunity through international cooperation that supports least developed and developing countries towards the accessibility of renewable energy, energy efficiency, clean energy technology and research and energy infrastructure investment will reduce the cost of renewable energy, eliminate barriers to energy efficiency (high discount rate) and promote new potentials towards climate change mitigation.

The study brought to light the opportunities associated with renewable energy sources; energy security, energy access, social and economic development and climate change mitigation and reduction of environmental and health impacts. There are challenges that tend to hinder the sustainability of renewable energy sources and its ability to mitigate climate change. These challenges are market failures, lack of information, access to raw materials for future renewable resource deployment, and most importantly our (humans) way of utilizing energy in an inefficient way.

From the findings, the following suggestions are made that can help improve the concerns of renewable energy being sustainable and also reduce the rate of the depletion of the ozone layer due to the emissions of GHG especially carbon dioxide (CO_2):

- Formulation of policies and discussions from all sectors towards the improvement of technologies in the renewable sector to sustain them.

- Changes in our use of energy in a more efficient way as individuals, countries and the world as a whole. Efforts that aim at increasing the share of renewable energy and clean fossil fuel technologies into global energy portfolio will help reduce climate change and its impacts. Energy efficiency programmes should be introduced globally, which give tax exemptions to firms who prove to provide energy efficiency initiatives (energy-efficient homes), product design (energy-efficient equipment) and services (industrial combined heat and power). Introducing the concept of usability, adaptability and accessibility into energy-dependent product design is a way of promoting energy efficient behaviours.

- Increase research in these areas, so that the fear of some renewables posing risks in the future is limited.

- Improve education, awareness-raising and human institutional capacity on climate change mitigation, adaptation, impact reduction and early warning. Developed countries should incorporate decarbonization policies and strategies into the industry, energy, agricultural, forest, health, transport, water resource, building and other sectors that have potential of increasing greenhouse gas emissions. Efforts in developing countries aimed at improving institutional training, strengthening institutions and improving capacity of research on climate change will increase awareness, promote adaptation and sustainable development. Least developed countries should develop and test tools and methods with a global support that direct policy and decision-making for climate change mitigation, adaptation and early warnings. Supporting a global dialogue through international cooperation and partnership with developed, developing and least developed countries will promote the development, dissemination and transfer of environmentally friendly technologies, innovation and technology, access to science, and among others which will increase the mutual agreement towards combating climate change and its impacts.

If these suggestions are implemented, the sustainability of renewable energy resources would be addressed as well as the seventh and thirteenth goal of sustainable development which seeks to ensure access to affordable, reliable, sustainable, modern energy for all and combat climate change and its impact.

Funding
The authors received no direct funding for this research.

Author details
Phebe Asantewaa Owusu[1]
E-mail: phebe.owusu@metu.edu.tr
Samuel Asumadu-Sarkodie[1]
E-mail: samuel.sarkodie@metu.edu.tr
[1] Sustainable Environment and Energy System, Middle East Technical University, Northern Cyprus Campus, Kalkanli, Guzelyurt 99738, TRNC, Turkey.

References
Abbasi, T., & Abbasi, S. (2010). *Renewable energy sources: Their impact on global warming and pollution*. PHI Learning.
Abbasi, T., Premalatha, M., & Abbasi, S. (2011). The return to renewables: Will it help in global warming control? *Renewable and Sustainable Energy Reviews, 15*, 891–894. http://dx.doi.org/10.1016/j.rser.2010.09.048

Ajanovic, A. (2011). Biofuels versus food production: Does biofuels production increase food prices? *Energy, 36*, 2070–2076. http://dx.doi.org/10.1016/j.energy.2010.05.019
Asumadu-Sarkodie, S., & Owusu, P. A. (2016a). Feasibility of biomass heating system in Middle East Technical University, Northern Cyprus campus. *Cogent Engineering, 3*. doi:10.1080/23311916.2015.1134304
Asumadu-Sarkodie, S., & Owusu, P. A. (2016b). A review of Ghana's energy sector national energy statistics and policy framework. *Cogent Engineering, 3*. doi:10.1080/233 11916.2016.1155274
Asumadu-Sarkodie, S., & Owusu, P. A. (2016c). Multivariate co-integration analysis of the Kaya factors in Ghana. *Environmental Science and Pollution Research*. doi:10.1007/s11356-016-6245-9
Asumadu-Sarkodie, S., & Owusu, P. A. (2016d). The potential and economic viability of solar photovoltaic in Ghana. *Energy Sources, Part A: Recovery, Utilization, and Environmental Effects*. doi:10.1080/15567036.2015.112 2682
Asumadu-Sarkodie, S., & Owusu, P. A. (2016e). The potential and economic viability of wind farms in Ghana *Energy Sources, Part A: Recovery, Utilization, and Environmental Effects*. doi:10.1080/15567036.2015.1122680
Asumadu-Sarkodie S, & Owusu, P. A. (2016f). The relationship between carbon dioxide and agriculture in Ghana, a comparison of VECM and ARDL model. *Environmental*

Science and Pollution Research. doi:10.1007/ s11356-016-6252-x

Asumadu-Sarkodie, S., & Owusu, P. A. (2016g). Carbon dioxide emissions, GDP, energy use and population growth: A multivariate and causality analysis for Ghana, 1971–2013. *Environmental Science and Pollution Research International.* doi:10.1007/s11356-016-6511-x

Asumadu-Sarkodie, S., Owusu, P. A., & Jayaweera, H. M. (2015). Flood risk management in Ghana: A case study in Accra. *Advances in Applied Science Research, 6,* 196–201.

Asumadu-Sarkodie, S., Owusu, P. A., & Rufangura, P. (2015). Impact analysis of flood in Accra, Ghana. *Advances in Applied Science Research, 6,* 53–78.

Asumadu-Sarkodie, S., Rufangura, P., Jayaweera, H. M., & Owusu, P. A. (2015). Situational analysis of flood and drought in Rwanda. *International Journal of Scientific and Engineering Research, 6,* 960–970. doi:10.14299/ ijser.2015.08.013

Ayoub, M., & Abdullah, A. Z. (2012). Critical review on the current scenario and significance of crude glycerol resulting from biodiesel industry towards more sustainable renewable energy industry. *Renewable and Sustainable Energy Reviews, 16,* 2671–2686. http://dx.doi.org/10.1016/j.rser.2012.01.054

Baños, R., Manzano-Agugliaro, F., Montoya, F., Gil, C., Alcayde, A., & Gómez, J. (2011). Optimization methods applied to renewable and sustainable energy: A review. *Renewable and Sustainable Energy Reviews, 15,* 1753–1766. http://dx.doi.org/10.1016/j.rser.2010.12.008

Barbier, E. (2002). Geothermal energy technology and current status: An overview. *Renewable and Sustainable Energy Reviews, 6,* 3–65. http://dx.doi.org/10.1016/S1364-0321(02)00002-3

Baum, C., Leinweber, P., Weih, M., Lamersdorf, N., & Dimitriou, I. (2009). Effects of short rotation coppice with willows and poplar on soil ecology. *Landbauforschung vTI Agriculture and Forestry Research, 59,* 09–2009.

Baum, S., Weih, M., Busch, G., Kroiher, F., & Bolte, A. (2009). The impact of short rotation coppice plantations on phytodiversity. *Landbauforschung vTI Agriculture and Forestry Research, 3,* 163–170.

Brew-Hammond, A. (2010). Energy access in Africa: Challenges ahead. *Energy Policy, 38,* 2291–2301. http://dx.doi.org/10.1016/j.enpol.2009.12.016

Demirbas, M. F., Balat, M., & Balat, H. (2009). Potential contribution of biomass to the sustainable energy development. *Energy Conversion and Management, 50,* 1746–1760.

Earth System Research Laboratory (2015) *The NOAA annual greenhouse gas index (AGGI).* Retrieved October 24, 2015, from http://www.esrl.noaa.gov/gmd/aggi/aggi.html

Edenhofer, O., Pichs-Madruga, R., Sokona, Y., Seyboth, K., Matschoss, P., Kadner, S., ... von Stechow, C. (2011). *Renewable Energy Sources and Climate Change Mitigation.* Cambridge : Cambridge University Press. http://dx.doi.org/10.1017/CBO9781139151153

EEA. (2016). Mitigating climate change, greenhouse gas emissions. Retrieved from http://www.eea. europa.eu/soer-2015/countries-comparison/ climate-change-mitigation

Esteban, M., & Leary, D. (2012). Current developments and future prospects of offshore wind and ocean energy. *Applied Energy, 90,* 128–136. http://dx.doi.org/10.1016/j.apenergy.2011.06.011

Førsund, F. R. (2015). *Hydropower economics* (Vol. 217). New York: Springer.

Fräss-Ehrfeld, C. (2009). *Renewable energy sources: A chance to combat climate change* (Vol 1). Kluwer Law International.

Hák, T., Janoušková, S., & Moldan, B. (2016). Sustainable development goals: A need for relevant indicators.

Ecological Indicators, 60, 565–573. http://dx.doi.org/10.1016/j.ecolind.2015.08.003

Hamann A. (2015). *Coordinated predictive control of a hydropower cascade.*

Headey, D., & Fan, S. (2008). Anatomy of a crisis: The causes and consequences of surging food prices. *Agricultural Economics, 39,* 375–391. http://dx.doi.org/10.1111/agec.2008.39.issue-s1

Hoogwijk, M., Faaij, A., Eickhout, B., de Vries, B., & Turkenburg, W. (2005). Potential of biomass energy out to 2100, for four IPCC SRES land-use scenarios. *Biomass and Bioenergy, 29,* 225–257. http://dx.doi.org/10.1016/j.biombioe.2005.05.002

International Energy Agency. (2014). *World Energy Outlook Special Report.* Retrieved August 17, 2015, from http:// www.iea.org/publications/freepublications/publication/ WEO2014_AfricaEnergyOutlook.pdf

Jacobson, M. Z., & Delucchi, M. A. (2011). Providing all global energy with wind, water, and solar power, Part I: Technologies, energy resources, quantities and areas of infrastructure, and materials. *Energy Policy, 39,* 1154–1169.

Kaygusuz, K. (2012). Energy for sustainable development: A case of developing countries. *Renewable and Sustainable Energy Reviews, 16,* 1116–1126. http://dx.doi.org/10.1016/j.rser.2011.11.013

Koh, L. P., & Ghazoul, J. (2008). Biofuels, biodiversity, and people: Understanding the conflicts and finding opportunities. *Biological Conservation, 141,* 2450–2460. http://dx.doi.org/10.1016/j.biocon.2008.08.005

Kruyt, B., van Vuuren, D. P., de Vries, H., & Groenenberg, H. (2009). Indicators for energy security. *Energy Policy, 37,* 2166–2181. http://dx.doi.org/10.1016/j.enpol.2009.02.006

Larsen H. H., Kristensen N. B., Sønderberg Petersen L., Kristensen H. O. H., Pedersen A. S., Jensen T. C., & Schramm J. (2009, March 17-18). *How do we convert the transport sector to renewable energy and improve the sector's interplay with the energy system?* Background paper for the workshop on transport-renewable energy in the transport sector and planning, Technical University of Denmark. Technical University of Denmark.

Lu, Y., Nakicenovic, N., Visbeck, M., & Stevance, A.-S. (2015). Policy: Five priorities for the UN sustainable development goals. *Nature, 520,* 432–433. http://dx.doi.org/10.1038/520432a

Manwell, J. F., McGowan, J. G., & Rogers, A. L. (2010). *Wind energy explained: Theory, design and application.* Wiley.

Owusu, P. A., Asumadu-Sarkodie, S., & Ameyo, P. (2016). A review of Ghana's water resource management and the future prospect. *Cogent Engineering, 3.* doi:10.1080/23311 916.2016.1164275

Panwar, N., Kaushik, S., & Kothari, S. (2011). Role of renewable energy sources in environmental protection: A review. *Renewable and Sustainable Energy Reviews, 15,* 1513– 1524. http://dx.doi.org/10.1016/j.rser.2010.11.037

Robertson, G., Dale, V. H., Doering, O. C., Hamburg, S. P., Melillo, J. M., Wander, M. M., ... Wilhelm, W. W. (2008). Sustainable biofuels redux. *Science, 322,* 49–50. doi:10.1126/ science.1161525

Rogelj, J., McCollum, D. L., Reisinger, A., Meinshausen, M., & Riahi, K. (2013). Probabilistic cost estimates for climate change mitigation. *Nature, 493,* 79–83. http://dx.doi.org/10.1038/nature11787

Schulz, U., Brauner, O., & Gruß, H. (2009). Animal diversity on short-rotation coppices–a review. *VTI Agriculture and Forestry Research, 3,* 171 181.

Tester J. W. (2005). *Sustainable energy: Choosing among options.* London: MIT Press.

Tilman, D., Hill, J., & Lehman, C. (2006). Carbon-negative biofuels from low-input high-diversity grassland biomass. *Science, 314,* 1598–1600. http://dx.doi.org/10.1126/science.1133306

Tiwari, G. N., & Mishra, R. K. (2011). *Advanced renewable energy sources*. Royal Society of Chemistry.

Twidell, J., & Weir, T. (2015). *Renewable energy resources*. Routledge.

U.S. Energy Information Administration. (2012). *International energy statistics*. Retrieved October 18, 2015, from http://www.eia.gov/cfapps/ipdbproject/IEDIndex3.cfm?tid=2&pid=2&aid=2

UNFCC. (2015). *Adoption of the Paris agreement*. Retrieved October 24, 2015, from http://unfccc.int/resource/docs/2015/cop21/eng/l09.pdf

United States Environmental Protection Agency. (2014). *Carbon dioxide emissions*. Retrieved December 2, 2015, from

http://www3.epa.gov/climatechange/ghgemissions/gases/co2.html

Urban, F., & Mitchell, T. (2011). *Climate change, disasters and electricity generation*.

Verbruggen, A., Fischedick, M., Moomaw, W., Weir, T., Nadaï, A., Nilsson, L. J., ... Sathaye, J. (2010). Renewable energy costs, potentials, barriers: Conceptual issues. *Energy Policy, 38*, 850–861. http://dx.doi.org/10.1016/j.enpol.2009.10.036

World Energy Council. (2013). *World Energy Resources: Hydro*. Retrieved January 26, 2016, from https://www.worldenergy.org/wp-content/uploads/2013/10/WER_2013_5_Hydro.pdf

5

Analysis of isotope element by electrolytic enrichment method for ground water and surface water in Saurashtra region, Gujarat, India

Sajal Singh[1], Athar Hussain[2]* and S.D. Khobragade[3]

*Corresponding author: Athar Hussain, Civil Engineering Department, Ch. Brahm Prakash Government Engineering College, Jaffarpur, Delhi, India

E-mail: athariitr@gmail.com

Reviewing editor: Shashi Dubey, Hindustan College of Engineering, India

Abstract: The present study has been aimed for the assessment of isotope element Tritium (^3H). It is a great threat to human health and environment for lengthy duration. The tritium exists in earth in diverse forms such as (1) small amounts of natural tritium are produced by alpha decay of lithium-7, (2) natural atmospheric tritium is also generated by secondary neutron cosmic ray bombardment of nitrogen, (3) atmospheric nuclear bomb testing in the 1950s, although the contribution from nuclear power plants is small. Tritium or ^3H is a radioactive isotope of hydrogen with a half-life of 12.32 ± 0.02 years. Water samples from ground water, surface water, and precipitation were collected from different locations in Gujarat area and were analyzed for the same. Distillation of samples was done to reduce the conductivity. Deuterium and Hydrogen were removed by the process of physico-chemical fractionation in the tritium enrichment unit. The basis of physico-chemical fractionation is the difference in the strength of bonds formed by the light vs. the heavier isotope of a given element. A total of 10 cycles (runs) were executed using Quintals process. Tritium concentration files were created with help of WinQ and Quick start software in Quintals process (Liquid Scintillation Spectrometer). The concentration of tritium in terms of tritium units (TU) of various samples has been determined. The TU values of the samples vary in the range of 0.90–6.62 TU.

Subjects: **Environmental Health; Pollution; Water Engineering; Water Science**

Keywords: **tritium; isotope; half-life; groundwater; deuterium; hydrogen; evironmental sustainability engineering**

ABOUT THE AUTHORS

Sajal Singh (M Tech Environmental Science & Engineering), research completed at National Institute of Hydrology (Ministry of Water Resources), Roorkee (IIT Campus). Research and development work of the monitoring to effluent treatment plant, sewage treatment plant, common effluent treatment plant, and other water and wastewater sources. His research interest includes the treatment of the typically contaminated water and wastewater treatment. Athar Hussain is currently an associate professor at Ch. B.P. Govt. Engineering College, Jaffarpur, Delhi, India. His research interest includes water and wastewater treatment. He is specially focused on biological treatment of wastewater.

PUBLIC INTEREST STATEMENT

Present study has been aimed to determine the concentration of the tritium from different sources of water. Because most of tritium is disseminated in the environment as water, it enters the hydrologic cycle as precipitation and eventually becomes concentrated in levels detectable in groundwater. It is a great threat to human health and environment for lengthy duration. Environmental tritium is a powerful tool for quantitative groundwater and surface water research, because it is a useful tracer for groundwater movement. Deuterium and Hydrogen were removed by the process of physico-chemical fractionation in the tritium enrichment unit. A total of 10 cycles (runs) were executed in Quintals process. Tritium concentration files were created with help of WinQ and Quick start software in Quintals process (Liquid Scintillation Spectrometer). Classification: Environmental Engineering, Water treatment.

1. Introduction

Tritium (T) or ^3H is a radioactive isotope of hydrogen (having two neutrons and one proton) with a half-life of 12.32 ± 0.02 years, decays to ^3He emitting a beta particle having a radiation energy of 0.0057 MeV. Tritium concentration is measured in tritium units (TU). One TU is defined as the presence of one tritium in 10^{18} atoms of hydrogen (H). In the earth, small amount of natural tritium is produced by alpha decay of lithium-7 (Kumar & Somashekar, 2011). Natural atmospheric tritium is also generated by secondary neutron cosmic ray bombardment of nitrogen. Tritium atoms then combine with oxygen, forming water that subsequently falls as precipitation. The extent of water/rock interaction, and hence the groundwater chemistry (Grahame, Manrico, & Offer, 1986), depends on the mineralogy of the aquifer rock and the residence time that the groundwater has been in contact with the rock. Prior to atmospheric nuclear bomb testing in the 1950s, tritium's natural average concentration ranged from approximately 2–8 TU. Since cessation of atmospheric nuclear tests, tritium concentrations have dropped between 12 and 15 TU, (Blavoux et al., 2013) although small contributions from nuclear power plants occur. Because most of tritium is disseminated in the environment as water, it enters the hydrologic cycle as precipitation and eventually becomes concentrated in levels detectable in groundwater.

Any possible health effects from tritium are the result of the beta radiation it emits. Because tritium's radiation cannot penetrate the skin, the only real exposure a person receives is the radiation received while tritium is inside the body. Exposure time and thus the possibility of health effects of the tritium depend on the form of tritium present, elemental tritium gas or tritium oxide. Tritium primarily enters the body when people drink tritiated water. People may also inhale tritium as a gas or absorb it through their skin, but exposure to harmful levels of tritium through these exposure pathways is very unlikely. Once tritium enters the body, it is quickly and uniformly distributed throughout the body, going directly into soft tissues and organs. The associated dose to these tissues is generally uniform and dependent on the tissues' water content. Regardless of the way it enters the body, tritium oxide immediately mixes with the body fluids and is eliminated like normal water (Stamoulis, Karamanis, & Ioannides, 2011). When it passes through a human body, it can produce permanent changes in cells. There are three principal potential health effects: cancer, genetic effects, and effects on fetuses. The body removes tritium naturally in the same way it removes water—by excreting it in the urine. As with all ionizing radiation, in theory, we assume exposure to tritium could increase the risk of developing cancer.

Tritium does not totally harmfully for human being, but it is also advantageous to water research. Environmental tritium is a powerful tool for quantitative groundwater and surface water research, because it is a useful tracer for groundwater movement. The age of water can be defined with help of tritium concentration (Iwatsuki, Xu, Itoh, Abe, & Watanabe, 2000). However, it is not possible to determine the correct age of groundwater by measuring the tritium concentration alone, but one should have other information about the aquifer in question and the flux of water. Tritium concentration in precipitation varied quite a lot during the last forty years due to atmospheric nuclear tests. Usually ground water will be mixture of many year input of tritium concentrations (Colville, 1984). If the tritium concentration is zero, then we can say that the water is old and has been cut off from the atmosphere for more than 50 years.

The tritium distribution in the groundwater is determined by (1) the input functions for the tritium concentration in recharge water, (2) the movement and mixing of water within the saturated zone (Cheng et al., 2010). Both the above factors have to be addressed when using the tritium concentration to estimate the age of water. The aim of this paper is to assessment of isotope element (tritium) of different sources of water. To meet these objectives, water samples from ground water, surface water, and precipitation were collected from different location in Gujarat area and were analyzed. The samples are provided by NIH.

2. Materials and methods

2.1. Sources of samples for tritium analysis

Typically, for environmental tritium analysis an aliquot of one liter is sufficient. The samples were collected as raw unfiltered water with no preservatives and stored either in a glass or high density polyethylene bottle with air-tight caps. Samples were given a unique sequential number and their description (i.e. sample, source type, location, etc.) and logged into a record book for reference.

A total number of 16 samples were collected and analyzed. The samples were collected from various sources such as hand pump, piezometer, and surface water from Saurashtra region. Saurashtra region is located in Gujarat state of India. The hand pump samples represent the shallow aquifer while piezometer samples represent deep aquifer. Initial analysis indicated that the samples had very high conductivity which is not enviable for tritium enrichment process. Conductivity of all samples to be analyzed should always be less than 100 µs/cm for carrying out tritium enrichment process. Therefore, distillation of the samples was done to reduce the conductivity. The distillation process was repeated three times till the desired conductivity levels were attained. Details of the various samples are given in Table 1.

The four steps involved in the process of tritium analysis are:

(1) Pre-distillation of sample, (2) Tritium Enrichment process, (3) Post-distillation of sample, (4) Quintals process. The various steps are discussed below.

2.2. Pre-distillation of sample

The samples were distilled (primary distillation) in the laboratory to remove all dissolved salts a pre-requirement before further processing. The various ions that are naturally found in water (Cl^-, SO_4^-, CO_3, Mg^{++}, Na^+, K^+, etc.) could interfere with the electrolysis process (i.e. produces gasses at either the

S. No.	Sample ID	Location	EC (I testing) (µS/cm)	EC (II testing) (µS/cm)	EC (III testing) (µS/cm)	Latitude	Longitude
1	JND-Pz-32	Khageshri	70	–	–	2,405,478	601,509
2	Pz-P-12	Adityana	300	60	–	2,399,341	567,817
3	JND-Pz-18	Ranavav	30	–	–	2,398,553	576,711
4	Pz-P-08	Ranavav	340	30	–	2,397,599	565,882
5	Pz-P-11	Mokar	1,310	80	–	2,387,885	580,535
6	Pz-P-02	Jambu	450	280	100	2,386,291	590,799
7	Pz-P-06	Kuchhadi	350	70	–	2,398,229	558,715
8	Pz-P-01	Mokar	420	170	70	2,389,970	581,307
9	JND-21	Rana kandorna	60	–	–	2,393,592	591,520
10	WS-2 creek near 3S-4	Tukda	320	60	–	2,82,173	574,511
11	Pz-P-10	Oddadar	220	100	–	2,386,552	572,478
12	Hand pump Khirasara	Khirasara	80	–	–	–	–
13	JND-Pz-17	Kantela	440	120	80	2,400,626	554,017
14	P5N-2	Kuchadi	50	–	–	2,396,509	557,523
15	Pz-P-09	Chaya	20	–	–	2,391,779	565,670
16	WS-4	Oddadar	30	–	–	2,383,771	576,831

Table 1. Details of all the samples collected along with different locations

anode or cathode other than oxygen or hydrogen) and also corrode the mild steel electrode. It is important to note that not all salts needs to be removed through primary distillation. If volatile salts or soluble gasses are present they could distill over with the water. When sample in distillation flask is nearly or completely distilled to dryness, turn off the heating unit check the conductivity of the distilled water sample. If it is up to 100 μS/cm then place (Clark & Fritz, 1997) the distilled samples in a tightly sealed container. In case of saline samples, the conductivity of the distilled sample may be more than 100 μS/cm. In such cases water samples should be distilled again.

2.3. Tritium enrichment unit

2.3.1. Tritium enrichment system
The electrolytic enrichment is the simplest and provides relatively high enrichment compared to other technique. In the electrolytic enrichment unit, tritium in sample water is concentrated by low temperature electrolysis. The electrolytic enrichment system contains: (1) electrolytic cell, (2) constant current (DC) supply unit, (3) cooling unit.

The process of electrolysis is very slow. The tritium measurement involves enrichment of standards, local water as a secondary standard, and a background sample. Long measurement duration and large number of samples analysis necessitate electrolytic system large enough to accommodate sufficient number of electrolytic cells. The system in use at National Institute of Hydrology (NIH), Roorkee, India (since 11–12 years) accommodates 20 electrolytic cells in the unit. The electrolytic reduction is done to concentrate the tritium. The samples contain mainly HHO and HTO molecules. By passing electric current through a conducting water solution, the bonds of the water molecules are broken with evolution of hydrogen and oxygen (Kluge et al., 2010). The temperature of the sample is maintained between 0 and 5 C in order to achieve the maximum fractionation or enrichment of HTO.

2.3.2. Electrolytic enrichment tritium theory
Electrolytic of water is a process in which water molecule dissociates into basic elements namely, hydrogen and oxygen on applying an electric field. During electrolysis, hydrogen appears at the cathode and oxygen at the anode. Similarly like hydrogen in case of HTO, tritium gets released at the cathode.

In the dissociation process lighter molecules get dissociated at a lower dissociation energy compared to that of heavier molecules (Gat, 1980). The molecules weight of HTO (20) is 2 units higher as compared to the molecules of H_2O (=18). Therefore on electrolysis the H_2O dissociates much rapidly compared to HTO molecule. This differential rate of dissociation leads to enrichment of HTO molecule in the water phase. The process therefore not only dissociates water into their respective gasses with the observed depletion in the quantity but also relatively enriches HTO molecules with respect to H_2O molecule in the liquid phase. In fact, the parameters like supplied electrical energy, reduction in the water amount, and enrichment of HTO in water phase are directly related.

The enrichment is also expressed in terms of isotopic ratio ($^3H/H$) in water to vapor phase (or released gasses in the case of electrolytic enrichment) as,

$$\beta\left(^3H\right)_{water-vapor} = \frac{\left(_{water}^3H/H\right)}{\left(_{vapor}^3H/H\right)} \tag{1}$$

Thus, the fractionation is the isotope concentration which is observed due to isotopic mass dependent control on the rate of physico-chemical reactions. The fractionation affects the individual isotope concentration in the reactant and product phase with no effect on the chemical nature of reactant or product of product species. The enrichment of deuterium and tritium is due to isotope effects that take place during electrolysis. Isotope fractionation occurs in any thermodynamic reaction due to differences in the rates of reaction for different molecular species. The result is a disproportionate concentration of one isotope over the other.

2.3.3. Isotopic fractionation (thermodynamic and quantum mechanical explanation)

The basis of physico-chemical fractionation is the difference in the strength of bonds formed by the light vs. the heavier isotope of a given element. Difference in the bond strength for an isotope of the same element is because of the differences in their reaction rates. The zero point energy is essentially a minimum potential energy of a molecular bond in a vibrating atom. The dissociation energy is more for heavy isotopes as compared to the lighter isotopes. The difference in the dissociation energy affects their reaction rates with lighter isotope reacting rapidly as compared to heavier isotope (Kitaoka, 1981). Isotope fractionation during physico-chemical reaction arises from this differential rate of chemical reaction. Isotope fractionation during physico-chemical reactions arises from the difference itself.

During the phase transition process, the dissociation energy relates to the partition function Q by the relation:

$$Q = \sum \sigma^{-1} m^{3/2} e^{-E/KT} \tag{2}$$

where σ is the asymmetry value, m is the isotopic, E is the energy state (in J/mole), Σ is the summation taken over all energy states from the point energy to the energy of the dissociated molecule, K is the Boltzmann's constant (gas constant per molecule), T is the temperature in Kelvin.

2.4. Distillation of enrichment samples (secondary distillation)

2.4.1. Use of lead chloride in post distillation

This is done to remove the Na^+ that got added through addition of Na_2O_2. The sample is transferred into a 50-ml round bottom flasks, then 8 g of $PbCl_2$ is added. Thereafter, condenser and collector are connected and the distillation is carried out. The flask is kept heated even after the drying of water to recover all of the hydrogen.

2.5. Quintals process

2.5.1. Liquid scintillation spectrometer

Quantulus is an ultra-low level spectrometer which provides stable measurement conditions, where no atmospheric pressure correction is needed even during long-term low radioactivity sample counting. The system includes PC software to allow complete data processing and the plotting of up to six spectra on screen. Liquid scintillation spectrometer is controlled by two PC software (WinQ and Quick start software).

Alpha, beta, and gamma radiation all fall into the category of ionizing radiation. Alpha and beta particles directly ionize the atoms with which they interact, adding or removing electrons. Gamma rays cause secondary electron emissions, which then ionize other atoms. The ionized particles left in the wake of a ray or particle can be detected as increasing conductivity in an otherwise insulating gas, which is done in electroscopes, ionization chambers or proportional counting chambers. These devices measure the pulse of conductivity between two electrodes when a particle or ray ionizes the gas between them. If a sufficiently high voltage is applied between the electrodes, an amplification of the signal can be obtained, and such counters can be quite sensitive. Their utility is severely limited by the fact that for most research applications only gas phase isotopes can be detected. This greatly complicates sample preparation and may preclude the analysis of some compounds entirely.

2.5.2. Observations

Counting is carried out in polyethylene scintillation vials. For this 8 g of enrichment sample/spike/ background is mixed with 13 g of Ultima Gold LLT. Sample was prepared for counting containing the standards STD-B and STD-C (two vials each), three vials of tritium free water and all the enriched samples (Saito, Shimamune, Nasik, Schimizu, & Hayashi, 1996). Special activity was being calculated as count per minute per gram of sample weight. The vials were placed in counter trays in the three pre-planned positions corresponding to the plan previously entered in the quantulus logbook. The standard and Tritium- Free vials were placed in a spatially distributed manner. The whole counting was being done in 10 cycles. Each vial has been counted for 40 min in each cycle thus each sample is counted for a total time of about 400 min.

The cycling helps in averaging any change that may occur such as lose of sample/cocktail weight, change in counting characteristics of the scintillator, change in counter stability, etc., which affects the sample count rate. This also helps in discarding any particular data, which is not statistically acceptable. Allow sample to remain in the counter for 24 h prior to the counting in order to adjust to counter temperature. The quantulus machine is operated through the WinQ and Quick start software.

2.5.3. Estimation of tritium concentration

The spiked samples estimation with tritium has been carried out. Out of all electrolytic cells, any three cells were used for enriching tap water samples spiked with tritium activity at another three for blank tap water sample. The spiked samples were meant for determining their tritium enrichment factor "Z" (final tritium concentration/initial tritium concentration). C.B. Taylor, I.A.E.A. has developed the following equation for the estimation of tritium enrichment factor of unknown samples.

$$\frac{\ln Z}{\ln N \cdot r_a} = \left(1 - \frac{1}{\beta}\right) + \frac{r_c}{r_a}\left(1 - d\right) \tag{3}$$

where r_a is the rate of electrolysis/total loss rate of hydrogen from cell, N is W_0/W_f (initial weight of the sample/final weight of the sample), β is instantaneous electrolytic separation factor for tritium, r_c is the rate of hydrogen loss as water vapor/total loss of hydrogen from cell, d is the fractionation factor for loss of tritium by evaporation.

On the right-hand side of Equation (3) the second term is of order of 0.5% of the first turn. The variation from cell to cell is negligible if it is based on a same set operated under the similar conditions (Michael, 2012). Equation (3) therefore indicates that a set of cells with uniform separation factor assigned as the symbol β when run simultaneously should show uniformity for the parameter $\ln Z/r_a$ $\ln N$. This parameter E should be used to evaluate enrichment factor, and to monitor the performance of a set of batch electrolysis cells. With slight modification in Equation (3), enrichment parameter E may be represented by Equation (4):

$$E = \left[\frac{W_0 - W_f}{W_e}\right]\frac{\ln Z}{\ln N \cdot r_a} = \left[\frac{W_0 - W_f}{W_e}\right]\left(1 - \frac{1}{\beta}\right) + \frac{r_c}{r_a}\left(1 - d\right) \tag{4}$$

where $(W_0 - W_f)$ is the total weight loss of the sample due to electrolysis evaporation and spray losses and "W_e" is the weight of electrolyzed water. If "W_e" cannot be measured accurately in the absence of an accurate ampere hour meter, then it can be made constant for all cells by passing series current for the same time. Because the second term on the right-hand side of Equation (4) is so small, the separation factor β is the parameter which mostly affects the magnitude and variability of the enrichment parameter E. In the case of spike samples, Z is measured and applied to determine E; the mean value "E_m" of E of the three samples, is then used to calculate Z for the unknown sample. If the total charge passes Q A h, then

$$W_e = \frac{Q}{2.97545}$$

(5)

Since the same amount of current is passed through all cells in series for the same length of time, the modified enrichment parameter, W_eE is used for the calculation of Z of unknown sample.

$$W_eE = \left[W_0 - W_f\right]\frac{\ln Z}{\ln N. \, r_a}$$

(6)

Therefore,

$$\ln Z = \left[W_0 - W_f\right]\cdot\left[W_eE_m \ln N\right]$$

(7)

where W_eE_m is the mean enrichment parameter of the three spiked cells, and N, Z and $(W_0 - W_f)$ are the mass reduction factor, tritium enrichment factor and total weight loss of the unknown sample (due to electrolysis, evaporation, and spray losses), respectively. The separation factor β for a spiked cell can be calculated using the equation,

$$B_{eff} = \ln\left(W_f \Big/ \ln\left(Z.\frac{W_f}{W_0}\right)\right)$$

(8)

The tap water sample is enriched mainly to determine the background count rate of spiked samples. This has become necessary as large quantity of tritium-free water is always not readily available to prepare the spike samples.

As the three spike cells are changed in rotation one after another in each run, one gets the enrichment parameter values of every cell which is found to be uniform if the cathode surfaces are well developed and unaffected by corrosion, chemical damage, mechanical damage, etc.

The tritium concentration of a sample on counting data is calculated as

$$C_s = C_{st}\cdot\left(\frac{N_s}{N_{st}}\right)\cdot\frac{1}{Z}$$

(9)

where C_s is the tritium concentration in TU of the sample, C_{st} is the tritium concentration in TU of the standard, N_s is the net count rate of the sample, N_{st} is the net count rate of the standard.

3. Results and discussion

A total of 10 cycles (runs) were completed in liquid scintillation spectrometer. The registry file of each cycle of samples was created with the help of WinQ software. The registry files so created are being run in quick start software. Using this software, the tritium concentration in Cpm (count per minute) unit was found. The tritium concentration file in Cpm was then run in WinQ software and the tritium concentrations in TU (Tritium Unit) were determined. Details of result of various samples are given in Table 2.

Data summarized in Table 3 indicate a high concentration of tritium in ground water in maximum samples. The tritium concentration values range between 0.90 and 6.62 TU (Figure 1). Ranavav region water contains a high concentration of tritium as compared to the other samples of different regions. The Jambu region water contains a lowest concentration of tritium as compared to other samples. While both regions located in Saurashtra are, of Gujarat. But tritium concentration of all samples is greater than exposure limit. Therefore, this water is very harmful for living organism.

The tritium concentration can be used as a powerful tool for quantitative groundwater research and rain water research. The radioactive element decaying time can be determined using half-life

	Table 2. Analysis details of samples collected at different locations					
S. No.	Sample	Code	Enrichment date	Cpm	TU	TU error (±)
1	JND-Pz-32	20	16-May-13	1.08	1.97	0.40
2	PZ-P-12	20	16-May-13	0.89	1.17	0.36
3	JND-PZ-18	20	16-May-13	1.18	2.48	0.43
4	PZ-P-08	20	16-May-13	2.08	6.62	0.46
5	PZ-P-11	20	16-May-13	1.12	2.85	0.54
6	PZ-P-02	20	16-May-13	0.80	0.90	0.37
7	PZ-P-06	20	16-May-13	1.26	3.20	0.51
8	PZ-P-01	20	16-May-13	1.12	2.16	0.41
9	JND-21	20	16-May-13	1.12	2.31	0.53
10	WS-2 creak near 3S-4	20	16-May-13	1.24	3.13	0.62
11	PZ-P-10	20	16-May-13	1.02	1.97	0.53
12	Hand pump Khirasara	20	16-May-13	0.92	1.50	0.51
13	JND-Pz-17	20	16-May-13	0.93	1.32	0.39
14	P5N-2	20	16-May-13	1.05	1.69	0.44
15	PZ-P-09	20	16-May-13	0.84	1.00	0.43
16	WS-4	20	16-May-13	1.24	2.60	0.52

formula used for radioactive elements. But one should have other information like aquifer (confined and unconfined). The decaying of parent isotope activity (related to radiation emission rate) takes place at a systematic rate governed by equation:

$$A = A_0 e^{-\lambda t} \tag{10}$$

where A_0 represents the maximum concentration of the tritium (for each sample) at initial period while A represents the concentration of the element at any time t. λ is the decay constant which is

	Table 3. Details of decaying time for various samples			
S. No.	Sample	Enrichment date	Tritium concentration in TU	Time (t) in years
1	JND-Pz-32	16-May-13	1.9	21.6
2	PZ-P-12	16-May-13	1.17	30.9
3	JND-PZ-18	16-May-13	2.48	17.5
4	PZ-P-08	16-May-13	6.62	0
5	PZ-P-11	16-May-13	2.85	15
6	PZ-P-02	16-May-13	0.90	35.6
7	PZ-P-06	16-May-13	3.20	12.9
8	PZ-P-01	16-May-13	2.16	20
9	JND-21	16-May-13	2.31	18.8
10	WS-2 creak near 3S-4	16-May-13	3.13	13.3
11	PZ-P-10	16-May-13	1.97	21.6
12	Hand pump Khirasara	16-May-13	1.50	26.5
13	JND-Pz-17	16-May-13	1.32	10.2
14	P5N-2	16-May-13	1.69	24.3
15	PZ-P-09	16-May-13	1.00	33.7
16	WS-4	16-May-13	2.60	16.7

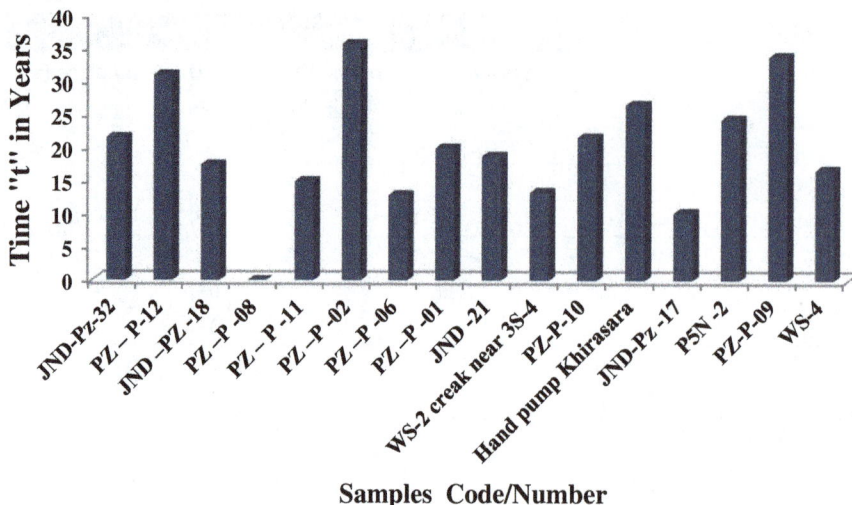

Figure 1. Tritium concentration variations of all samples from different locations.

characteristic of an isotope. Other information should be available such as ground water flex, aquifer situation, etc. However, it is not possible to determine the time for decay of its initial strength of tritium concentration by measuring the tritium concentration alone (Motzer, 2007), but one should have other information like aquifer (confined and unconfined). Here, A_0 denotes the maximum concentration of tritium depending on the aquifer condition. The study on ground water aquifer of Saurashtra region has already been carried out by the NIH. Based on that study it can be assumed that the A_0 is the maximum concentration of the tritium within all samples.

$$\lambda = \frac{0.693}{t^{\frac{1}{2}}}$$

(11)

where λ is the decay constant which is characteristic of an isotope and is definite in terms of its half-life is denoted by $t^{1/2}$. This $t^{1/2}$ is defined as the time in which the radioactive element decays to one half (1/2) of its initial strength. Half-life of the tritium is calculated as 12.35 ± 0.02.

From the present study, the time values of groundwater samples as calculated in the study area ranges between 0 and 35.6 year (Figure 2). The lower concentration of tritium indicates the old water or multiple birth of the present water and higher concentration of tritium indicates recently

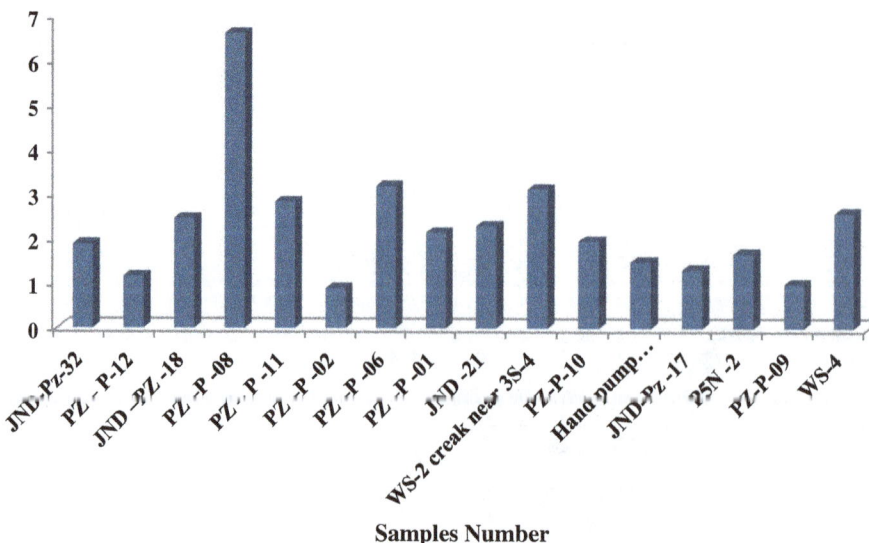

Figure 2. Decaying time variations of tritium in samples from different locations.

generated water (Morgenstern & Daughney, 2012). It can be inferred that the tritium concentration is decreasing with respect to time in all the samples of present study area.

In Equation (12), A is determined experimentally through measurement of sample activity, A_0, the initial activity is estimated from combination of measured data and field data. On solving the Equation (12), the age can be calculated which is the time elapsed to decay the parent activity from A_0 to A. For the case of ^3H Equation (12) reduces to age equations as:

$$^3H \, age(Years) = \frac{lnA_o}{A} \tag{12}$$

4. Conclusion

The samples from different sources such as hand pump and piezometer belonging to Saurashtra region of Gujarat were analyzed for tritium. The samples analysis was carried out in Nuclear Hydrology Laboratory at NIH, Roorkee. The experimental analysis of tritium involves four steps namely pre-distillation, tritium enrichment, post-distillation, and quintals analysis. The concentrations of tritium in all samples has been analyzed and quoted in terms of tritium unit (TU). The tritium concentration in all samples lies in the range from 0.90 to 6.62 TU. The obtained concentration values can be used to determine the age of the source of water. However, it appears that sample number 6 has added values of the samples.

Funding
The authors received no direct funding for this research.

Author details
Sajal Singh[1]
E-mail: chodharysajal5@gmail.com
Athar Hussain[2]
E-mail: athariitr@gmail.com
S.D. Khobragade[3]
E-mail: sdkhobragade@yahoo.com
[1] School of Engineering, Gautam Buddha University, Greater Noida 201310, Uttar Pradesh, India.
[2] Civil Engineering Department, Ch. Brahm Prakash Government Engineering College, Jaffarpur, Delhi, India.
[3] Hydrological Investigations Division, National Institute of Hydrology, Roorkee 247667, Uttarakhand, India.

References
Blavoux, B., Lachassagne, P., Henriot, A., Ladouche, B., Marc, V., Beley, J., ... Olive, P. (2013). A fifty-year chronicle of tritium data for characterising the functioning of the Evian and Thonon (France) glacial aquifers. *Journal of Hydrology, 494,* 116–133. http://dx.doi.org/10.1016/j.jhydrol.2013.04.029

Cheng, W., Kuo, T., Su, C., Chen, C., Fan, K., Liang, H., & Han, Y. (2010). Evaluation of natural recharge of Chingshui geothermal reservoir using tritium as a tracer. *Radiation Measurements, 45,* 110–117. http://dx.doi.org/10.1016/j.radmeas.2009.08.002

Clark, I. D., & Fritz, P. (1997). Environmental isotopes in hydrogeology (328 p). Boca Raton, FL: CRC Press/Lewis.

Colville, J. S. (1984). Estimation of aquifer recharge and flow from natural tritium content of groundwater, CSIRO. *Journal of Hydrology, 67,* 195–222.

Gat S. R. (1980). *The isotope of hydrogen and oxygen in precipitation, hand book of environmental isotope geochemistry* (Vol. 1, pp. 21–47). Amsterdam: Elsevier.

Grahame, J. L., Manrico, R. D., & Offer, S. (1986). *Application of the tritium interface method for determining recharge rates to unconfined drift aquifers.* East Lansing: Department of Geological Sciences, Michigan State University.

Iwatsuki, T., Xu, S., Itoh, S., Abe, M., & Watanabe, M. (2000). Estimation of relative groundwater age in the granite at the Tono research site, Tono Geoscience Center. *Japan Nuclear Cycle Development Institute (JNC), 959-3,* 509–5102.

Kitaoka, K. (1981). The electrolytic separation factor of Tritium Japan. *Radioisotopes, 30,* 247–252. http://dx.doi.org/10.3769/radioisotopes.30.5_247

Kluge, T., Riechelmann, D. F. C., Wieser, M., Spötl, C., Sültenfuß, J. D., Schröder-Ritzrau, A., ... Aeschbach-Hertig, W. (2010). Dating cave drip water by tritium. *Journal of Hydrology, 394,* 396–406. http://dx.doi.org/10.1016/j.jhydrol.2010.09.015

Kumar, P. R., & Somashekar, R. K. (2011). Environmental tritium and hydrochemical investigations to evaluate groundwater in Varahi and Markandeya river basins, Karnataka, India. *Journal of Environmental Radioactivity, 102,* 153–162.

Michael, K. S. (2012). A 40-year record of carbon-14 and tritium in the Christchurch groundwater system, New Zealand: Dating of young samples with carbon-14, New Zealand. *Journal of Hydrology, 430–431,* 50–68.

Morgenstern, U., & Daughney, C. J. (2012). Groundwater age for identification of baseline groundwater quality and impacts of land-use intensification – The national groundwater monitoring programme of New Zealand. *Journal of Hydrology, 456–457,* 79–93. http://dx.doi.org/10.1016/j.jhydrol.2012.06.010

Motzer, W. E. (2007). *Age dating groundwater.* Emeryville, CA. Retrieved from bmotzer@toddengineers

Saito, M. S. T., Shimamune, T., Nasik, Y., Schimizu, H., & Hayashi, T. (1996). Tritium enrichment by electrolysis using solid polymer electrolyte Japan. *Radioisotope, 45,* 3–10.

Stamoulis, K. C., Karamanis, D. B., & Ioannides, K. G. (2011). Assessment of tritium levels in rivers and precipitation in north-western Greece before the ITER operation. *Fusion Engineering and Design, 86,* 206–213. http://dx.doi.org/10.1016/j.fusengdes.2010.12.056

6

Electricity production potential and social benefits from rice husk

Obaidullah Mohiuddin[1], Abdullah Mohiuddin[1], Madina Obaidullah[1], Humayun Ahmed[1] and Samuel Asumadu-Sarkodie[1]*

*Corresponding author: Samuel Asumadu-Sarkodie, Sustainable Environment and Energy Systems, Middle East Technical University, Northern Cyprus Campus, Mersin 10, Turkey
E-mails: samuel.sarkodie@metu.edu.tr, samuelsarkodie@yahoo.com, asumadusarkodiesamuel@yahoo.com
Reviewing editor: Shashi Dubey, Hindustan College of Engineering, India

Abstract: Pakistan has been experiencing energy crisis owing to its sole dependence on fossil fuels. Reduction in local fossil fuel reserves has led to an increase in their prices, thereby increasing the cost of electricity. Since the tariff remains the same, Pakistan is over-burdened with circular debts and observes a daily power shortfall of about 12–14 h. Being an Agra-economic country, many major and minor crops are produced and exported in large quantities. This results in a bulk of the agricultural waste which are not utilized. The waste can be utilized to meet the country's energy demand while mitigating climate change and its impact. The study examines the electricity production potential and social benefits of rice husk in Pakistan. It is estimated in this study that if 70% of rice husk residues are utilized, there will be annual electricity production of 1,328 GWh and the cost of per unit electricity by rice husk is found at 47.36 cents/kWh as compared to 55.22 cents/kWh of electricity generated by coal. Importantly, the study will increase the awareness of the benefits of utilizing agricultural waste for useful products such as silica, with several social and environmental benefits such as a reduction of 36,042 tCO_{2e}/yr of methane, reducing

ABOUT THE AUTHORS

Obaidullah Mohiuddin is a chemical engineer and currently enrolled as a master's student in Sustainable Environment and Energy System in Middle East Technical University, Northern Cyprus Campus with Graduate Teaching assistant in Chemistry Department. His research interest includes but not limited to: Renewable energy, sustainable environment, and green energy/green chemistry.

Madina Obaidullah is a chemical engineer and currently enrolled as a master's student in Sustainable Environment and Energy System in Middle East Technical University Northern Cyprus Campus. Her research interest includes sustainability, renewable energy, and nanotechnology.

Samuel Asumadu-Sarkodie is a multidisciplinary researcher who currently studies master's in Sustainable Environment and Energy Systems at Middle East Technical University, Northern Cyprus Campus where he is also a graduate assistant in the Chemistry department. His research interest includes but not limited to: Renewable energy, econometrics, energy economics, climate change, and sustainable development.

PUBLIC INTEREST STATEMENT

Increasing population has resulted in increasing demand for energy which has a role to play in economic growth and human development. Nevertheless, there is a global crusade for universal access to clean and affordable energy and its related services while mitigating climate change and its impact. Pakistan as an agrarian country has being battling with agricultural waste disposal which has resulted in poor air quality due to burning of such residues. However, residues from agriculture have a potential of producing electricity and other social benefits. The study examines the electricity production potential and social benefits of rice husk in Pakistan. The study will increase the awareness of the benefits of utilizing agricultural waste (rice husk) for useful products with several social and environmental benefits such as reducing carbon dioxide emissions, improving air quality, and decreasing unemployment rate.

carbon dioxide emissions, improving the air quality, and providing 4.5 k new jobs. The paper concludes with the policy recommendations based on this study.

Subjects: Bio Energy; Clean Tech; Clean Technologies; Environmental; Renewable Energy; Traditional Industries - Clean & Green Advancements

Keywords: biomass; rice husk; renewable energy; energy demand; power generation; agricultural waste; Pakistan; climate change; fossil-fuel reduction; environmental sustainability engineering

1. Introduction

There is a growing need for clean energy technologies throughout the world to meet energy demand due to a global decline in fossil fuel reserves within the last decades (S. Asumadu-Sarkodie & P. A. Owusu, 2016e). The need for energy and its related services to satisfy human need; social and economic development, welfare and health is increasing (Owusu & Asumadu-Sarkodie, 2016; Owusu, Asumadu-Sarkodie, & Ameyo, 2016). Energy development is closely linked with the economic development of a country. Supplying the energy demand for residential use, industrial use, and commercial use will directly affect the economic growth of a country (S. Asumadu-Sarkodie & P. Owusu, 2016). However, the inadequate energy supply, power outages, and load shedding have become challenging in developing and least developing countries due to increasing population and energy demand (S. Asumadu-Sarkodie & P. A. Owusu, 2016d). Pakistan's current generation capacity is not enough to keep up with its ever-increasing demand, causing an electricity shortage of up to 4,500–5,500 MW. This has caused the supply and demand gap to increase dramatically over the past 5 years (Ministry of Water & Power, 2013). Figure 1 shows the main sources of electricity production in Pakistan. The main sources of electricity production are hydro, thermal, and nuclear with 6,858, 15,440, and 750 MW installed capacity, respectively. During July–March 2013–2014, the installed capacity of electricity was 23,048 MW. Thus, the hydropower capacity accounts for 29.7%, thermal 67.0%, and nuclear 3.3%. Nonetheless, the power production is practically 50% of installed capacity (Ministry of Finance, 2013).

Owing to the major reliance on thermal fuel sources, the cost of electricity production is extremely high (i.e. PKR 12/unit). The cost of supplying electricity per unit to the customers has been set at Rs. 14.70 by the NEPRA. This shows that PKR 2.70 per unit must be added as a subsidy to meet the cost of electricity production (Ministry of Water & Power, 2013). According to Yasmeen and Sharif (2014)

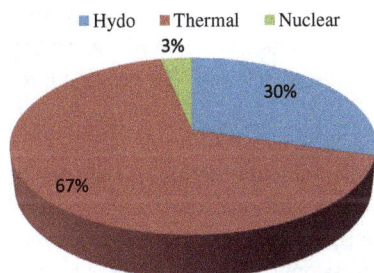

Figure 1. Electricity production in Pakistan (Ministry of Finance, 2013).

| Table 1. Circular debt due to expensive power generation ||
Date	PKR (billion)
1 June 2009	214
30 June 2009	216
18 May 2010	120
2014–2015	250

the present power crunch has affected the entire country including the industrial sector such as: small-scale to large-scale enterprises, exports and employments causing business to lose almost 157 billion rupees while 400,000 employees losing their jobs (Yasmeen & Sharif, 2014). Due to a rise in international oil price in 2008, the condition became more problematic. As the subsidy element (difference between cost and tariff) increased, enormous quantity of circular debt was created whereby power generation companies could not disburse payment to the fuel suppliers as a result of non-payments from supplier-companies (Ministry of Finance). Table 1 shows the circular debt due to expensive power generation.

This circular debt resulted in disruption of power supply causing an increase in demand and supply gap. Utilization of Biomass such as rice husk can provide a solution to this problem. Pakistan as an agrarian country has been battling with agricultural waste disposal which has resulted in poor air quality due to burning of such residues. However, residues from agriculture have a potential of producing electricity and other social benefits. As suggested by some studies, heat and power generation from combustion of Biomass has now become mature technology offering the solution of sustainable fuel and waste disposal (Amer & Daim, 2011). This combustion of biomass is an efficient and environmental friendly approach of disposing of the public waste which is gathered in huge amounts each day in different towns and cities of the country (Ashraf Chaudhry, Raza, & Hayat, 2009). According to S. Asumadu-Sarkodie and P.A. Owusu (2016b) the development and adoption of a renewable energy technology like biomass power plant will save costs on buying the conventional type of fuel (fossil-based) and result in a large techno-economic potential for climate change mitigation which will satisfy the sustainable development goals (S. Asumadu-Sarkodie & P. Owusu, 2016; S. Asumadu-Sarkodie & P. A. Owusu, 2016c).

There are a few existing studies in literature that have investigated biomass energy potential particularly in Pakistan. Mirza, Ahmad, and Majeed (2008) investigated biomass energy utilization in Pakistan, discussed the different dimensions of producing electricity in the rural areas through biomass. They concluded that biomass is a clean and cost-effective fuel option with tremendous potential for application in Pakistan. However, there is a need to allocate necessary resources for improving these technologies through a widespread dissemination plan. Bhutto, Bazmi, and Zahedi (2011) highlighted the issues and challenges in the efficient and effective utilization of biomass as an energy source in Pakistan. They focused on electricity production, industrial, and domestic fuels. They identified areas that require attention for establishing and improving biomass energy production and delivery system. Zuberi, Hasany, Tariq, and Fahrioglu (2013) examined the potential of two major biomass energy resources available in Pakistan: Livestock and Bagasse. They determined that biomass resources in Pakistan are capable of contributing 42% of the power portfolio of the country. They also suggested that utilizing this technology can contribute immensely toward raising the economy, reducing not only the oil and natural gas imports and carbon emissions but also the fraction of unemployment by 17.1%.

This study, however, narrows down the focus to rice husk in particular and examines the electricity production potential and social benefits of rice husk in Pakistan. The study will increase the awareness on the benefits of utilizing agricultural waste (rice husk) for useful products with several social and environmental benefits such as reducing carbon dioxide emissions, improving air quality, and decreasing unemployment rate.

2. Energy generation methods from rice husk

Although carbon dioxide (CO_2) is produced by combustion of biomass such as rice husk, the carbon produced is absorbed by the plants by the process of photosynthesis. Hence, the combustion of biomass and biogas reduces the global warming effect since CO_2 net emission becomes zero. While combustion of conventional fossil fuels adds an extra amount of CO_2, causing global warming.

Several methods have been formulated for energy extraction from biomass, which includes liquefaction, anaerobic digestion, alcoholic fermentation, trans-esterification, pyrolysis, direct combustion, and gasification. Each technology has its own advantages, depending on the economics of biomass availability and the end product desired (Mirza et al., 2008). The end product could be either power/ heat, transportation fuel, or chemical feedstock. The various options of biomass utilization are shown in Figure 2 (Cheng, 2009). Currently there are two methods used for producing heat and electrical power from rice husk (waste of rice mills) namely gasification and direct combustion. S. Asumadu-Sarkodie and P. A. Owusu (2016b) analyzed the feasibility of biomass heating system in Middle East Technical University, Northern Cyprus campus using rice straw as the main source of fuel. They reported the economic feasibility, GHG emission reduction, and the advantages of using the proposed technology. It is reported that direct combustion (Steam turbine) consumes 1.3 kg of rice husk per kWh electricity while 1.86 kg of rice husk consumes per kWh of electricity by gasification (Islam & Ahiduzzaman, 2013).

Several other methods have also been introduced for the conversion of agricultural byproducts such as rice husk into electrical energy. Straw-fired power station started its operation in 2000 at Sutton, Ely in Cambridge shire with an output capacity of 36 MW, and the cost was £60 M. Yearly 200 thousand tons of straw are utilized. The annual electric output by conventional steam turbine and generator was over 270 GWh/year (Boyle, 2004).

A rice husk power system was installed in India, which required sack loads of rice husk or other biomass residues poured into the gasifier hopper for every 30–45 min. The biomass then burn in a restricted supply of air to produce an energy-rich gas. The gas passes through a series of filter for cleaning, after which the gas is used as the fuel engine that drives the electricity generator. Electricity is then distributed to customers via insulated overhead cables (Pandey, 2011).

EGCO Group (2006) carried out a Green Power Project in Rio Et that used rice husk as the primary fuel for an SPP-Renewable Power plant to generate electricity. This pilot project had a 10 MW capacity, which uses 225 tons of rice husk and 1,400 tons of water in a day. The plant has a power requirement of 1 MW and its net power output is about 8.8 MW (Chungsangunsit, Gheewala, & Patumsawad, 2005).

Denmark has the largest biomass-fired combined heat and power plant with a capacity of 26,000 MWh of electrical power and 50,000 MWh of heat each year at very favorable prices. In this

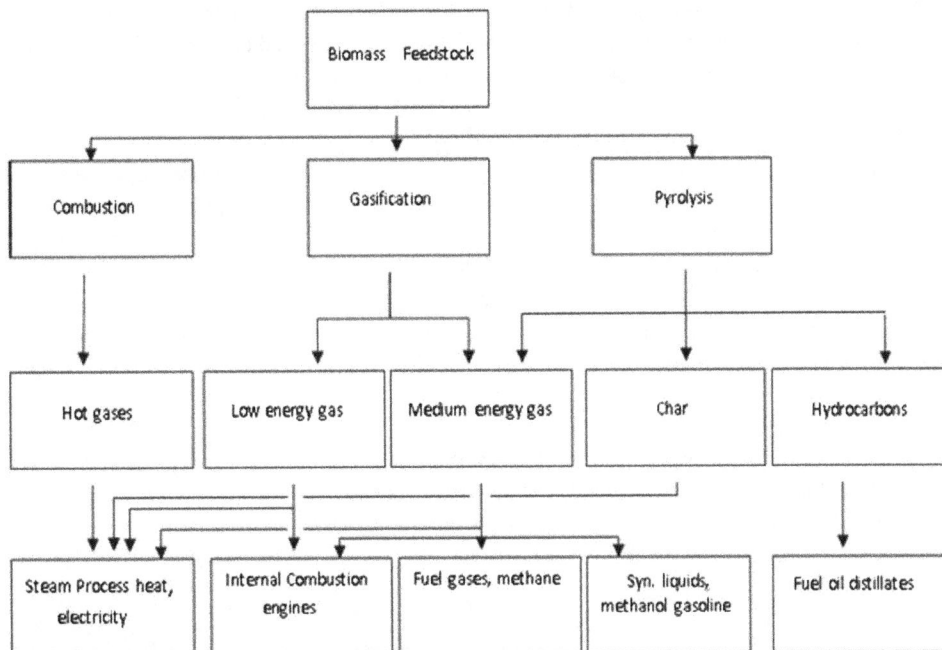

Figure 2. Biomass thermochemical conversion pathway (Cheng, 2009).

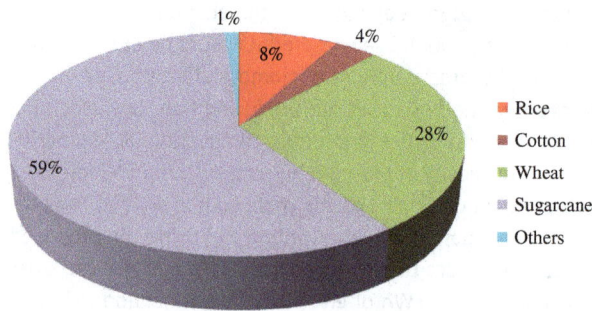

Figure 3. Total crop production of Pakistan in 2011 (Bhutto et al., 2011).

Table 2. Rice husk availability in Pakistan (Mirani, Ahmad, Kalwar, & Ahmad, 2013)		
Year	**Paddy production (1,000 tons)**	**Rice husk (1,000 tons)**
2008	6,952	1,390.4
2009	6,883	2,064.9
2010	4,823	1,446.9
2011	6,160	1,848
2012	5,536	1,660.8
2013	6,798	2,039.4
2014	7,005	2,101.5
2015	6,900	2,070
Average	6,382	1,828

process, much of the fuel fed to the boiler is combusted while flying through the chamber, and the large parts burn in the grate. The heat from the boiler is used to produce steam at a high temperature of 525°C. It has an electrical power efficiency of 27% (Morten Tony Hansen, 2014).

A biomass gasifier was installed in 2007 in Bangladesh, which consisted of a gasifier and a dual fuel internal combustion engine, with the capacity of 250 kW. Biomass are burned in a controlled way with the partial oxidation technique to form syngas, the gas filter for purification and is then burned in the engine with an initial start-up from diesel and when desired amount of syngas is produced, then the ratio of syngas to diesel becomes 70:30 (IDCOL, 2007).

3. Rice husk availability and price in Pakistan

Pakistan is an agriculture country producing all major crops like wheat, maize, rice, sugarcane, and cotton. In 2014–2015 production of rice is 6,900 thousand tons, which was 7,005 thousand tons in 2013–2014 (Ministry of Finance, 2013). Figure 3 represents the total crop production in Pakistan (Bhutto et al., 2011). There is a considerable potential of generating electricity from biomass in the form of crop residues, animal waste (manure), and municipal waste in Pakistan. Presently 50,000 tons of solid waste and approximately 1,500 m³ of forestry firewood are generated daily. 225,000 tons of crop residue calculated with 1.7% rice husk and 16% bagasse (residue from sugar mills) daily and animal manure is estimated 1 million tons every day (Uqaili, Harijan, & Memon, 2007).

Rice is cultivated mainly in interior Sindh and Punjab provinces of Pakistan. Rice husk is obtained after separating the rice grain from its husk in the rice mills as a byproduct and it consists of 20% by weight of rice. Availability of rice husk in Pakistan per annum for the past 8 years is shown in Table 2. It can be seen that in last four years, average production of rice husk was around 1,828 ton per annum.

Rice husk obtained as a byproduct is considered as a waste and usually huge quantities of the rice husk are dumped as waste causing waste disposal problems and emissions of methane. The

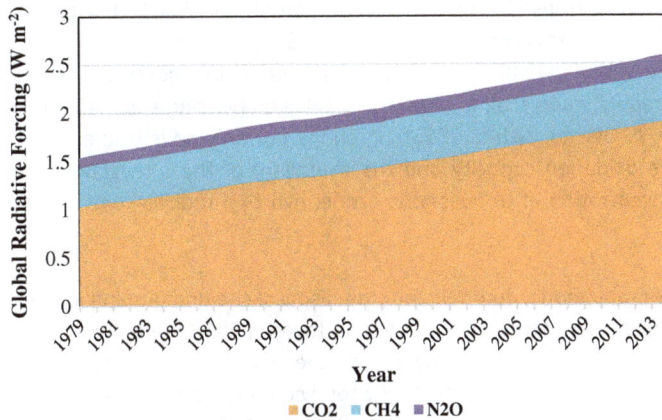

Figure 4. Global radiative forcing.

Source: Adapted from Earth System Research Laboratory (2015).

Table 3. Ultimate and proximate analysis of rice husk (Yin et al., 2002)

Ultimate analysis (%)		Proximate analysis (%)	
Carbon	36.74	Volatile matters	53.1
Hydrogen	5.51	Fixed carbons	20.4
Oxygen	42.55	Moisture	11.7
Nitrogen	0.28	Ash	14.8
Sulfur	0.55		

Table 4. Comparison of coal and rice husk (Mirani et al., 2013)

Fuel	Heating value (MJ/kg)	Cost (Rs./Ton)	Energy cost (Rs./MJ)
Rice husk	15	3,750	56.25
Coal	36	10,260	369.36

particulate nature of certain portions of the rice husk when inhaled affect air quality which may lead to lung cancer and heart conditions. If instead of disposing rice husk, it is used to generate electricity, the problem of the incorrect disposal method could be solved. Rice husk is a valuable source of renewable energy which should be utilized and converted to a beneficial form of energy to fulfill the thermal and mechanical energy requirement for the mills (Chungsangunsit, Gheewala, & Patumsawad, 2009).

Energy produced from rice husk mainly depends on its composition; ultimate and proximate analysis of the rice husk as shown in Table 3. Rice husk contains 16–23% ash contents which are mainly silica as its composition is approximately 95% in ash (Yin, Wu, Zheng, & Chen, 2002). Carbon and Hydrogen make around 42% of the total composition which actually take part in combustion.

Rice husk has the calorific value of approximately15 MJ/kg that means its heating value is 41% lesser than coal, however its cost is 36% cheaper than coal. Table 4 shows the amount of energy produced per kg by coal and rice husk and the cost per MJ of energy. It can be seen that energy obtained from rice husk is cheaper than coal and is carbon neutral. Although coal still is the most efficient solid fuel in the world, but its operational cost is higher because of hazardous materials present in it. In addition, it produces the highest amount of additional emissions causing global warming.

4. Methane emission from conventional rice husk disposal methods

Methane (CH_4) is one of the environmental pollutants that is considered as a greenhouse gas. Nevertheless its concentration is lower compared to carbon dioxide emissions, which increased from

275 to 345 ppm, but one molecule of CH_4 traps 30 times more heat than carbon dioxide. In Figure 4, the global radiative forcing is given. As expected, data from Earth System Research Laboratory (2015) shows that carbon dioxide emissions still dominant global greenhouse gasses, however, methane and nitrous oxide are gradually decreasing in pace yet, still experiencing exponential increase (S. Asumadu-Sarkodie & P. A. Owusu, 2016a, 2016f). Methane not only adds to the greenhouse effect, but also affects the oxidation capacity and the chemistry of the atmosphere, for instance by manipulating the concentration of tropospheric ozone, hydroxyl radicals, and carbon monoxide (Neue, 1993).

Disposing rice husk has become problematic, dumping into the disposal site and burning in open air releases methane into the atmosphere, thereby contributing to the greenhouse effect and poor air quality. The amount of CO_2 released into the open air by burning the rice husk mainly depends on the amount of rice husk and the carbon fraction of rice husk. The total carbon and methane released from burning of rice husks can be calculated using Equations (1–3), respectively.

$$TCR = \text{Amount of Rice Husk} \times \text{CF of Rice husk} \tag{1}$$

where, TCR is a total carbon released with tons of C/yr and CF is carbon fraction of rice husk, which is taken as 0.3674.

$$AMR = TCR \times CR \times MCF \times GWP \tag{2}$$

where AMR is the annual methane released with tons of CO_{2e}/yr, CR is carbon released as CH_4 in open air taken as 0.005%, MCF is mass conversion factor and its value is taken as $\frac{12}{16}$ GWP is a global warming potential which is 21 for methane. The parameters employed in Equations (1 and 2) are adopted from the clean energy finance committee of bio-power rice husk power project (Mitsubishi Securities, 2003). Assuming 70% of rice husk is considered, following results were obtained.

Total carbon released = 457,672.38 tC/yr

Annual methane release = 36,042 tCO_{2e}/yr

It is evident from the above results that the current practice of disposing rice husk produces large amount of CH4, which leads to serious health conditions in human population and contribute to the global increase of greenhouse gasses thereby leading to climate change. Therefore, this study proposes that even if 70% of total rice husk produced in Pakistan is used to produce electricity, a reduction of 36,042 tCO_{2e}/yr of methane is possible.

5. Total potential of rice husk as electricity source in Pakistan

The study followed the experimental parameters of a pilot project in Rio Et (EGCO Group, 2006) for analyzing the total potential of electricity from rice husk. For 10 MW of SPP-Renewable Power plant, the net output is 9 MW (Chungsangunsit et al., 2005) therefore the amount of electricity generation by rice husk can be calculated using Equation (3). Rice husk availability in Pakistan is $1,779.58 \times 10^3$ tons per year. 225 tons of rice husk produce 10 MW of electricity per day. Assuming 70% of the total rice husk (1,245,706 tons per year) is utilized for electricity production then:

$$1,779,580 \text{ tons} \times 0.7 = \frac{10}{225}\left(\frac{MW}{tons}\right) \times 0.7 \times 1,779,580\left(\frac{tons}{Year}\right) = 55,364\left(\frac{MW}{Year}\right) \tag{3}$$

Thus:

1,245,706 tons of rice husk will produce = $55,364 \times 24$ MWh of electricity per year.

Annual electricity production potential = 13.28×10^2 GWh.

Table 5. Carbon dioxide emission from different fuels (APEIS, 2003)		
Fuel	CO_2 emission (kg/MW)	CO_2 emission (kg/55,364 MWh electricity)
Coal	1,269.524	7.03×10^7
Oil	812.608	4.50×10^7
Gas	568.878	3.15×10^7
Rice husk	Neutral	Neutral

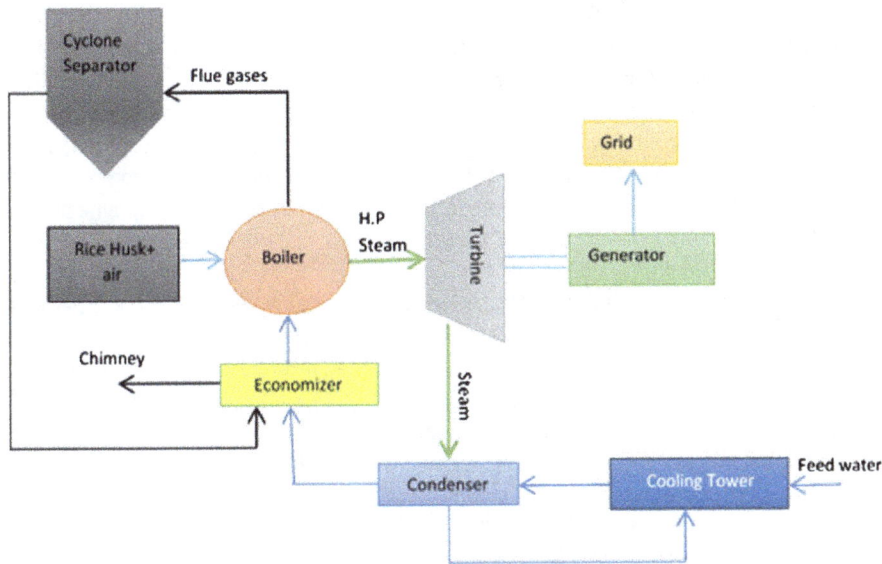

Figure 5. Pathway for electricity production from the rice husk power plant.

6. Total reduction potential for greenhouse gasses

For any amount of electricity production by coal there is some emission of CO_2. If the same amount of electricity is produced with the rice husk, the amount of carbon dioxide emitted will be recycled with the cultivation of new crops, and the emission of methane as a result of air pollution from open burning rice husk will be controlled leading to better air quality and healthy environment. Table 5 shows the comparison of carbon dioxide released from coal, oil, and gas with rice husk. From Equations (1 and 2), the annual CH_4 released by openly burning rice husk was estimated as 36,042 tCO_{2e}/yr and if the same amount of electricity is produced by coal than the total emissions would be 70,285.9 tCO_2/yr as shown in Table 5. The total savings of emission would be 106,327.93 tCO_{2e}/yr.

7. Estimation of cost of unit electricity

It is suggested to install rice husk electrical power plant in the district of Badin for various reasons. Badin district has an area of 6,726 km² and more than 60 rice mills are in operation and labor availability is guaranteed (Government of Sindh, 2013; Shalim Kamran Dost, 2007). Trucks are always available for transportation of rice husk. A large area is required for the storage of rice husk and its handling, which should be in-door. The storage and transportation costs are included in the cost of rice husk as fuel price as shown in Table 2.

In order to estimate, per unit cost, a simple steam cycle plant is assumed. Figure 5 shows the pathway and schematic of the assumed rice husk power plant. By using a conveyor belt the rice husk is transferred to the boiler for combustion, the husk will be burned in the boiler and produce a high pressurized steam, which will rotate the steam turbine connected to a shaft that turns the generator. The burned gas (flue gas) will first go through the multi-cyclone separator and later pass through the economizer to heat the water and finally emitted into the atmosphere. Carbon dioxide produced from the rice husk power plant could be viewed as carbon neutral since emissions would eventually

Table 6. Job opportunities from rice husk power plant

Task	Man–day/ton
Raw material collection	0.75
Transportation process	0.25
Trading	0.33
Total	1.33

be sequestered in future biological growth. In a steady-state biological system, gross emissions would be equal to gross sequestration, thus net emissions would be zero (Della, Kühn, & Hotza, 2002). The burning of rice husk in the air always leads to the formation of ash, which is rich in silica, reported as 96.34% (Della et al., 2002; Srivastava, Mall, & Mishra, 2006). Silica is considered as a by-product which is an essential requirement of Portland cement (EPA, 2012b). It can be sold out to the cement industries or be used as a fertilizer to boost agricultural production. By using combine heat and power system (CHP) the energy can be recovered from the steam after the steam turbine makes the system more feasible and economical (EPA, 2012a).

Following the analysis of cost estimation of electricity per kWh energy conversion takes place in two stages. The first stage deals with the conversion efficiency of the boiler and combustion process. Compared to a boiler efficiency of 55–60% achievable with stepped grate furnace, the fluidized bed combustion technology gives a high efficiency of the order of 75–80%. In this case, we take 75% efficiency (IEA, 2010). Second stage deals with the steam cycle efficiency. Modern Rankine cycle adopted in coal-fired power plants has efficiencies that vary from 32 to 42% (IEA, 2010). This depends mainly on the steam parameters. Higher steam pressure and temperatures in the range of 600°C–230 bars have efficiencies around 42%. A value of 38% is assumed in this case. This results in an overall conversion efficiency of 0.28%.

$$\text{Amount of fuel (Ton/kWh)} = \frac{3600\frac{kJ}{hr}}{0.28 \times CV_{fuel}} \times PC \times 1000 \tag{4}$$

where CV_{fuel} is the calorific value of coal and PC is the proposed plant capacity. The cost of electricity can thus be found by the following equation.

$$\text{Cost of electricity}_{PC} = \text{Fuel price}\left(\frac{cents}{Ton}\right) \times \text{Amount of fuel}\left(\frac{Ton}{kWh}\right) \tag{5}$$

Assuming a calorific value of coal as 20,000 J/kg and coal price as 10.26 cents the cost of electricity production for a plant capacity of plant capacity of 100 MW is found as 55.22 cents/kWh, while the cost of electricity production by rice husk is found as 47.36 cents/kWh.

8. Total potential of employment generation
The employment generation opportunities from rice husk are from the raw material collection, transportation, trading, and power plant working personnel. Table 6 which is obtained by a survey performed by Ahiduzzaman (2007) shows the requirements of man-day/ton for the raw material collection, transportation, and trading of rice husk. An 8-h work day was assumed in this survey.

Equation (6) is used to calculate the total number of possible employment slots.

$$\text{Vacancies} = \frac{\frac{Man-day}{ton} \times 0.7 \times \text{Total rice husk}}{365} \tag{6}$$

Since the unemployment rate is on the rise, taking advantage of this renewable energy alternative has a potential of solving this social issue to some extent. In this regard, it is important to know that almost 4,530 job opportunities will be created from raw material collection, transportation, and rice husk trading in Pakistan.

9. Total potential of silica production

As it is discussed in Section 7 of this article, some amount of silica will be produced during the utilization of rice husk for power generation. The total potential of silica production can be found using Equation (7),

$$\text{Silica production}\left(\frac{\text{tons}}{\text{year}}\right) = \text{Total amount of rice husk}(\frac{\text{tons}}{\text{year}}) \times \text{Fraction of ash} \times 0.95 \quad (7)$$

Using Equation (7) and the fraction of ash from Table 3, the total possible silica extraction from rice husk ash is found to be 175,146.3 tons/yr.

10. Conclusion and policy recommendations

For several years, Pakistan has been experiencing energy crisis owing to its sole dependence on fossil fuels. The decline of fossil fuel reserves has led to an increase in their prices, thereby increasing the cost of electricity. Since the tariff remains the same, Pakistan is over-burdened with circular debts and observes a daily power shortfall of about 12–14 h. Being an Agra-economic country, many major and minor crops are produced and exported in large quantities. However, a large amount of it forms a bulk of the agricultural waste which are not utilized. This waste can be utilized to meet the country's energy demand while mitigating climate change and its impact. Technological advancement makes it easy to convert the agricultural waste into thermal energy which is considered as carbon neutral. Adopting these technologies would help bridge the demand and supply gap, decreasing the reliance on fossil fuel, thereby increasing the energy security and economic growth. The study examines the electricity production potential and social benefits of rice husk in Pakistan. It is estimated that if 70% of rice husk residues are utilized, there will be annual electricity production of 1,328 GWh. The following policy recommendations emanate from the study:

- Efforts by the government to implement policies that prevent waste generation through preventive measures, reduction management, recycling and encouraging reuse will be essential in combating climate change and its impact.

- The government should institute policies that encourage public and private sector to adopt sustainable practices and incorporate information on sustainability into their production, labeling, and reporting cycles.

- Efforts by a government that provide relevant and create awareness on reducing the carbon footprint and sustainable lifestyles in Pakistan is worthwhile.

- Government's effort toward the creation of an enabling environment and support for renewable energy technologies like subsidies, removal of market distortions, tax reductions, and incentives will boost the country's effort toward patronizing clean energy whiles mitigating climate change.

- Efforts by the government to sponsor scientific research on renewable energy technologies in Institutions through the provision of grants, loans, and merit-based awards will boost the scientific and technological advancement of the country.

It is believed that exploiting the renewable energy potential and utilizing agricultural waste will be beneficial to the country's energy portfolio while adding several social and environmental benefits such as reducing carbon dioxide emissions, improving air quality, and decreasing the unemployment rate. Future studies should examine how different agricultural residues in Pakistan can serve as an alternative source of electricity generation and/or social benefit based on a sensitivity and risk analysis.

Funding
The authors received no direct funding for this research.

Author details
Obaidullah Mohiuddin[1]
E-mail: obaidullah.mohiuddin@metu.edu.tr
ORCID ID: http://orcid.org/0000-0003-1484-0521
Abdullah Mohiuddin[1]
E-mail: abdullah.mohiuddin@metu.edu.tr
ORCID ID: http://orcid.org/0000-0003-1370-4210
Madina Obaidullah[1]
E-mail: madina.obaidullah@metu.edu.tr
ORCID ID: http://orcid.org/0000-0003-1488-1139
Humayun Ahmed[1]
E-mail: humayun.ahmed@metu.edu.tr
ORCID ID: http://orcid.org/0000-0003-2741-5794
Samuel Asumadu-Sarkodie[1]
E-mails: samuel.sarkodie@metu.edu.tr, samuelsarkodie@
yahoo.com, asumadusarkodiesamuel@yahoo.com
ORCID ID: http://orcid.org/0000-0001-5035-5983

[1] Sustainable Environment and Energy Systems, Middle East
 Technical University, Northern Cyprus Campus, Mersin 10,
 Turkey.

References
Ahiduzzaman, M. (2007). *Rice husk energy technologies in Bangladesh*.
Amer, M., & Daim, T. U. (2011). Selection of renewable energy technologies for a developing county: A case of Pakistan. *Energy for Sustainable Development, 15*, 420–435.
APEIS. (2003). *Use of rice husks as fuel in process steam boilers*. Retrieved November 17, 2014, from http://enviroscope.iges.or.jp/contents/APEIS/RISPO/inventory/db/pdf/0004.pdf
Ashraf Chaudhry, M., Raza, R., & Hayat, S. (2009). Renewable energy technologies in Pakistan: Prospects and challenges. *Renewable and Sustainable Energy Reviews, 13*, 1657–1662. http://dx.doi.org/10.1016/j.rser.2008.09.025
Asumadu-Sarkodie, S., & Owusu, P. (2016). A review of Ghana's energy sector national energy statistics and policy framework. *Cogent Engineering*. doi:10.1080/23311916.2016.1155274
Asumadu-Sarkodie, S., & Owusu, P. A. (2016a). Carbon dioxide emissions, GDP, energy use and population growth: A multivariate and causality analysis for Ghana, 1971–2013. *Environmental Science and Pollution Research International*. doi:10.1007/s11356-016-6511-x
Asumadu-Sarkodie, S., & Owusu, P. A. (2016b). Feasibility of biomass heating system in Middle East Technical University, Northern Cyprus Campus. *Cogent Engineering, 3*, 1134304. doi:10.1080/23311916.2015.1134304
Asumadu-Sarkodie, S., & Owusu, P. A. (2016c). Multivariate co-integration analysis of the Kaya factors in Ghana. *Environmental Science and Pollution Research International*. doi:10.1007/s11356-016-6245-9
Asumadu-Sarkodie, S., & Owusu, P. A. (2016d). The potential and economic viability of solar photovoltaic in Ghana. *Energy Sources, Part A: Recovery, Utilization, and Environmental Effects*. doi:10.1080/15567036.2015.1122682

Asumadu-Sarkodie, S., & Owusu, P. A. (2016e). The potential and economic viability of wind farms in Ghana. *Energy Sources, Part A: Recovery, Utilization, and Environmental Effects*. doi:10.1080/15567036.2015.1122680
Asumadu-Sarkodie, S., & Owusu, P. A. (2016f). The relationship between carbon dioxide and agriculture in Ghana: A comparison of VECM and ARDL model. *Environmental Science and Pollution Research*. doi:10.1007/s11356-016-6252-x
Bhutto, A. W., Bazmi, A. A., & Zahedi, G. (2011). Greener energy: Issues and challenges for Pakistan—Biomass energy prospective. *Renewable and Sustainable Energy Reviews, 15*, 3207–3219. http://dx.doi.org/10.1016/j.rser.2011.04.015
Boyle, G. (2004). *Renewable energy*. Oxford University Press.
Cheng, J. (2009). *Biomass to renewable energy processes*. CRC Press.
Chungsangunsit, T., Gheewala, S. H., & Patumsawad, S. (2005). Environmental assessment of electricity production from rice husk: A case study in Thailand. *International Energy Journal, 6*, 347–356.
Chungsangunsit, T., Gheewala, S. H., & Patumsawad, S. (2009). Emission assessment of rice husk combustion for power production. *World Academy of Science: Engineering and Technology, 53*, 1070.
Della, V., Kühn, I., & Hotza, D. (2002). Rice husk ash as an alternate source for active silica production. *Materials Letters, 57*, 818–821. http://dx.doi.org/10.1016/S0167-577X(02)00879-0
Earth System Research Laboratory. (2015). *The NOAA annual greenhouse gas index (AGGI)*. Retrieved October 24, 2015, from http://www.esrl.noaa.gov/gmd/aggi/aggi.html
EGCO Group. (2006). *Roi-Et Green Power Project*. Retrieved November 17, 2014, from http://www.jst.go.jp/asts/asts_j/files/ppt/22_ppt.pdf
EPA. (2012a). *Combined heat and power*. Retrieved December 4, 2014, from http://www.epa.gov/chp/documents/faq.pdf
EPA. (2012b). *Overview of Portland cement and concrete*. Retrieved November 17, 2014, from http://www3.epa.gov/epawaste/conserve/tools/cpg/pdf/app-a.pdf
Government of Sindh. (2013). *List of rice mills & owners–district Badin rice mill*. Retrieved November 17, 2014, from http://sindhagri.gov.pk/rice%20mills/dist%20badin%20rice%20mill.pdf
IDCOL. (2007). *Biomass technologies–Biogas gasification in Bangladesh*. Retrieved November 15, 2014, from https://practicalaction.org/media/download/7167
IEA. (2010). *Power generation from coal*. Retrieved November 15, 2014, from https://www.iea.org/ciab/papers/power_generation_from_coal.pdf
Islam, A. S., & Ahiduzzaman, M. (2013). Green electricity from rice husk: A model for Bangladesh. *Thermal Power Plants-Advanced Applications*, 127–144.
Ministry of Finance. (2013). *Highlights of Pakistan economic survey 2013–14*. Retrieved November 15, 2015, from http://finance.gov.pk/survey/chapters_14/Highlights_ES_201314.pdf
Ministry of Water & Power. (2013). *National Power Policy, 2013 Government of Pakistan*. Retrieved November 15, 2014, from http://www.ppib.gov.pk/National%20Power%20Policy%202013.pdf
Mirani, A. A., Ahmad, M., Kalwar, S. A., & Ahmad, T. (2013). A rice husk gasifier for paddy drying. *Science and Technology Development, 32*, 120–125.
Mirza, U. K., Ahmad, N., & Majeed, T. (2008). An overview of biomass energy utilization in Pakistan. *Renewable and Sustainable Energy Reviews, 12*, 1988–1996. http://dx.doi.org/10.1016/j.rser.2007.04.001
Mitsubishi Securities. (2003). *Clean Energy Finance Committee*. Retrieved November 17, 2014, from http://www.sc.mufg.jp/english/company/sustainability/cef/index.html

Morten Tony Hansen. (2014). *Biomass based combined heat and power generation*. Retrieved November 15, 2014, from http://www.videncenter.dk/exportcat/combined_heat_and_power.pdf

Neue, H.-U. (1993). Methane emission from rice fields. *Bioscience, 43*, 466–474.

Owusu, P., & Asumadu-Sarkodie, S. (2016). A review of renewable energy sources, sustainability issues and climate change mitigation. *Cogent Engineering*. doi:10.1080/23311916.2016.1167990

Owusu, P. A., Asumadu-Sarkodie, S., & Ameyo, P. (2016). A review of Ghana's water resource management and the future prospect. *Cogent Engineering*. doi:10.1080/23311916.2016.1164275

Pandey, G. (2011). *Case study summary husk power systems India*. Retrieved November 15, 2014, from https://www.ashden.org/files/Husk%20winner.pdf

Shalim Kamran Dost. (2007). *Disaster risk management plan district Badin*. District Disaster Management Authority. Retrieved November 15, 2014, from http://bdro.org/wp-content/uploads/2013/04/District-Disaster-Management-Plan-Badin.pdf

Srivastava, V. C., Mall, I. D., & Mishra, I. M. (2006). Characterization of mesoporous rice husk ash (RHA) and adsorption kinetics of metal ions from aqueous solution onto RHA. *Journal of Hazardous Materials, 134*, 257–267. http://dx.doi.org/10.1016/j.jhazmat.2005.11.052

Uqaili, M. A., Harijan, K., & Memon, M. (2007). Prospects of renewable energy for meeting growing electricity demand in Pakistan. In *Renewable energy for sustainable development in the Asia Pacific region* (Vol. 1., pp. 53–61). AIP Publishing.

Yasmeen, F., & Sharif, M. (2014). Forecasting electricity consumption for Pakistan. *International Journal of Emerging Technology and Advanced Engineering, 4*, 496–503.

Yin, X. L., Wu, C. Z., Zheng, S. P., & Chen, Y. (2002). Design and operation of a CFB gasification and power generation system for rice husk. *Biomass and Bioenergy, 23*, 181–187. http://dx.doi.org/10.1016/S0961-9534(02)00042-9

Zuberi, M. J. S., Hasany, S. Z., Tariq, M. A., & Fahrioglu, M. (2013). Assessment of biomass energy resources potential in Pakistan for power generation. In *2013 Fourth International Conference on Power engineering, energy and electrical drives (POWERENG)* (pp. 1301–1306). IEEE.

7

Flood frequency analysis using method of moments and L-moments of probability distributions

N. Vivekanandan[1]*

*Corresponding author: N. Vivekanandan, Central Water and Power Research Station, Pune, Maharashtra 411024, India

E-mail: anandaan@rediffmail.com

Reviewing editor: Sanjay Shukla, Edith Cowan University, Australia

Abstract: Estimation of maximum flood discharge (MFD) at a desired location on a river is important for planning, design and management of hydraulic structures. This can be achieved using deterministic models with extreme storm events or through frequency analysis by fitting of probability distributions to the recorded annual maximum discharge data. In the latter approach, suitable probability distributions and associated parameter estimation methods are applied. In the present study, method of moments and L-moments (LMO) are used for determination of parameters of six probability distributions. Goodness-of-Fit tests such as Chi-square and Kolmogorov–Smirnov are applied for checking the adequacy of fitting of probability distributions to the recorded data. Diagnostic test of D-index is used for the selection of a suitable distribution for estimation of MFD. The study reveals that the Extreme Value Type-1 distribution (using LMO) is better suited amongst six distributions used in the estimation of MFD at Malakkara and Neeleswaram gauging stations in Pampa and Periyar river basins, respectively.

Subjects: Statistics & Probability; Civil, Environmental and Geotechnical Engineering; Engineering Mathematics; Engineering Management

Keywords: chi-square; D-index; Kolmogorov–Smirnov; L-moments; method of moments; maximum flood; probability distribution

ABOUT THE AUTHOR

N. Vivekanandan post graduated in mathematics from Madurai Kamaraj University in 1991. He obtained Master of Engineering in hydrology from University of Roorkee in 2000. He obtained Master of Philosophy in mathematics from Bharathiar University in 2006 and MBA (Human Resources) from Manonmaniam Sundaranar University in 2013. From May 1993 to March 2006, he worked as a research assistant in Central Water and Power Research Station (CWPRS), Pune. From April 2006 to till date, he is working as an assistant research officer in CWPRS, wherein conducting hydrological and hydrometeorological studies using probabilistic and stochastic models for various water resources projects.

PUBLIC INTEREST STATEMENT

Estimation of maximum flood discharge (MFD) at a desired location on a river is important for planning, design and management of hydraulic structures. This can be achieved using deterministic models with extreme storm events or through frequency analysis by fitting of probability distributions to the recorded annual maximum discharge data. In the latter approach, suitable probability distributions and associated parameter estimation methods are applied. This paper details the procedures adopted in method of moments and L-moments that are used for determination of parameters of probability distributions. Goodness-of-fit tests such as Chi-square and Kolmogorov–Smirnov are applied for checking the adequacy of fitting of probability distributions to the recorded data. However, the diverging results based on GoF tests lead to adopt qualitative test to aid the selection of a suitable distribution for estimation of MFD.

1. Introduction

Estimation of maximum flood discharge (MFD) with a specified return period is crucial for the design of hydraulic structures such as bridges, barrages, culverts, dams and drainage systems. Since the hydrologic phenomena governing the MFD is highly stochastic in nature, the MFD can be effectively determined by fitting of probability distributions to the series of recorded annual maximum discharge (AMD) data. An AMD is the highest instantaneous discharge value at a definite cross-section of a natural stream throughout an entire hydrologic year (water year). The longer the period of observation, the greater would be the length of the recorded series that may offer better results of the flood frequency analysis (FFA).

A number of probability distributions such as Exponential (EXP), Extreme Value Type-1 (EV1), Extreme Value Type-2 (EV2), Generalized Extreme Value (GEV), Generalized Pareto (GPA) and Normal (NOR) are used in FFA (Haktanir & Horlacher, 1993). The distributions viz., EV1, EV2, GEV and GPA are classified as extreme value family of distributions. Likewise, EXP and NOR distributions are classified as Gamma and Normal family of distributions. Generally, method of moments (MOM) is used for determination of parameters of the probability distributions. Sometimes, it is difficult to assess exact information about the shape of a distribution that is conveyed by its third and higher order moments. Also, when the sample size is small, the numerical values of sample moments can be very different from those of the probability distribution from which the sample was drawn. It is also reported that the estimated parameters of distributions fitted by the MOM are often less accurate than those obtained by other parameter estimation procedures such as maximum likelihood method, method of least squares and probability weighted moments. To address the aforesaid shortcomings, the application of alternative approach, namely L-moments (LMO) discussed in this paper is used for FFA (Hosking, 1990).

In the recent past, a number of studies have been carried out by different researchers on adoption of probability distributions for FFA. Kjeldsen, Smithers, and Schulze (2002) applied LMO in regional flood frequency analysis (RFFA) for KwaZulu-Natal province of South Africa. Kumar, Chatterjee, Kumar, Lohani, and Singh (2003) carried out RFFA adopting 12 frequency distributions (using LMO) and found that the GEV distribution is better suited distribution for estimation of MFD. Yue and Wang (2004) applied LMO to identify the suitable probability distribution for modelling annual stream flow in different climatic regions of Canada. Kumar and Chatterjee (2005) employed the LMO to define homogenous regions within 13 gauging sites of the north Brahmaputra region of India. Atiem and Harmancio˜lu (2006) carried out RFFA using the index flood LMO approach for 14 gauged sites on the Nile River tributaries. Saf (2009) observed that the Pearson Type-III distribution is better suited for modelling extreme values in Antalya and lower west Mediterranean sub-regions and the generalized logistic for the upper west Mediterranean sub-region.

Bhuyan, Borah, and Kumar (2010) applied LH-moments (generalized version of LMO) to carry out RFFA for river Brahmaputra. They found that RFFA based on the GEV distribution by using level one LH-moment is superior to the use of LMO. Malekinezhad, Nachtnebel, and Klik (2011) concluded that the GEV (using LMO) is better suited amongst five distributions studied for modelling AMD of three different regions in Iran. Badreldin and Feng (2012) carried out RFFA for the Luanhe basin using LMO and cluster techniques. Haberlandt and Radtke (2014) carried out FFA for three mesoscale catchments in northern Germany. Thus, the studies reported didn't suggest applying a particular distribution for FFA for different region or country. This apart, when different distributions are used for estimation of MFD, a common problem is encountered as regards the issue of best model fits for a given set of data. This can be answered by formal statistical procedures involving Goodness-of-Fit (GoF) and diagnostic tests; and the results are quantifiable and reliable (Zhang, 2002). Qualitative assessment is made from the plots of the recorded and estimated MFD. For quantitative assessment on MFD within in the recorded range, Chi-square (χ^2) and Kolmogorov–Smirnov (KS) tests are applied. A diagnostic test of D-index is used for the selection of suitable probability distribution for FFA (United States Water Resources Council [USWRC], 1981). The study compares the performance of six probability distributions that were employed for FFA, and illustrates the applicability of GoF and diagnostic tests procedures in identifying which distribution is the best suited for estimation of MFD for Malakkara and Neeleswaram.

2. Methodology

The study is to assess the probability distribution function (PDF) for FFA. Thus, it is required to process and validate the data for application such as: (1) select the PDFs for FFA (say, EXP, EV1, EV2, GEV, GPA and NOR); (2) select parameter estimation methods (say, MOM and LMO); (3) select quantitative GoF and diagnostic tests; and (4) conduct FFA and analyse the results obtained thereof.

2.1. Theoretical description of MOM

MOM is a technique for constructing estimators of the parameters that is based on matching the sample moments with the corresponding distribution moments (Ghorbani, Ruskeep, Singh, & Sivakumar, 2010). The rth central moment (μ_r) about the mean ($\bar{Q}$) of a random variable Q is defined by:

$$\mu_r = E\left(Q - \bar{Q}\right)^r = \int \left(Q - \bar{Q}\right)^r f(Q)dQ, \text{ if } Q \text{ is continuous variable} \qquad (1)$$

where $f(Q)$ is PDF of a random variable Q. The second moment (μ_2) about $\bar{Q}$ is called as variance. Similarly, third and fourth moments (μ_3 and μ_4) about $\bar{Q}$ are used to define skewness (C_S) and kurtosis (C_K), which are as follows:

$$C_S = \mu_3 / \mu_2^{3/2} \text{ and } C_K = \left(\mu_4 / \mu_2^2\right) - 3 \qquad (2)$$

2.2. Theoretical description of LMO

LMOs are summary statistics for probability distributions and data samples. They are analogous to ordinary moments, which provide measures of location, dispersion, skewness, kurtosis and other aspects of the shape of probability distributions or data samples. But, LMOs are computed from linear combinations of the ordered data values (Vogel & Wilson, 1996). LMO can be used as the basis of a unified approach to the statistical analysis adopting probability distributions. According to Central Water Commission (2010), LMOs has the following advantages:

(i)   LMO characterize wider range of probability distributions than conventional moments.

(ii) LMOs are less sensitive to outliers in the data.

(iii) LMOs approximate their asymptotic normal distribution more closely.

(iv) LMOs are nearly unbiased for all combinations of sample sizes and populations.

LMO will thus particularly useful in providing accurate quantile estimates of hydrological data in developing counties, where small sample size typically exist. LMO is a linear combination of probability weighted moments. Let $Q_1, Q_2, ..., Q_N$ be a conceptual random sample of size N and $Q_{1N} \leq Q_{2N} \leq, ..., \leq Q_{NN}$ denote the corresponding order statistics. The $r + 1$th LMO defined by Hosking and Wallis (1993) is:

$$l_{r+1} = \sum_{k=0}^{r} \frac{(-1)^{r-k}(r+k)!}{(k!)^2(r-k)!} b_k \qquad (3)$$

where l_{r+1} is the $r + 1$th sample moment and b_k is an unbiased estimator with

$$b_k = N^{-1} \sum_{i=k+1}^{N} \frac{(i-1)(i-2)\ldots(i-k)}{(N-1)(N-2)\ldots(N-k)} Q_{iN} \qquad (4)$$

The first two sample LMOs are expressed by:

$$l_1 = b_0 \text{ and } l_2 = 2b_0 - b_0 \qquad (5)$$

Table 1 gives the details of quantile functions and parameters of six probability distributions considered in the study.

Table 1. Quantile functions and parameters of probability distributions

Serial number	Distribution	Quantile function (Q)	Parameters	
			MOM	**LMO**
1	EXP	$Q_T = \xi - \alpha \ln (1 - F)$	$\xi = \bar{Q} - \alpha;\ \alpha = S_Q$	$\xi = l_1 - 2l_2;\ l_2 = \alpha/2$
2	EV1	$Q_T = \xi - \alpha \ln (-\ln (1 - F))$	$\xi = \bar{Q} - 0.5772157\alpha$	$\xi = l_1 - 0.5772157\alpha$
			$\alpha = \left(\sqrt{6/\pi}\right) S_Q$	$\alpha = l_2/\log 2$
3	EV2	$Q_T = \alpha e^{(-\ln(-\ln(F))/k}$	By using the logarithmic transformation of the recorded data, parameters of EV1 are initially obtained by MOM and LMO; and further used to determine the parameters of EV2 from $\alpha = e^{\xi l}$ and $k = 1/$(scale parameter of EV1)	
4	GEV	$Q_T = \xi + \left(\alpha(1 - (-\ln F)^k)/k\right)$	$\bar{Q} = \xi + \left(\alpha(\Gamma(1+k) - 1)/k\right)$	$z = (2/(3 + t_3)) - (\ln 2/\ln 3)$
			$S_Q = (\alpha/k)\left\{\Gamma(1 + 2k) - \Gamma(1 + k)^2\right\}^{1/2}$	$t_3 = (2(1 - 3^{-k})/(1 - 2^{-k})) - 3$
			$\psi = (\text{sign}\,k)\frac{\Gamma(1+3k)+3\Gamma(1+k)(1+2k)-2\Gamma^3(1+k)}{\left\{\Gamma(1+k)-\Gamma^2(1+k)\right\}^{1/2}}$	$k = 7.8590z + 2.9554z^2$
				$\alpha = l_2 k/(1 - 2^{-k})\Gamma(1 + k)$
				$\xi = l_1 + (\alpha(\Gamma(1 + k) - 1)/k)$
5	GPA	$Q_T = \xi + (\alpha(1 - (1 - F)^k)/k)$	$\bar{Q} = \xi + (\alpha/(1 + k))$	$\xi = l_1 + l_2(k + 2)$
			$S_Q = \alpha^2/(1 + 2k)(1 + k)^2$	$k = (4/(t_3 + 1)) - 3$
			$C_s = 2(1 - k)(1 + 2k)^{1/2}/(1 + 3k)$	$t_3 = (1 - k)/(3 + k)$
				$\alpha = (1 + k)(2 + k)l_2$
6	NOR	$Q_T = \mu + \sigma\phi^{-1}(F)$	$\mu = \bar{Q};\ \sigma = S_Q$	$\mu = l_1;\ \sigma = l_2\sqrt{\pi}$

Here, $F(Q)$ (or F) is the cumulative distribution function (CDF) of Q; ϕ^{-1} is the inverse of the standard normal distribution function and $\phi^{-1} = (P^{0.135} - (1 - P)^{0.135})/0.1975$; ξ, α, k are the location, scale and shape parameters, respectively; μ (or), σ (or S_Q) and C_s (or ψ) are the average, standard deviation and coefficient of skewness of the recorded AMD data; sign(k) is plus or minus 1 depending on the sign of k; Q_T is the estimated MFD by probability distribution for a return period T.

2.3. Goodness-of-Fit tests

GoF tests such as χ^2 and KS are applied for checking the adequacy of fitting of probability distributions to the series of recorded AMD data.

χ^2 statistic is defined by:

$$\chi^2 = \sum_{j=1}^{NC} \frac{\left(O_j(Q) - E_j(Q)\right)^2}{E_j(Q)} \tag{6}$$

where $O_j(Q)$ is the observed frequency value of jth class, $E_j(Q)$ is the expected frequency value of jth class and NC is the number of frequency classes. The rejection region of χ^2 statistic at the desired significance level (η) is $\chi^2 \geq \chi^2_{1-\eta,\,NC-m-1}$. Here, m denotes the number of parameters of the distribution.

KS statistic is defined by:

$$KS = \underset{i=1}{\overset{N}{\text{Max}}} \left(F_e(Q_i) - F_D(Q_i)\right) \tag{7}$$

where $F_e(Q_i) = (i - 0.44)/(N + 0.12)$ is the empirical CDF of Q_i and $F_D(Q_i)$ is the computed CDF of Q_i (Zhang, 2002). If the computed values of GoF tests statistic given by the distribution are lower than that of the theoretical values at the desired significance level, then the distribution is considered to be acceptable for estimation of MFD.

2.4. Diagnostic test

The selection of a suitable probability distribution for estimation of MFD is performed through D-index, which is defined by:

$$\text{D-index} = \left(1/\bar{Q}\right) \sum_{i=1}^{6} |Q_i - Q_i^*| \tag{8}$$

Here, $\bar{Q}$ is the average (or mean) of the recorded AMD, Q_i's ($i = 1$ to 6) are the first six highest sample values in the series and Q_i^* is the estimated value by the probability distribution. The distribution having the least D-index is identified as the better suited distribution for estimation of MFD (USWRC, 1981).

3. Application

An attempt has been made to estimate the MFD by six probability distributions (using MOM and LMO) at Malakkara and Neeleswaram gauging stations. These gauging stations are located in the Pamba and Periyar river basins, respectively. The catchment areas of these gauging stations are 1,713 and 4,234 km². Based on the water year (June to May), stream flow data for the period 1985–1986 to 2012–2013 for Malakkara and 1971–1972 to 2012–2013 for Neeleswaram is used. The series of AMD is derived from the daily stream flow data and further used in FFA. The summary statistics of the AMD values are presented in Table 2.

4. Results and discussions

A computer program was developed and used for performing FFA. The program computes the parameters of the six probability distributions (using MOM and LMO), flood estimates for different return periods, GoF tests statistic and D-index values for the stations under study.

4.1. Analysis based on GoF tests

In the present study, the degree of freedom ($NC - m - 1$) was considered as one for 3-parameter distributions (GEV and GPA) and two for 2-parameter distributions (EXP, EV1, EV2 and NOR) while computing the χ^2 statistic values for Malakkara and Neeleswaram. GoF tests statistics are computed through Equations 6 and 7, and presented in Table 3 for the stations under study.

Table 2. Summary statistics of the AMD

Gauging station	Statistical parameters				
	Mean (m³/s)	SD (m³/s)	CV (%)	Skewness	Kurtosis
Malakkara	1,007.2	326.3	32.4	1.064	1.632
Neeleswaram	2,026.2	663.2	32.7	0.247	−0.864

Notes: SD: Standard deviation; CV: Coefficient of variation.

Table 3. Computed values of GoF tests statistic

Distribution	Malakkara				Neeleswaram			
	χ^2		KS		χ^2		KS	
	MOM	LMO	MOM	LMO	MOM	LMO	MOM	LMO
EXP	2.000	2.714	0.108	0.127	8.857	7.143	0.163	0.146
EV1	0.571	1.643	0.059	0.055	3.143	2.571	0.134	0.124
EV2	2.714	2.000	0.145	0.118	21.714	8.000	0.187	0.171
GEV	2.357	1.643	0.055	0.058	5.143	6.286	0.102	0.096
GPA	1.643	1.643	0.109	0.061	5.429	5.429	0.060	0.055
NOR	1.643	3.071	0.081	0.083	3.714	3.714	0.113	0.108

From Table 3, it may be noted that the computed values of χ^2 statistic for the EXP, EV2, GEV and GPA distributions (using MOM and LMO) are greater than the theoretical values ($\chi^2_{0.05,1}$ = 3.84 for GEV and GPA; and $\chi^2_{0.05,2}$ = 5.99 for EXP, EV1, EV2 and NOR) at 5% significance level, and at this level, these four distributions are not acceptable for estimation of MFD in Neeleswaram. On the other hand, the computed values of χ^2 statistic for these distributions (using MOM and LMO) are lower than the theoretical values at 5% significance level, and at this level, the six distributions are acceptable for estimation of MFD in Malakkara. Also, from Table 3, it may be noted that the computed values of KS statistic for the six probability distributions (using MOM and LMO) are lower than the theoretical values (0.250 for Malakkara and 0.205 for Neeleswaram) at 5% significance level and at this level, the six distributions are acceptable for estimation of MFD in Malakkara and Neeleswaram.

4.2. Estimation of MFD by probability distributions

The parameters of the six probability distributions were determined by MOM and LMO; and further used for estimation of the MFD at Malakkara and Neeleswaram. The results are presented in Tables 4 and 5.

4.3. Flood frequency curves

The MFD estimates computed by the six probability distributions (using MOM and LMO) for Malakkara and Neeleswaram, as given in Tables 4 and 5, are used to develop flood frequency curves and these are presented in Figures 1 and 2, respectively.

Table 4. MFD estimates for different return periods for Malakkara

Return period (year)	Estimated MFD (m³/s)											
	MOM						LMO					
	EXP	EV1	EV2	GEV	GPA	NOR	EXP	EV1	EV2	GEV	GPA	NOR
2	907.1	953.6	928.2	955.8	925.0	1,007.3	896.4	952.3	908.4	950.2	940.9	1,007.3
5	1,206.1	1,242.1	1,170.8	1,245.5	1,268.4	1,281.9	1,227.5	1,247.8	1,189.8	1,245.7	1,293.8	1,276.8
10	1,432.3	1,433.1	1,365.4	1,435.0	1,481.5	1,425.4	1,478.0	1,443.4	1,422.5	1,443.4	1,487.3	1,417.7
20	1,658.5	1,616.3	1,582.3	1,614.9	1,661.5	1,544.0	1,728.5	1,631.1	1,688.4	1,634.7	1,634.1	1,534.1
50	1,957.4	1,853.4	1,915.1	1,845.3	1,857.0	1,677.4	2,059.7	1,874.0	2,107.7	1,884.6	1,775.3	1,665.0
100	2,183.6	2,031.1	2,209.6	2,016.1	1,978.4	1,766.3	2,310.2	2,056.0	2,488.8	2,073.6	1,852.8	1,752.3
200	2,409.8	2,208.2	2,548.0	2,184.7	2,080.8	1,847.7	2,560.7	2,237.4	2,937.1	2,263.4	1,911.5	1,832.2
500	2,708.8	2,441.8	3,075.1	2,404.8	2,192.2	1,946.4	2,891.8	2,476.6	3,654.3	2,516.1	1,968.0	1,929.1
1,000	2,934.9	2,618.3	3,544.6	2,569.3	2,261.3	2,015.6	3,142.3	2,657.5	4,310.4	2,708.9	1,999.0	1,997.0

Table 5. MFD estimates for different return periods for Neeleswaram

Return period (year)	Estimated MFD (m³/s)											
	MOM						LMO					
	EXP	EV1	EV2	GEV	GPA	NOR	EXP	EV1	EV2	GEV	GPA	NOR
2	1,822.7	1,917.2	1,865.5	1,992.0	1,968.9	2,026.2	1,790.9	1,909.7	1,812.1	1,984.0	1,978.1	2,026.2
5	2,430.4	2,503.6	2,357.3	2,586.1	2,692.2	2,584.4	2,493.6	2,536.7	2,405.9	2,594.8	2,700.5	2,598.3
10	2,890.1	2,891.7	2,752.3	2,910.8	2,989.7	2,876.1	3,025.2	2,951.9	2,902.4	2,934.3	2,987.7	2,897.3
20	3,349.8	3,264.1	3,193.2	3,179.1	3,166.9	3,117.1	3,556.8	3,350.1	3,474.8	3,218.7	3,154.0	3,144.2
50	3,957.4	3,746.1	3,870.5	3,473.2	3,296.4	3,388.2	4,259.6	3,865.6	4,386.5	3,535.0	3,271.6	3,422.1
100	4,417.1	4,107.2	4,470.5	3,659.9	3,349.7	3,569.0	4,791.2	4,251.8	5,223.3	3,738.8	3,318.4	3,607.4
200	4,876.8	4,467.1	5,161.0	3,821.4	3,381.4	3,734.5	5,322.8	4,636.7	6,215.7	3,917.4	3,345.5	3,777.0
500	5,484.5	4,941.9	6,237.6	4,002.5	3,404.6	3,935.0	6,025.5	5,144.4	7,819.1	4,120.6	3,364.6	3,982.4
1,000	5,944.2	5,300.7	7,197.9	4,118.8	3,414.1	4,075.6	6,557.1	5,528.2	9,300.1	4,252.9	3,372.3	4,126.6

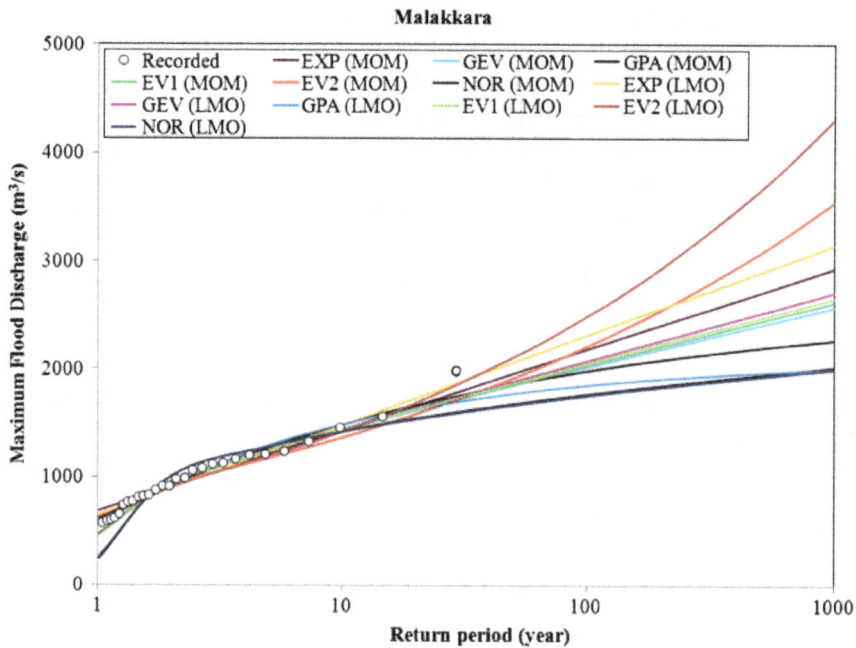

Figure 1. Plots of recorded and estimated MFD by six probability distributions (using MOM and LMO) for Malakkara.

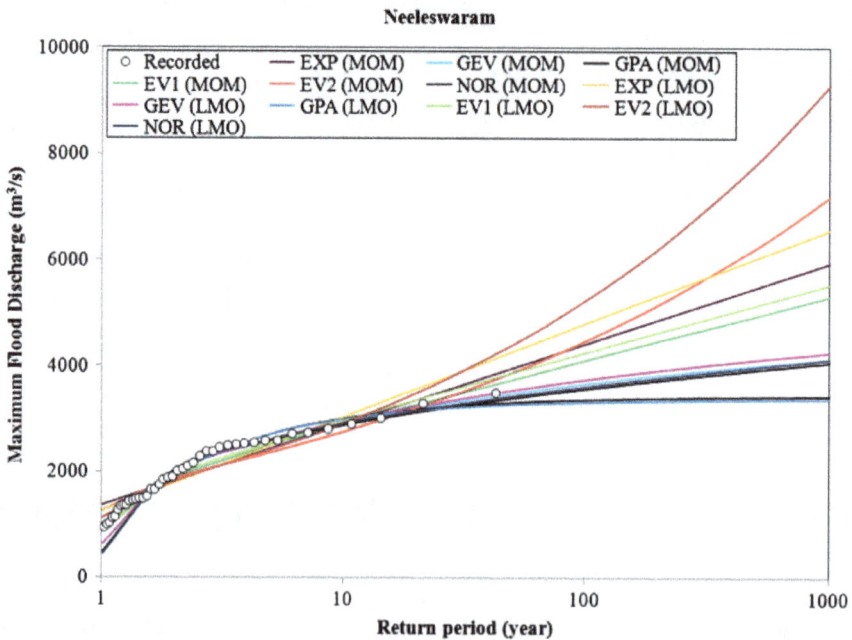

Figure 2. Plots of recorded and estimated MFD by six probability distributions (using MOM and LMO) for Neeleswaram.

From Figures 1 and 2, it can be seen that the estimated MFD by EV2 (LMO) was higher estimates when compared to the corresponding values of other distributions (using MOM and LMO) for the return period of 20 years and above. Also, from Figures 1 and 2, it can be seen that the fitted curves by EV1 distribution (using MOM and LMO) are linear for the stations under study.

4.4. Analysis based on diagnostic test

For the selection of the best suitable distribution for estimation of MFD, the D-index values of the six probability distributions are computed by Equation 8 and the results are presented in Table 6.

Gauging station	Indices of D-index											
	MOM						LMO					
	EXP	EV1	EV2	GEV	GPA	NOR	EXP	EV1	EV2	GEV	GPA	NOR
Malakkara	3.997	3.143	4.057	3.062	2.931	2.117	4.723	3.287	5.614	3.383	2.477	2.063
Neeleswaram	3.865	2.984	3.959	1.727	1.010	1.459	4.946	3.403	6.091	1.955	0.950	1.571

Table 6. D-index values of probability distributions (using MOM and LMO)

From Table 6, it may be noted that (1) the values of D-index viz., 2.063 of NOR (using LMO) for Malakkara and 0.950 of GPA (using LMO) for Neeleswaram are minimum when compared to the corresponding values of other probability distributions; (2) χ^2 test results don't support the use of EXP, EV2, GEV and GPA distributions (using MOM and LMO); and therefore, these four distributions are not found to be acceptable for estimation of MFD in Neeleswaram; and (3) D-index value of 1.571 (using NOR) is the second minimum next to the GPA when LMO is applied for determination of parameters for Neeleswaram. Therefore, NOR distribution (using LMO) is found to be better suited for estimation of MFD for Malakkara whereas NOR distribution (using MOM) for Neeleswaram.

Because of the diverging results obtained from GoF and diagnostic tests, qualitative assessment is made to identify the suitable probability distribution for estimation of MFD. From Figures 1 and 2, it can be seen that the fitted lines by estimated MFD values adopting NOR (using MOM and LMO) and GPA (using LMO) are not confined with the line of agreement of the observed MFD values. By considering the trend lines of the fitted curves using estimated MFD values, the study identifies the EV1 distribution (using LMO) is found to be a good choice for estimation of MFD at Malakkara though the D-index value of EV1 is higher than that of GPA. For Neeleswaram, EV1 distribution (using LMO) is also found to be a good choice for estimation of MFD.

5. Conclusions

The paper describes briefly the study carried out for estimation of MFD by adopting FFA using a computer aided procedure for determination of parameters of six probability distributions (using MOM and LMO) for Malakkara and Neeleswaram. The following conclusions are drawn from the study:

(i)   For the return period of 10 years and above, it was found that the estimated MFD by LMO of EXP, GEV, EV1 and EV2 distributions are higher than the corresponding values of MOM of these four distributions for Malakkara and Neeleswaram.

(ii) The study presents the selection of suitable distribution evaluated by GoF (using χ^2 and KS) and diagnostic (using D-index) tests.

(iii) The χ^2 test results showed that the EXP, EV1, EV2, GEV, GPA and NOR distributions (using MOM and LMO) are acceptable for estimation of MFD at Malakkara.

(iv) The χ^2 test results showed that the EXP, EV2, GEV and GPA distributions (using MOM and LMO) are not acceptable for estimation of MFD at Neeleswaram.

(v)   The KS test results indicated that the six probability distributions (using MOM and LMO) are acceptable for estimation of MFD at Malakkara and Neeleswaram.

(vi) By considering the trend lines of the fitted curves using estimated MFD values, the study presented that the EV1 distribution (using LMO) is better suited amongst six distributions studied for estimation of MFD at Malakkara and Neeleswaram.

(vii) The study suggested that the MFD values for different return periods computed by EV1 distribution (using LMO) could be considered as the design parameter for planning and designs of irrigation and flood protection; and also for hydraulic structures on these rivers near Malakkara and Neeleswaram.

Acknowledgements

The author is grateful to Shri S. Govindan, Director, Central Water and Power Research Station (CWPRS), Pune, for providing the research facilities to carry out the study. The author is thankful to the Executive Engineer, Central Water Commission, Kerala, and Dr C. Ramesh, Senior Research Officer, CWPRS, Pune, for supply of streamflow data used in the study.

Funding

The authors received no direct funding for this research.

Author details

N. Vivekanandan[1]

E-mail: anandaan@rediffmail.com

[1] Central Water and Power Research Station, Pune, Maharashtra 411024, India.

References

Atiem, I. A., & Harmancio˘lu, N. B. (2006). Assessment of regional floods using L-moments approach: The case of the river Nile. *Water Resources Management, 20*, 723–747. http://dx.doi.org/10.1007/s11269-005-9004-0

Badreldin, G. H. H., & Feng, P. (2012, January 5–7). Regional rainfall frequency analysis for the Luanhe Basin using L-moments and cluster techniques. *International Conference on Environmental Science and Development (ICESD 2012)* (Vol. 1, pp. 126–135). Hong Kong.

Bhuyan, A., Borah, M., & Kumar, R. (2010). Regional flood frequency analysis of north-bank of the river Brahmaputra by using LH-moments. *Water Resources Management, 24*, 1779–1790. http://dx.doi.org/10.1007/s11269-009-9524-0

Central Water Commission. (2010). *Development of hydrological design aids (surface water) under hydrology project II: State of the art report.* Consulting Engineering Services (India) in Association with HR Wallingford. New Delhi.

Ghorbani, A. M., Ruskeep, A. H., Singh, V. P., & Sivakumar, B. (2010). Flood Frequency analysis using mathematica. *Turkish Journal of Engineering and Environmental Sciences, 34*, 171–188.

Haberlandt, U., & Radtke, I. (2014). Hydrological model calibration for derived flood frequency analysis using stochastic rainfall and probability distributions of peak flows. *Hydrology and Earth System Sciences, 18*, 353–365. http://dx.doi.org/10.5194/hess-18-353-2014

Haktanir, T., & Horlacher, H. B. (1993). Evaluation of various distributions for flood frequency analysis. *Hydrological Sciences Journal, 38*, 15–32. http://dx.doi.org/10.1080/02626669309492637

Hosking, J. R. M. (1990). L-moments: Analysis and estimation of distributions using linear combinations of order statistics. *Royal Statistical Society (Series B), 52*, 105–124.

Hosking, J. R. M., & Wallis, J. R. (1993). Some statistics useful in regional frequency analysis. *Water Resources Research, 29*, 271–281. http://dx.doi.org/10.1029/92WR01980

Kjeldsen, T. R., Smithers, J. C., Schulze, R. E. (2002). Regional flood frequency analysis in the KwaZulu-Natal province, South Africa using the index-flood method. *Journal of Hydrology, 255*, 194–211. http://dx.doi.org/10.1016/S0022-1694(01)00520-0

Kumar, R., & Chatterjee, C. (2005). Regional flood frequency analysis using L-moments for north Brahmaputra region of India. *Journal of Hydrologic Engineering, 10*(1), 1–7. http://dx.doi.org/10.1061/(ASCE)1084-0699(2005)10:1(1)

Kumar, R., Chatterjee, C., Kumar, S., Lohani, A. K., & Singh, R. D. (2003). Development of regional flood frequency relationships using L-moments for Middle Ganga Plains Subzone 1(f) of India. *Water Resources Management, 17*, 243–257. http://dx.doi.org/10.1023/A:1024770124523

Malekinezhad, H., Nachtnebel, H. P., Klik, A. (2011). Regionalization approach for extreme flood analysis using L-moments. *Agricultural Science and Technology* (Iran), *13*, 1183–1196.

Saf, B. (2009). Regional flood frequency analysis using L-moments for the west Mediterranean region of Turkey. *Water Resources Management, 23*, 531–551. http://dx.doi.org/10.1007/s11269-008-9287-z

United States Water Resources Council. (1981). *Guidelines for determining flood flow frequency* (Bulletin No. 17B). Washington, DC.

Vogel, R. M., & Wilson, I. (1996). Probability distribution of annual maximum, mean, and minimum streamflows in the United States. *Journal of Hydrologic Engineering, 1*, 69–76. http://dx.doi.org/10.1061/(ASCE)1084-0699(1996)1:2(69)

Yue, S., & Wang, C. Y. (2004). Possible regional probability distribution type of Canadian annual streamflow by L-moments. *Water Resources Management, 18*, 425–438. http://dx.doi.org/10.1023/B:WARM.0000049145.37577.87

Zhang, J. (2002). Powerful goodness-of-fit tests based on the likelihood ratio. *Journal of the Royal Statistical Society: Series B (Statistical Methodology), 64*, 281–294. http://dx.doi.org/10.1111/rssb.2002.64.issue-2

8

Forecasting monthly groundwater level fluctuations in coastal aquifers using hybrid Wavelet packet–Support vector regression

Sujay Raghavendra. N[1]* and Paresh Chandra Deka[1]

*Corresponding author: Sujay Raghavendra. N, Department of Applied Mechanics and Hydraulics, National Institute of Technology Karnataka, Surathkal, Mangalore 575025, India
E-mail: sujayraghavendran@ymail.com
Reviewing editor: Sanjay Shukla, Edith Cowan University, Australia

Abstract: This research demonstrates the state-of-the-art capability of Wavelet packet analysis in improving the forecasting efficiency of Support vector regression (SVR) through the development of a novel hybrid Wavelet packet–Support vector regression (WP–SVR) model for forecasting monthly groundwater level fluctuations observed in three shallow unconfined coastal aquifers. The Sequential Minimal Optimization Algorithm-based SVR model is also employed for comparative study with WP–SVR model. The input variables used for modeling were monthly time series of total rainfall, average temperature, mean tide level, and past groundwater level observations recorded during the period 1996–2006 at three observation wells located near Mangalore, India. The Radial Basis function is employed as a kernel function during SVR modeling. Model parameters are calibrated using the first seven years of data, and the remaining three years data are used for model validation using various input combinations. The performance of both the SVR and WP–SVR models is assessed using different statistical indices. From the comparative result analysis of the developed models, it can be seen that WP–SVR model outperforms the classic SVR model in predicting groundwater levels at all the three well locations (e.g. $NRMSE_{(WP-SVR)} = 7.14$, $NRMSE_{(SVR)} = 12.27$; $NSE_{(WP-SVR)} = 0.91$, $NSE_{(SVR)} = 0.8$ during the test phase with respect to

ABOUT THE AUTHOR

Paresh Chandra Deka, currently an associate professor in the Department of Applied Mechanics & Hydraulics, National Institute of Technology Karnataka, has over 20 years of experience in teaching undergraduate and postgraduate students. He earned his PhD from the prestigious IIT Guwahati. His current research interests are Hydrological Modeling, applications of Artificial Neural Network, Fuzzy logic, Genetic algorithm, Wavelet transformation, and Support Vector Machine in water resources system modeling. He has published over 21 journal articles and numerous national and international conference papers. He has successfully guided three PhD scholars and presently, seven research scholars are working under his esteemed guidance. He is member of several profession bodies like Indian Society for Hydraulics, International Association of Hydraulic Research, and Indian Society for Technical Education, etc. He was visiting faculty at AIT Bangkok during the year 2012.

PUBLIC INTEREST STATEMENT

In general, Groundwater systems are characterized by non-stationary and nonlinear features. Modeling of these systems and predicting their future scenarios requires identification and capture of the rudimentary characteristics of the variables affecting them. The prolonged drought, population growth and emerging climate change impacts have increased demand for groundwater and sustainable management of groundwater resources is imperative to the agricultural, industrial, urban, rural and environmental viability. Such management requires not only a robust monitoring of groundwater levels, but also sufficient knowledge on future groundwater level trends. Hence a novel hybrid Wavelet packet–Support vector regression (WP–SVR) technique is proposed with a main objective to develop a reliable monthly groundwater level fluctuation forecasting system.

well location at Surathkal). Therefore, using the WP–SVR model is highly acceptable for modeling and forecasting of groundwater level fluctuations.

Subjects: Computation; Data Preparation & Mining; Civil, Environmental and Geotechnical Engineering; Water Engineering; Water Science

Keywords: groundwater systems; support vector machines; Wavelet packets; radial basis function (RBF); Wavelet packet–Support vector regression

1. Introduction

Groundwater is one of the major sources of supply for domestic, industrial, and agricultural purposes. In some areas, groundwater is the only dependable source of supply, while in some other regions, it is chosen because of its ready availability. The shallow water table depths have significant impacts on crop growth, vegetation development, and contaminant transport. Groundwater has many advantages over surface water for water supply. It is reliable in dry seasons or droughts because of the large storage (Hiscock, Rivett, & Davison, 2002). The consequences of aquifer depletion can lead to local water rationing, excessive reductions in yields, wells going dry or producing erratic ground water quality changes, changes in flow patterns of groundwater. Normal groundwater recharge to creeks and streams during low-flow periods could result in reduced supplies for surface water sources. To ensure the ecological sustainability of a watershed, the management of groundwater resources in conjunction with surface waters is very much substantial. So this necessitates the constant monitoring of the groundwater levels (Batu, 1998).

Assessment of groundwater levels from observation wells provides a principal source of information regarding the hydrological stresses acting over aquifers and how those stresses influence over groundwater recharge, storage, and discharge. It also allows water managers, engineers, and policy-makers to develop better strategies of groundwater management and protection. A better conception of the groundwater dynamics and underlying factors that affect groundwater levels paves way to investigate the requirement of agricultural, industrial, urban, and other demands and assess the benefits and costs of water conservation. The water levels if forecasted well in advance may help the administrators to plan for better utilization of groundwater resources (Healy & Cook, 2002).

In the past, several models have been proposed for forecasting groundwater levels using various stochastic, analytical, and soft computing techniques. In the recent times, the application of Artificial neural networks (ANN) and its hybrid models are found to be efficient in forecasting groundwater levels at different time scales (Chitsazan, Rahmani, & Neyamadpour, 2013; Daliakopoulos, Coulibaly, & Tsanis, 2005; Jalalkamali, Sedghi, & Manshouri, 2011; Nayak, Rao, & Sudheer, 2006; Nourani, Mogaddam, & Nadiri, 2008; Shi & Zhu, 2009; Sreekanth et al., 2009; Sreenivasulu, Deka, & Nagaraj, 2012; Wanakule & Aly, 2005). ANN, due to its "black box" nature, immense computational burden, prone to overfitting, and the empirical nature of model development somehow paved way for hydrological modeling using Support Vector Machines (SVMs). The current standard SVM algorithm based on statistical learning theory proposed by Cortes and Vapnik (1995) is an approximation implementation of the method of structural risk minimization with a good generalization capability. SVMs have also been successfully applied in groundwater level estimation studies. Liu, Chang, and Zhang (2009) proposed least squares-support vector machine (LS-SVM) arithmetic based on chaos optimization for dynamic prediction of groundwater levels, taking into account of groundwater level dynamic series length and peak mutation characters. Their simulation results revealed that the developed LS-SVM model was very effective in reflecting the dynamic evolution process of groundwater levels in a well. Behzad, Asghari, and Coppola (2010) made an attempt with SVMs and ANNs for predicting transient groundwater levels in a complex groundwater system under variable pumping and weather conditions. They predicted water-level elevations at different time horizons i.e. daily, weekly, biweekly, monthly, and bimonthly. Here, particularly for longer prediction horizons, SVM outperformed ANN even when fewer data events were available for model development. Yoon, Jun, Hyun, Bae, and Lee (2011) demonstrated the efficacy of SVM model in predicting short-term groundwater level fluctuations in a coastal aquifer giving significance to recharge from precipitation and tidal effect. The dilemma of salt water intrusion into coastal aquifer is accounted by considering the tidal effect. Sudheer, Shrivastava, Panigrahi, and Mathur (2011) proposed a hybrid Quantum behaved Particle Swarm Optimization (QPSO)-based SVM model for estimating the groundwater levels. QPSO function was adopted in their study in order to determine the optimal SVM parameters. Tapak, Rahmani, and Moghimbeigi (2014) developed a time series prediction model based on SVM using monthly peizometric groundwater level data. Their proposed SVM model outperformed the classic time series model in predicting groundwater levels. Raghavendra and Deka (2014a) investigated the potential of Sequential Minimal Optimization Algorithm-based Support Vector Regression (SMO-SVR) model for forecasting monthly groundwater level fluctuations observed in two shallow unconfined coastal aquifers. The relative performance of SMO-SVR models developed using three different kernel functions—Polynomial, RBF, and Pearson VII function-based universal kernel (PuK) was comparatively evaluated.

Discrete data pose a problem in SVR analysis. Even after the optimal choice of kernel and regularization parameter means, it would end up with all data being support vectors. The loss function of SVR doesn't possess clear statistical interpretation; henceforth, expert knowledge of the problem cannot be efficiently constructed. For this, the representation and quality of data are of foremost importance. The knowledge discovery during training phase is more difficult if there is much of noisy and unreliable data. Consequently, this provides way for data pre-processing before using it in any regression models. This study investigates the applicability of the Wavelet packet transform (WPT) as a data pre-processing technique and is hybridized with SVR to forecast groundwater level fluctuations in coastal aquifers.

Wavelets are mathematical functions that cut up data into different frequency components, and then study each component with a resolution matched to its scale. Wavelet packet analysis can be used to denoise the highly non-linear, non-stationary time series (such as groundwater levels, rainfall, tidal levels, temperature) so that the new denoised time series can be used as inputs for an SVR model. Wavelet transformation has numerous applications in the field of hydrology and water resource engineering such as rainfall run-off modeling (Nourani, Kisi, & Komasi, 2011), forecasting of precipitation (Kisi & Shiri, 2011), daily pan evaporation prediction (Abghari, Ahmadi, Besharat, & Rezaverdinejad, 2012), urban water demand forecasting (Campisi-Pinto, Adamowski, & Oron, 2012). Wavelet packet-based SVM applications can be witnessed in various other domains of engineering (Hu, Zhu, & Ren, 2008; Manimala, Selvi, & Ahila, 2012; Subramanian & Henry, 2010; Tong, Song, Lin, & Zhao, 2006). Adamowski and Chan (2011) proposed a novel method of coupling discrete wavelet transforms (WA) and artificial neural networks (ANN) for groundwater level forecasting. Moosavi, Vafakhah, Shirmohammadi, and Behnia (2013) developed a Wavelet-ANFIS hybrid model for groundwater level forecasting with different prediction times. Ping, Qiang, and Xixia (2013) developed a combined model of chaos theory, wavelet and SVM to overcome the limitations, including challenges with determination of orders of non-linear models and low prediction accuracy in groundwater level forecasting. Suryanarayana, Sudheer, Mahammood, and Panigrahi (2014) emphasizes the application of integrated Wavelet analysis–Support vector regression (WA–SVR) model for predicting the groundwater level variations in three observation wells. They comparatively evaluate the WA–SVR model with ANN and Auto Regressive Integrated Moving Average models and found the superior performance of the WA–SVR model in predicting groundwater levels. Only few research works are carried out in groundwater level forecasting using simple discrete Wavelet analysis hybridized with SVM technique. No research has been published yet which explores, coupling Wavelet packet analysis with SVM for groundwater level forecasting with multiple hydrological input parameters. Hence, a novel hybrid Wavelet packet–Support vector regression (WP–SVR) technique is proposed with a main objective to develop a reliable monthly groundwater level fluctuation forecasting system. The performance of SVR models with multiple hydrological input scenarios is tested and the combination yielding higher efficiency is adopted in hybrid WP–SVR modeling. A comparative analysis of results obtained by WP–SVR and basic SVR is also presented.

2. Theory of Support vector regression

Consider a simple linear regression problem trained on data-set $\chi = \{u_i, v_i; i = 1, \ldots, n\}$ with input vectors u_i and linked targets v_i. In order to associate the inherited relations between the data-sets, a function $g(u)$ has to be formulated approximately, and thereby it can be used in future to infer the output "v" for a new input data "u".

Standard SVM regression employs a loss function $L_\varepsilon(v, g(u))$ which explains the deviation of the estimated function from the original one. Various forms of loss functions, namely, linear, quadratic, exponential, Huber's loss function, etc. can be mined in the literature. The standard Vapnik's ε insensitive loss function is used in this context which is defined as:

$$L_\varepsilon(v, g(u)) = \begin{cases} 0 & \text{for } |v - g(u)| \le \varepsilon \\ |v - g(u)| - \varepsilon & \text{otherwise} \end{cases} \tag{1}$$

The ε-insensitive loss function, aid in determining $g(u)$ which can better approximate the actual output vector "v". It also provides at most error tolerance "ε" from the actual incurred targets "v_i" for all training data as flat as possible. Consider the regression function defined by

$$g(u) = w \cdot u + b \tag{2}$$

where $w \in \chi$, χ is the input space; $b \in R$ is a bias term and $(w \cdot u)$ is dot product of vectors w and u. Flatness in Equation 2 refers to a smaller value of parameter vector w. By minimizing the norm $\|w\|^2$, flatness can be ascertained along with model complexity. Thus, regression problem can be stated as the following convex optimization problem:

$$\min_{w, b, \xi, \xi^*} \quad \frac{1}{2}\|w\|^2 + C \sum_{i=1}^{n} \left(\xi_i + \xi_i^*\right)$$
$$\text{subject to} \quad \begin{array}{l} v_i - (w \cdot u_i + b) \le \varepsilon + \xi_i \\ (w \cdot u_i + b) - v_i \le \varepsilon + \xi_i^* \\ \xi_i, \xi_i^* \ge 0, \quad i = 1, 2, \ldots, n \end{array} \tag{3}$$

where ξ_i and ξ_i^* are slack variables introduced to evaluate the deviation of training samples outside ε-insensitive zone. The trade-off between the flatness of g and the quantity up to which deviations greater than ε are tolerated is depicted by $C > 0$. Whenever a training error occurs, C is a positive constant influencing the degree of penalizing. Underfitting and overfitting of training data can be prevented by minimization of the regularization term $w^2/2$ along with the training error term $C \sum_{i=1}^{n} \left(\xi_i - \xi_i^*\right)$ in Equation 3. The minimization problem in Equation 3 represents the primal objective function. A dual set of variables, $\underline{\alpha}_i$ and $\overline{\alpha}_i$ is introduced for the corresponding constraints and the problem is dealt by constructing a Lagrange function from the primal objective function. Optimality conditions are utilized at the saddle points of a Lagrange function, steering to the formulation of the dual optimization problem:

$$\max_{\underline{\alpha}_i, \overline{\alpha}_i} -\frac{1}{2} \sum_{i,j=1}^{n} \left(\underline{\alpha}_i - \overline{\alpha}_i\right)\left(\underline{\alpha}_j - \overline{\alpha}_j\right)\langle u_i \cdot u_j \rangle - \varepsilon \sum_{i=1}^{n} \left(\underline{\alpha}_i + \overline{\alpha}_i\right) + \sum_{i=1}^{n} v_i \left(\underline{\alpha}_i - \overline{\alpha}_i\right)$$
$$\sum_{i=1}^{n} \left(\underline{\alpha}_i - \overline{\alpha}_i\right) = 0$$
$$\text{subject to} \quad \begin{array}{l} 0 \le \underline{\alpha}_i \le 0 \quad i = 1, 2, \ldots, n \\ 0 \le \overline{\alpha}_i \le 0 \quad i = 1, 2, \ldots, n \end{array} \tag{4}$$

After determining Lagrange multipliers, $\underline{\alpha}_i$ and $\overline{\alpha}_i$, the parameter vectors w and b can be evaluated under Karush–Kuhn–Tucker (KKT) complementarity conditions (Fletcher, 2000) which are not discussed herein. Therefore, the prediction is a linear regression function that can be expressed as:

$$g(u) = \sum_{i=1}^{n} \left(\underline{\alpha}_i - \overline{\alpha}_i\right)\langle u_i \cdot u \rangle + b \tag{5}$$

Thus, SVM regression expansion is derived; where "w" is represented as a linear combination of the training patterns "v_i" and "b" can be found using primary constraints. For $|g(u)| \geq \varepsilon$, Lagrange multipliers may be non-zero for all the samples inside the ε-tube and these remaining coefficients are labeled as support vectors.

In order to craft SVM regression to deal with non-linear cases, pre-processing of training patterns "u_i" has to done by mapping the input space χ into some feature space $\mathfrak{I}$ using non-linear hypothesis function $\varphi = \chi \rightarrow \mathfrak{I}$ and then applied to the standard support vector algorithm. Let u_i be mapped into the feature space by non-linear function $\varphi(u)$ and hence, the decision function is given by:

$$g(w, b) = w \cdot \varphi(u) + b \tag{6}$$

This non-linear regression problem can be expressed as the following optimization problem:

$$\min_{w,b,\xi,\xi^*} \quad \frac{1}{2} \|w\|^2 + C \sum_{i=1}^{n} \left(\xi_i + \xi_i^* \right)$$
$$\text{subject to} \quad \begin{array}{l} v_i - \left(w \cdot \varphi(u_i) + b \right) \leq \varepsilon + \xi_i \\ \left(w \cdot \varphi(u_i) + b \right) - v_i \leq \varepsilon + \xi_i^* \\ \xi_i, \xi_i^* \geq 0, \quad i = 1, 2, \dots, n \end{array} \tag{7}$$

where w is the weight vector of coefficients, ξ_i and ξ_i^* are the upper and lower training bounds of the region where the errors less than ε are ignored and b is a constant. The index i labels the "n" training cases. The $v \in \pm 1$ is the class label and u_i is the independent variable.

Then, the dual form of the non-linear SVR can be expressed as:

$$\max_{\underline{\alpha}_i, \overline{\alpha}_i} -\frac{1}{2} \sum_{i,j=1}^{n} \left(\underline{\alpha}_i - \overline{\alpha}_i \right) \left(\underline{\alpha}_j - \overline{\alpha}_j \right) \left\langle \varphi(u_i) \cdot \varphi(u_j) \right\rangle - \varepsilon \sum_{i=1}^{n} \left(\underline{\alpha}_i + \overline{\alpha}_i \right) + \sum_{i=1}^{n} v_i \left(\underline{\alpha}_i - \overline{\alpha}_i \right)$$
$$\sum_{i=1}^{n} \left(\underline{\alpha}_i - \overline{\alpha}_i \right) = 0$$
$$\text{subject to} \quad \begin{array}{l} 0 \leq \underline{\alpha}_i \leq 0 \quad i = 1, 2, \dots, n \\ 0 \leq \overline{\alpha}_i \leq 0 \quad i = 1, 2, \dots, n \end{array} \tag{8}$$

The "kernel trick" $K(u_i, u_j) = \left\langle \varphi(u_i), \varphi(u_j) \right\rangle$ is used for computations in input space χ to fetch the inner products into feature space $\mathfrak{I}$. Any function satisfying Mercer's theorem (Vapnik, 1999) can be used as kernels.

Finally, the decision function of non-linear SVR with the allowance of the kernel trick is expressed as follows:

$$g(u) = \sum_{i,j=1}^{1} \left(\underline{\alpha}_i - \overline{\alpha}_i \right) K \langle u_i \cdot u \rangle + b \tag{9}$$

The parameters that impact over the effectiveness of the non-linear SVR are the cost constant C, the radius of the insensitive tube ε, and the kernel parameters. These parameters are mutually dependent over one another and hence, altering the value of one parameter affects the other linked parameters also (Raghavendra & Deka, 2014b).

Platt (1998) proposed an algorithm called sequential minimal optimization (SMO) for solving the problem of regression with SVMs. SMO breaks the large quadratic programming (QP) problem into a series of smallest possible QP problems. For the evaluation of the decision function, SMO devotes significant time, rather than performing QP. The SMO algorithm has a much naïve formulation and is trouble-free to implement. Shevade, Keerthi, Bhattacharyya, and Murthy (2000) proposed an improvement that enhances the algorithm to perform significantly faster by avoiding the use of threshold "b" in checking KKT conditions.

3. Theory of Wavelet packet transform (WPT)

Wavelets are mathematical functions that slice up data into distinct frequency domains, and then examine each domain with a resolution matched to its scale. They have benefits over traditional Fourier methods in analyzing physical situations where the signal contains discontinuities and sharp spikes (Shinde & Hou, 2004).

The WPT is a generalization of the discrete wavelet transform (DWT), and it is convenient to introduce the WPT from the DWT. The DWT is a basis transformation, i.e. it estimates the coordinates of a data vector (spectrum or signal) in the so-called wavelet basis (Coifman & Wickerhauser, 1992). A wavelet is a function that appears like a small wave, a ripple of the baseline, thus its name. The wavelet basis is generated by stretching out the wavelet to fit different scales of the signal and by moving it to cover all parts of the signal. The DWT is said to give a time-scale, or time-frequency, analysis of signals. In wavelet transform, signals split into a detail and an approximation. The approximation obtained from the first level is split into new detail and approximation, and this process is repeated. Because of the fact that DWT decomposes only the approximations of the signal, it may pose problems while applying DWT in certain applications where the valuable information is localized in higher frequency components. The major dissimilarity between DWT and WPT is that WPT splits not only approximations but also details. The peak level of the WPT is the time representation of the signal, whereas, the base level has a better frequency resolution. Thus, with the aid of WPT, a better frequency resolution can be achieved from the decomposed signal. In addition, the use of WPT mines much more features about the signal. Consequently, the Wavelet packet analysis provides better check of frequency resolution and more features about signal than DWT (Lei, Meyer, & Ryan, 1994).

Each component in the Wavelet packet tree can be regarded as a filtered component with a bandwidth of a filter, decreasing with increasing level of decomposition and the whole tree can be considered as a filter bank. At the crown of the tree, the time resolution of the WP components is reasonable, but at the cost of poor frequency resolution, whereas at the tail end with the use of Wavelet packet analysis, the frequency resolution of the decomposed component with high-frequency content can be increased. As a result, the Wavelet packet analysis provides better check of frequency resolution for the decomposition of the signal (Learned & Willsky, 1995). A Wavelet packet is represented as a function, $\psi_{j,t}^i(t)$, where "i" is the modulation parameter, "j" is the dilation parameter, and "k" is the translation parameter.

$$\psi_{j,k}^i(t) = 2^{-j/2} \psi^i \left(2^{-j} \cdot t - k\right) \tag{10}$$

where $i = 1, 2, \dots, j_n$ and "n" is the level of decomposition in the Wavelet packet tree. The wavelet is obtained by the following recursive relationships:

$$\psi^{2i}(t) = \frac{1}{\sqrt{2}} \sum_{k=-\infty}^{\infty} h(k) \cdot \psi^i \left(\frac{t}{2} - k\right) \tag{11}$$

$$\psi^{2i+1}(t) = \frac{1}{\sqrt{2}} \sum_{k=-\infty}^{\infty} g(k) \cdot \psi^i \left(\frac{t}{2} - k\right) \tag{12}$$

where $\psi^i(t)$ is called as a mother wavelet and the discrete filters $h(k)$ and $g(k)$ are quadrature mirror filters associated with the scaling function and the mother wavelet function. These two filters, $h(k)$ and $g(k)$, are also called group conjugated orthogonal filters.

The Wavelet packet coefficients $C_{j,k}^i$ corresponding to the signal $f(t)$ can be obtained by

$$C_{j,k}^i = \int_{-\infty}^{\infty} f(t) \cdot \psi_{j,k}^i(t) \cdot dt \tag{13}$$

Provided the wavelet coefficients satisfy the orthogonality condition (Fan & Zuo, 2006). The Wavelet packet component of the signal at a particular node can be obtained as:

$$f_j^i(t) = \sum_{k=-\infty}^{\infty} C_{j,k}^i \cdot \psi_{j,k}^j(t)\, dt \tag{14}$$

After performing Wavelet packet decomposition up to jth level, the original signal can be represented as a summation of all Wavelet packet components at jth level as shown in equation:

$$f(t) = \sum_{i=1}^{2i} f_j^i(t) \tag{15}$$

In Wavelet packet, standard structure composed of low- and high-pass filters is used for perfect reconstruction (Wickerhauser, 1994). Wavelet packets are particular linear combinations of wavelets which form bases that hold back many of the orthogonality, smoothness, and localization properties of their parent wavelets. The coefficients in the linear combinations are computed by a recursive algorithm, making each newly computed Wavelet packet coefficient sequence, the root of its own analysis tree (Graps, 1995).

4. Study area and data analysis

Dakshina Kannada is a maritime district located in the southwestern part of Karnataka adjoining the Arabian Sea. The geographical area is 4770 km^2 extending from 12°30'00" to 13°11'00" north latitudes and 74°35'00" and 75°33'30" east longitudes. The observation wells selected for the current research are located in two adjacent micro-watersheds that come under the Pavanje and Gurpura river catchments. The observation well located at Surathkal lies at 12°59'00" north latitude and 74°48'10" east longitude, the well location at Mangalore lies at 12°53'23" north latitude and 74°51'36" east longitude, and lastly, the well near Ganjimatta is located at 12°59'02" north latitude and 74°57'15" east longitude as shown in Figure 1. The study area is dominated by the southwest monsoon (June–September) and non-monsoon period (October–May). The average annual rainfall over the watershed is around 3,500 mm.

In the study area, Lateritic soil is the predominant with highly porous and permeable nature. Due to this lateritic soil property, the infiltration rate is high and the shallow wells give very quick response to rainfall with water table rising immediately, but subsequent drastic draw down is also observed within a short period of time.

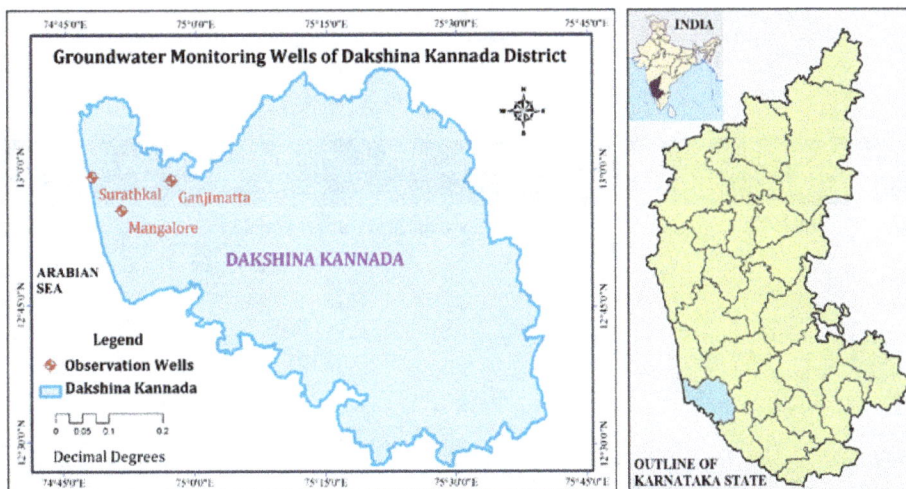

Figure 1. Study area (location of observation wells).

Main input to the groundwater in the micro-watershed is from the monsoon rains. The point rain-fall observed at the rain gauge stations established at National Institute of Technology Karnataka (NITK) campus, Surathkal; Bajpe and Mangalore (Kavoor) are being utilized in this study. The rainfall data of these stations for the years 1996–2006 have been used in this study. The monthly total rain-fall, average temperature, and mean tidal-level data are used to correlate the measurement of groundwater levels in the observation wells on a monthly basis.

The water-level data of the observation wells located at Surathkal, Mangalore, and Ganjimatta for the years 1996–2006 used in the study were obtained from Department of Mines and Geology, Dakshina Kannada Dist. The tidal-level data of the study period are obtained from "Tide Tables" published by the Geological Survey of India.

From the data analysis, the maximum, minimum, mean, standard deviation S_d, variance, and correlation values of all the variables influencing groundwater table fluctuation is being tabulated in Tables 1–3. The mean and variance of all the variables do not stay steady over time and this proves that time series data employed to be non-stationary. The predictive relationship between two dependent variables is represented in terms of correlation. The tidal levels of the sea have also exhibited a fair correlation with the monthly groundwater level time series, which needles its use during modeling. In this context, it is seen that rainfall is negatively correlated with groundwater level time series.

Table 1. Statistical parameters of groundwater level, rainfall, temperature, and tidal level data with respect to well location at Surathkal

Data-set	Variable	Statistical parameters					
		X_{mean}	X_{max}	X_{min}	S_d	Variance	CC with GWT
Training	GWT	8.33	11.04	3.14	1.65	2.71	1
	Rainfall	314.70	1,523.00	0.00	427.94	183,130.20	−0.87
	Temperature	27.68	35.12	25.15	1.54	2.36	0.45
	Tidal Level	7.08	7.25	6.78	0.10	0.01	0.49
Testing	GWT	8.52	10.98	5.52	1.18	1.39	1
	Rainfall	261.50	885.20	0.00	322.61	104,074.09	−0.81
	Temperature	27.85	31.38	26.19	1.27	1.60	0.52
	Tidal Level	7.08	7.22	6.89	0.09	0.01	0.47

Table 2. Statistical parameters of groundwater level, rainfall, temperature, and tidal level data with respect to well location at Mangalore

Data-set	Variable	Statistical parameters					
		X_{mean}	X_{max}	X_{min}	S_d	Variance	CC with GWT
Training	GWT	10.43	12.13	6.26	1.11	1.23	1
	Rainfall	310.28	1489.20	0.00	423.87	179,662.76	−0.59
	Temperature	27.53	31.21	23.81	1.51	2.29	0.43
	Tidal Level	7.08	7.25	6.78	0.10	0.01	0.41
Testing	GWT	10.24	11.91	7.94	1.03	1.07	1
	Rainfall	296.58	1,129.80	0.00	369.43	136,479.98	−0.68
	Temperature	27.41	30.10	25.90	1.12	1.25	0.39
	Tidal Level	7.08	7.22	6.89	0.09	0.01	0.35

Table 3. Statistical parameters of groundwater level, rainfall, temperature, and tidal level data with respect to well location at Ganjimatta

Data-set	Variable	Statistical parameters					
		X_{mean}	X_{max}	X_{min}	S_d	Variance	CC with GWT
Training	GWT	4.41	9.45	0.95	2.25	5.04	1.00
	Rainfall	305.32	1466.30	0.00	424.95	180,584.57	−0.67
	Temperature	27.53	31.21	23.81	1.51	2.29	0.81
	Tidal Level	7.08	7.25	6.78	0.10	0.01	0.58
Testing	GWT	4.42	9.55	1.32	1.96	3.85	1.00
	Rainfall	269.71	992.80	0.00	349.54	122,180.58	−0.75
	Temperature	27.41	30.10	25.90	1.12	1.25	0.68
	Tidal Level	7.08	7.22	6.89	0.09	0.01	0.45

5. Methodology

5.1. Development of basic SVR model

In this study, the SMO-based SVR model is adopted for the prediction of monthly groundwater table using various input combinations. Monthly total rainfall, average temperature, mean tidal levels, and previous groundwater levels are used as input variables. The model is trained utilizing seven years (1996–2003) of data and tested with three years (2003–2006) of data. The SVR models were developed using the WEKA software (Hall et al., 2009). There is no way in advance to know which kernel function will be best for an application; henceforth, the RBF kernel is being adopted for prediction of groundwater levels considering its effective generalization capability. The accuracy of an RBF kernel-based SVR model is principally dependent on the selection of the model parameters such as C—regularization parameter, Gamma (γ)—kernel parameter, and epsilon parameter (ε). WEKA provides two methods for finding optimal parameter values, firstly, a grid search and the other one is cross-validation parameter selection. Grid search tries to find values of each parameter across the specified search range using geometric steps (Cherkassky & Ma, 2004). Usually, grid searches are computationally expensive as the model is evaluated at various points within the grid for each parameter. If cross-validation parameter selection is opted, a V-fold cross-validation is used by the search to calculate the optimal parameters using the error computed from the training data. The input–output combination of the SMO-SVR model is given below:

⇒ GW Level + Rainfall = 1 month lead GW Level forecast

⇒ GW Level + Rainfall + Temperature = 1 month lead GW Level forecast

⇒ GW Level + Rainfall + Temperature + Tidal levels = 1 month lead GW Level forecast

Initially, the time series data of input variables are normalized in SVR model and the data are fitted using RBF kernel function. The best RBF-based SVR model is obtained from a grid search of hyperparameters—(C, γ, and ε). The developed SVR model is trained using the optimal values of hyperparameters of RBF kernel. The model is then simulated again with testing data-set and evaluated. The predicted values obtained from the SMO-SVR model is then compared with the known observed values.

5.2. Development of hybrid WP–SVR model

The time series data of all input variables, namely, monthly total rainfall, average temperature, mean tidal levels, and previous groundwater levels are considered for Wavelet packet denoising. These time series signals are highly non-stationary and non-linear. They possess high peaks and sudden downfalls. This affects the predicting capability of an SVR model. Henceforth, the application of denoising of input signals serves as a suitable data pre-processing technique to address the non-linear time series data. All the input signals are denoised using WPT and the resulting Wavelet packet coefficients from the denoised signals are used as inputs for the prediction of the target variable. Note, the target variable (i.e. one month lead GW level) is unchanged i.e. original observed time series is retained without any transformation. To perform wavelet packet analysis, haar, db2, db3, and db4 wavelets have been selected as mother Wavelets and various decomposition levels have been evaluated. Figure 2 depicts the flowchart of Wavelet packet analysis. Each input variable is denoised separately using Wavelet packet analysis in order to obtain Wavelet packet coefficients from the denoised signal. MATLAB-Wavelet Design and Analysis Tool (Misiti & Misiti, 1996) was used for Wavelet packet analysis. After obtaining output in the form of coefficients from the discrete Wavelet packet transformation, the next step is to train the SVR model as explained in the above Section 5.1 using WP coefficients of all input variables of the training phase. Once the model is trained, the WP coefficients of test phase are used to simulate the model and thereby tested. Figure 3 represents the methodology of the hybrid WP–SVR model. Wavelet packets serve as a powerful tool for the task of signal denoising. The ability to decompose a signal into different scales is very important for denoising, and it improves the analysis of the signal significantly.

6. Performance evaluation

The statistical indices assess the degree of confidence one can have in the predictions of the model and whether the model displays any bias which could lead to "fail-dangerous" predictions. Models usually involve prediction or measurement of a number of variables. The usual practice in texts is to develop the concept of error propagation in connection with precision (random) errors. To evaluate the performance of SVR and hybrid WP–SVR models, the following statistical indices were adopted.

(1) Normalized Root Mean Square Error (NRMSE)

$$NRMSE = \frac{RMSE}{(X_{max} - X_{min})} \times 100 \text{ where, } RMSE = \sqrt{\frac{\sum\limits_{i=1}^{N}(X_i - Y_i)}{N}} \qquad (16)$$

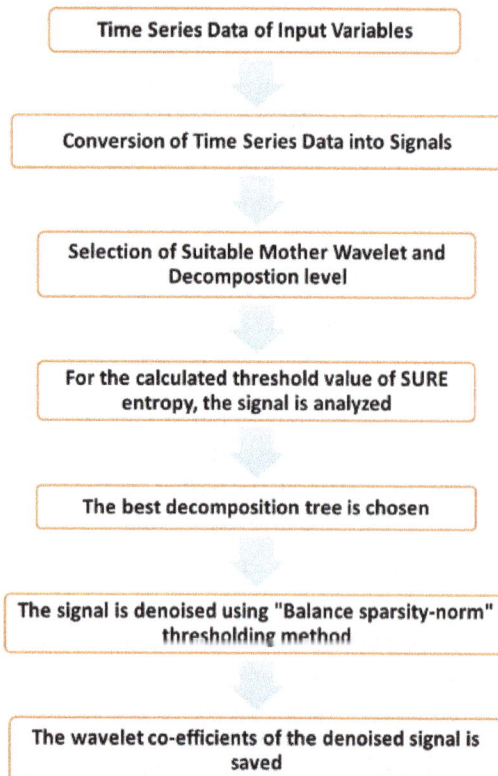

| Time Series Data of Input Variables |

| Conversion of Time Series Data into Signals |

| Selection of Suitable Mother Wavelet and Decompostion level |

| For the calculated threshold value of SURE entropy, the signal is analyzed |

| The best decomposition tree is chosen |

| The signal is denoised using "Balance sparsity-norm" thresholding method |

| The wavelet co-efficients of the denoised signal is saved |

Figure 2. Flowchart of Wavelet packet analysis.

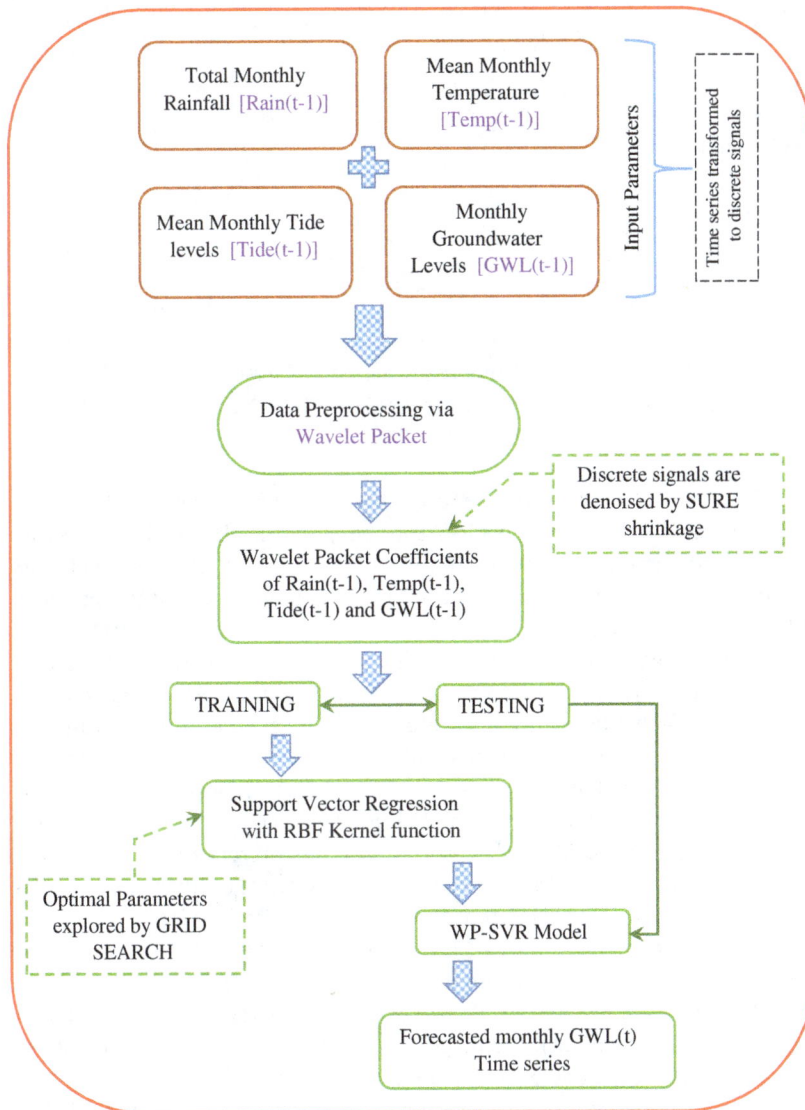

Figure 3. Schematic portrait depicting WP–SVR methodology.

(2) Normalized Mean Bias

$$NMB = \sum_{i=1}^{N} \frac{(Y_i - X_i)}{X_i} = \left(\frac{\overline{Y}_i}{\overline{X}_i} - 1 \right) \qquad (17)$$

(3) Absolute Relative Error (ARE) and Average Absolute Relative Error (AARE)

$$ARE = \sum_{i=1}^{N} \left| \frac{Y_i - X_i}{X_i} \right| \times 100; \quad AARE = \frac{1}{N} \sum_{i=1}^{N} \left| \frac{Y_i - X_i}{X_i} \right| \times 100 \qquad (18)$$

(4) Nash–Sutcliffe coefficient (NSE)

$$NSE = 1 - \frac{\sum_{i=1}^{N} \left(X_i - Y_i \right)^2}{\sum_{i=1}^{N} \left(X_i - \overline{x} \right)^2} \qquad (19)$$

(5) Threshold Statistics (TS)

$$TS_x = \frac{n_x}{N} \times 100$$

(20)

(6) Correlation Coefficient (CC)

$$CC = \frac{\sum_{i=1}^{N} \left(X_i - \overline{X} \right) \cdot \left(Y_i - \overline{Y} \right)}{\sqrt{\sum_{i=1}^{N} \left(X_i - \overline{X} \right)^2 \cdot \sum_{i=1}^{N} \left(Y_i - \overline{Y} \right)^2}}$$

(21)

where X is the Observed/Actual Values; Y is the Computed Values; $\overline{X}$ is the Mean of Actual Data; n_x is the Number of Data points whose ARE value is less than x%; N is the Total Number of Data Points.

7. Results and discussions

The aim of using the SMO-SVR is to test its ability to predict groundwater level fluctuation in observation wells located at Surathkal, Mangalore, and Ganjimatta. The analysis is being performed for one month lead time prediction of groundwater levels for the above-mentioned input combinations (Section 5.1) at all the three well locations. The optimal parameters obtained after tuning the SVR model is as tabulated in Table 4. The number of support vectors (nsv) provides the information regarding the efficacy of the model developed i.e. whether the model is underfitting or overfitting. The SVR models developed above have "nsv" in 30–40% range and this indicates that the models are neither underfitting nor overfitting. Table 5 presents the statistical results of the SVR models developed for all the three input cases. The magnitude of the NMB and AARE computations infers that the SVR model trained with four input variables closely predicts the observed groundwater level time series than the SVR models trained with two and three input variables. The negative and positive NMB values can be interpreted as the model underestimates and overestimates, respectively. The performance in terms of Threshold Statistics of ARE computed for 5 and 10%, also indicates that the SVR model with four input variables is performing better. During the test phase, the TS < 5% and TS < 10% for an SVR model with four input variables are 66 and 86.11%, respectively with respect to well at Surathkal. The NRMSE and NSE statistics of the SVR models with different input cases is presented in Figures 4 and 5, respectively. The degree of residual variance portrayed using NRMSE statistic shows the global goodness of fit between the computed and observed groundwater levels for input combination with four variables. Since the observation wells considered in the study are near to the coast, the effect of tidal levels can have a considerable effect on the shoreline aquifers. The results obtained during this modeling demonstrate the relevance of inclusion of mean tidal level data as an input variable.

Table 4. Optimal SVR parameters for different input combinations

Well location	Input combinations	Optimal values			
		C	γ	ε	nsv
Surathkal	GWL + RAIN + TEMP + TIDE	6.25	3.163	0.1	34/84
	GWL + RAIN + TEMP	12	2.633	0.1	35/84
	GWL + RAIN	15	3.366	0.1	35/84
Mangalore	GWL + RAIN + TEMP + TIDE	15	2	0.1	39/84
	GWL + RAIN + TEMP	10	2.5	0.1	39/84
	GWL + RAIN	7.5	2.25	0.1	38/84
Ganjimatta	GWL + RAIN + TEMP + TIDE	7.25	3.163	0.1	36/84
	GWL + RAIN + TEMP	7.3	1.5858	0.1	38/84
	GWL + RAIN	6.5	1.7783	0.1	37/84

Table 5. Statistical results of SVR models with different input combinations

Input combination	Statistical indices	Well location					
		Surathkal		Mangalore		Ganjimatta	
		TRAIN	TEST	TRAIN	TEST	TRAIN	TEST
GWL + RAIN + TEMP + TIDE	NRMSE	6.96	12.27	7.32	11.58	8.47	9.59
	NMB	−0.41	−0.75	−0.54	−1.01	0.19	−0.38
	AARE	4.21	5.29	3.66	4.18	4.61	5.65
	NSE	0.89	0.8	0.82	0.79	0.89	0.83
	TS < 5%	75	66	72.61	72.22	70.23	66.86
	TS < 10%	92.85	86.11	96.42	94.44	90.47	88.88
GWL + RAIN + TEMP	NRMSE	10.0	15.38	9.71	15.11	10.11	10.81
	NMB	0.52	−0.79	−0.61	−1.19	−1.08	1.1
	AARE	6.12	6.97	4.51	5.14	5.32	6.51
	NSE	0.77	0.73	0.73	0.68	0.73	0.63
	TS < 5%	54.76	52.77	61.9	58.33	63.09	61.11
	TS < 10%	83.33	72.22	94.04	91.66	86.9	83.33
GWL + RAIN	NRMSE	10.25	15.93	11.92	18.63	14.23	15.91
	NMB	−0.84	−0.91	−0.69	0.96	−1.16	−1.02
	AARE	6.83	7.92	4.96	5.7	5.83	6.78
	NSE	0.71	0.51	0.56	0.47	0.65	0.6
	TS < 5%	42.85	41.66	60.71	50	59.52	55.55
	TS < 10%	80.95	72.22	92.85	83.33	85.71	80.55

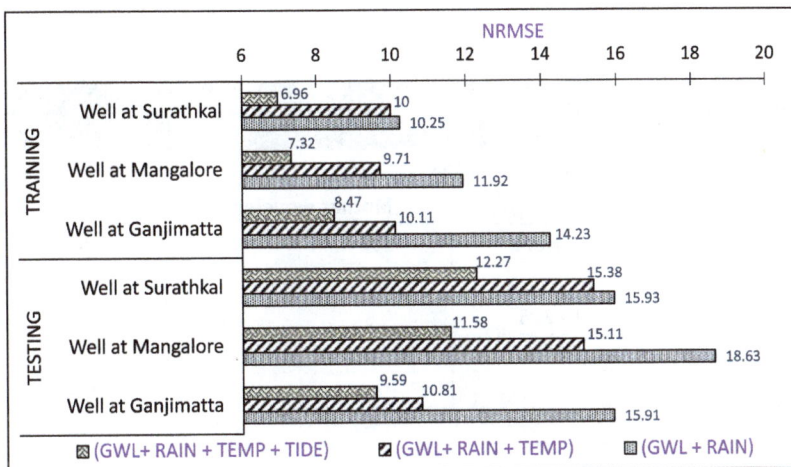

Figure 4. NRMSE of SVR models with different input combinations.

During Wavelet packet analysis, Haar, Daubechies 2 (db2), db3, db4 wavelets were selected as mother wavelet. Due to the satisfactory performance of SVR model with four input variables, namely, monthly time series of groundwater level, total rainfall, average temperature and mean tidal levels, the same input combination is retained and considered for Wavelet packet analysis. The optimal SVR parameters obtained while modeling using Wavelet packet coefficients as input, for selected mother wavelets with respect to well locations at Surathkal, Mangalore, and Ganjimatta are presented in Tables 6–8, respectively. The individual performance of different mother wavelets is evaluated using correlation coefficient (CC) statistic. The correlation (CC) statistic infers that the db4 mother wavelet performed more efficiently than other mother wavelets at level 4. Hence, the db4 Wavelet packet

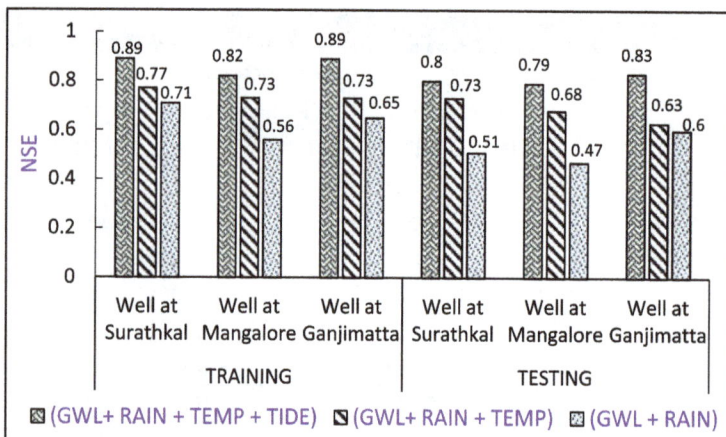

Figure 5. NSE of SVR models with different input combinations.

Table 6. Optimal SVR parameters for forecasting using Wavelet packet coefficients with respect to well location at Surathkal

	Well at Surathkal			
Optimal SVR parameters	**Mother wavelet**			
	Haar	**db2**	**db3**	**db4**
C	14	18	7.25	9.25
Gamma (γ)	0.25	1.633	2.25	3.5
Epsilon (ε)	0.1	0.1	0.1	0.1
Train CC	0.75	0.77	0.82	0.97
Test CC	0.63	0.68	0.75	0.95

Table 7. Optimal SVR parameters for forecasting using Wavelet packet coefficients with respect to well location at Mangalore

	Well at Mangalore			
Optimal SVR parameters	**Mother wavelet**			
	Haar	**db2**	**db3**	**db4**
C	9	12	21	7.75
Gamma (γ)	2.25	3.8	4.25	5.25
Epsilon (ε)	0.1	0.1	0.1	0.1
Train CC	0.55	0.67	0.84	0.96
Test CC	0.51	0.59	0.79	0.96

Table 8. Optimal SVR parameters for forecasting using Wavelet packet coefficients with respect to well location at Ganjimatta

	Well at Ganjimatta			
Optimal SVR parameters	**Mother wavelet**			
	Haar	**db2**	**db3**	**db4**
C	12	14	20	12.25
Gamma (γ)	2.276	2.88	3.75	1.75
Epsilon (ε)	0.1	0.1	0.1	0.1
Train CC	0.69	0.74	0.8	0.97
Test CC	0.61	0.62	0.73	0.97

Table 9. Optimal SVR parameters for forecasting using db4-Wavelet packet coefficients

Well location	Input combinations	Optimal SVR parameters			
		C	Gamma (γ)	Epsilon (ε)	nsv
Surathkal	GWL + RAIN + TEMP + TIDE	9.25	3.5	0.1	38/84
Mangalore	GWL + RAIN + TEMP + TIDE	7.75	5.25	0.1	39/84
Ganjimatta	GWL + RAIN + TEMP + TIDE	12.25	1.75	0.1	39/84

Table 10. Statistical results of WP–SVR and SVR models

Well location	Model		NRMSE	NMB	AARE	NSE	TS < 5%	TS < 10%
Surathkal	WP–SVR	TRAIN	4.43	−0.22	3.82	0.94	77.38	95.23
		TEST	7.14	−0.70	3.58	0.91	75	97.22
	SVR	TRAIN	6.96	−0.41	4.21	0.89	75	92.85
		TEST	12.27	−0.75	5.29	0.8	66	86.11
Mangalore	WP–SVR	TRAIN	4.94	−0.49	2.63	0.92	79.76	100
		TEST	7.55	−0.58	2.45	0.91	75	100
	SVR	TRAIN	7.32	−0.54	3.66	0.82	72.61	96.42
		TEST	11.58	−1.01	4.18	0.79	72.22	94.44
Ganjimatta	WP–SVR	TRAIN	5.52	0.13	3.65	0.94	82.14	96.42
		TEST	6.43	−0.29	4.68	0.93	77.77	94.44
	SVR	TRAIN	8.47	0.19	4.61	0.89	70.23	90.47
		TEST	9.59	−0.38	5.65	0.83	66.86	88.88

coefficients of input variables are considered for the development of hybrid WP–SVR model in order to generate one month ahead groundwater level forecasts. Table 9 presents the optimal SVR parameters of hybrid WP–SVR model.

The comparative evaluation of results obtained from the WP–SVR and SVR models for one month lead GW-level prediction is presented in the form of various performance indices in Table 10 and in the form of graphs (Figures 6–8). The advantage and robustness of WP–SVR model are being checked and the results were found to be quite satisfactory over the basic SVR model.

Study area 1 – Well location at Surathkal: From the time series graph and scatter plot as presented in Figures 9 and 12 for 1 month lead GW-level prediction, it is observed that both SVR and WP–SVR model results are closely following the observed time series. The proposed WP–SVR model is almost

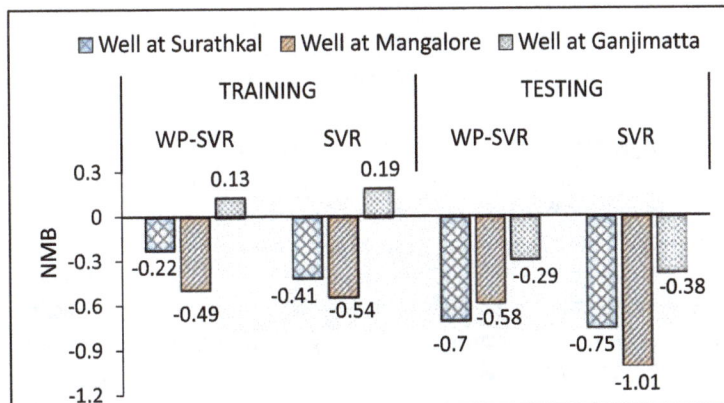

Figure 6. Normalized mean bias of WP–SVR and SVR forecast.

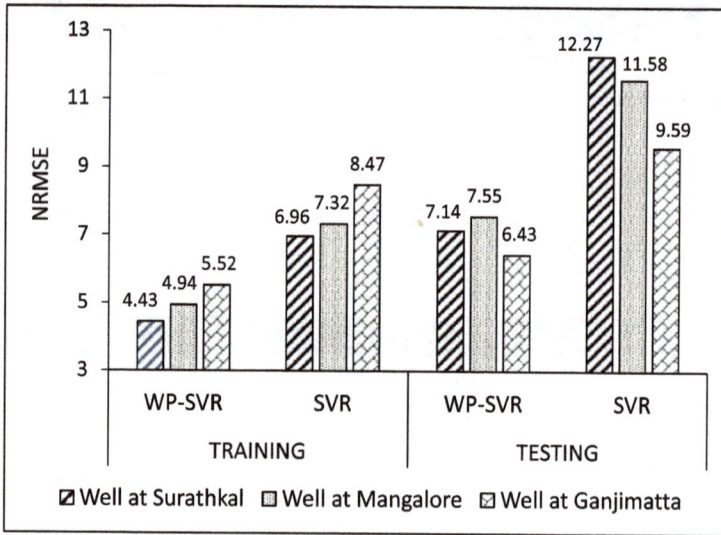

Figure 7. NRMSE of WP–SVR and SVR forecast.

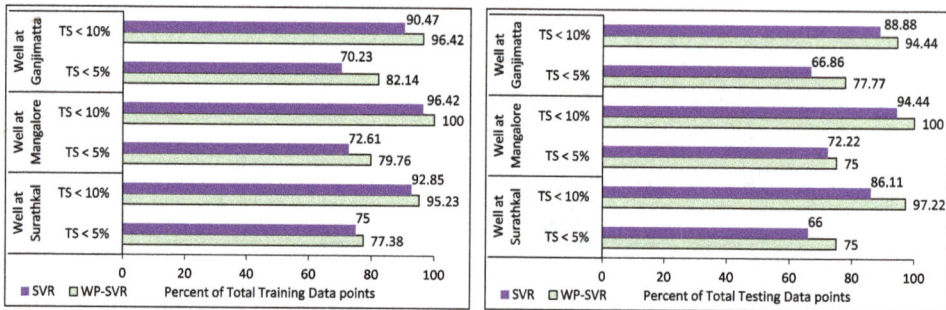

Figure 8. Comparative performance evaluation of WP–SVR and SVR models using absolute relative error (%) as threshold statistic (TS) during training and testing phase.

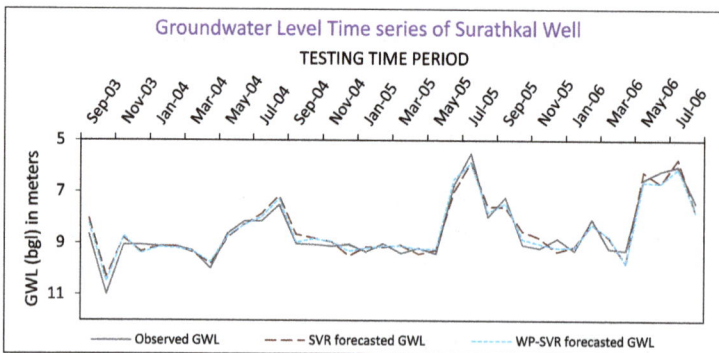

Figure 9. Forecasted groundwater level using SVR and WP–SVR models during testing with respect to well location at Surathkal.

following the trend of observed time series, while the SVR foresee slightly deviates at peaks and base, not properly catching the higher and lower values. The WP–SVR model performance appears to have accepted accuracy during training and testing phase as observed from different statistical indices presented in Table 10 and from the scatter plot of test phase as shown in Figure 12. Here, the TS < 5% and TS < 10% of WP–SVR model being 75 and 97.22%, respectively, during test phase verifies the close agreement of WP–SVR model with the observed groundwater levels. The other statistical evaluations of NMB and NRMSE as presented in Figures 6 and 7, respectively, demonstrate the efficacy of WP–SVR prediction to that of the SVR model.

Study area 2—Well location at Mangalore: Figure 10 displays the time series plot of WP–SVR and SVR model outputs in test stage, wherein it can be visualized that the WP–SVR prediction closely

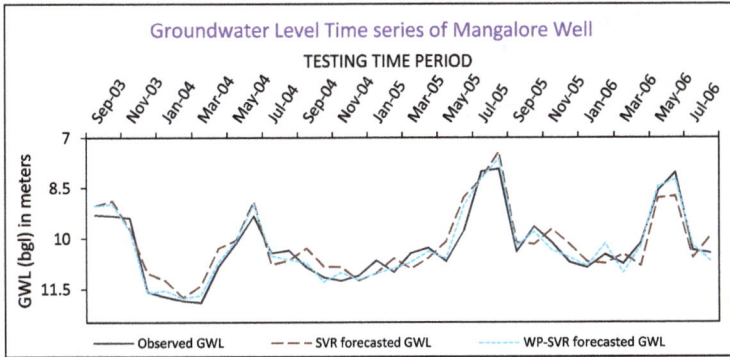

Figure 10. Forecasted groundwater level using SVR and WP–SVR models during testing with respect to well location at Mangalore.

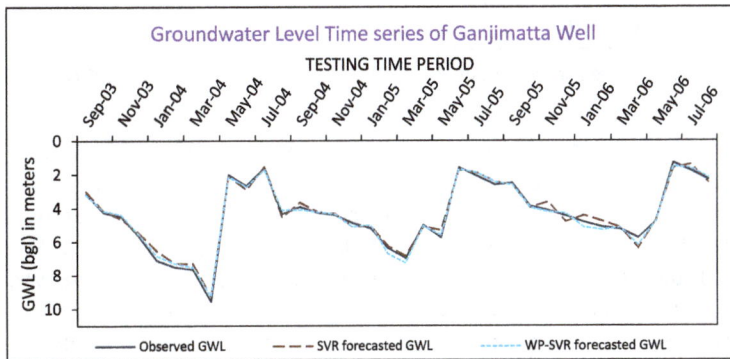

Figure 11. Forecasted groundwater level using SVR and WP–SVR models during testing with respect to well location at Ganjimatta.

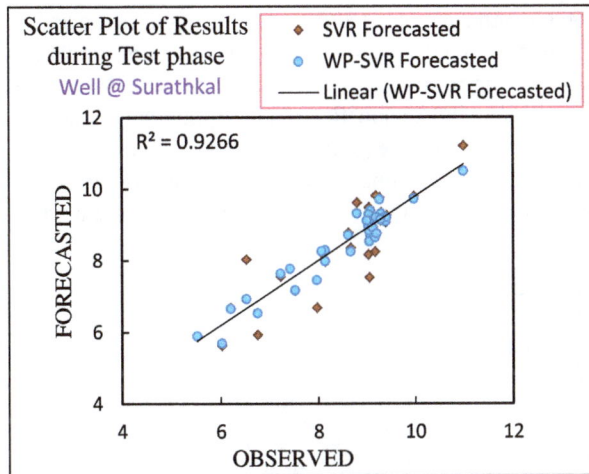

Figure 12. Scatter plot of observed versus forecasted of WP–SVR and SVR models with respect to well location at Surathkal.

follows the observed groundwater level time series. Here, NRMSE of 4.94 and 7.55%, NSE of 0.92 and 0.91 observed during training and test phase, respectively, from WP–SVR model depicts the significance of the hybrid model developed. Figure 13 shows the closely spaced scatters of the computed and observed groundwater levels of WP–SVR and SVR models during the test phase. It can be observed that the SVR predictions have more number of outliers than that of WP–SVR predictions during the test phase. Referring to Figure 8, it can be seen that the WP–SVR yields 75% forecast lower than 5% absolute relative error (ARE) and 100% forecast less than 10% ARE during test phase, thereby clearly ascertaining the consistency and robust performance of the hybrid model that is developed.

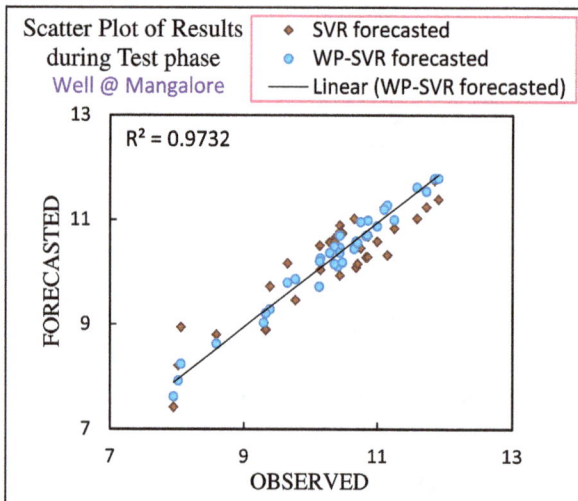

Figure 13. Scatter plot of observed versus forecasted of WP–SVR and SVR models with respect to well location at Mangalore.

Study area 3—Well location at Ganjimatta: From the time series plot of WP–SVR and SVR model outputs of the test phase as represented in Figure 11, it can be inferred that the WP–SVR prediction is capable of catching the lower peak values efficiently and it also diligently shadows the observed groundwater level time series. During training phase, positive NMB values are seen in both WP–SVR and SVR forecasts, thereby indicating the overestimation of observed data values (Figure 6). However, in the test phase, both the models exhibit underestimation of observed values. The SVR predictions show slightly weak correlation with observed groundwater levels when compared with WP–SVR predictions during the test phase as depicted from the scatter plot (Figure 14).

It can be observed that the performance is satisfactory with RBF Kernel-based SVR models and the WP–SVR model forecasted the water levels with superior accuracy in terms of all the statistical indices during calibration and validation period at all well locations. It is obvious from the fit line equations and R^2 values that the WP–SVR model performs much better than the basic SVR model (Figures 12–14). In overall, the Wavelet packet and SVR conjunction model framed by integrating two methods, Discrete WPTs and SVR, seems to be more adequate than the single SVR model for forecasting monthly groundwater levels. The denoised periodic components obtained from the Wavelet packet technique is found to be most effective in yielding more accurate forecast when used as inputs in the SVR models.

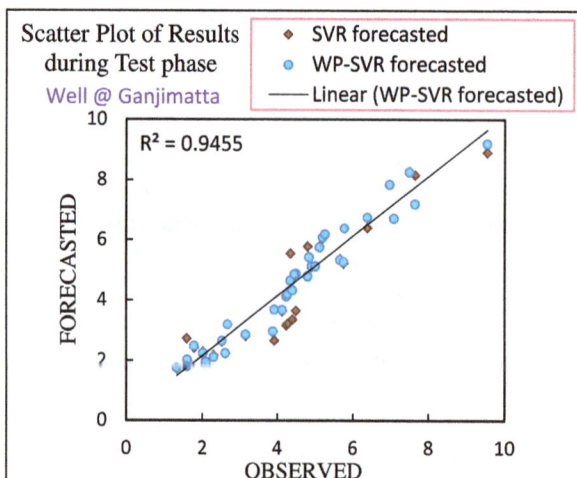

Figure 14. Scatter plot of observed versus forecasted of WP–SVR and SVR models with respect to well location at Ganjimatta.

8. Summary and conclusions

In this study, a novel hybrid model of WP–SVR has been developed to forecast monthly groundwater levels of three different wells located in Dakshina Kannada district, southwest coast of India. From the advantages of discrete Wavelet packet transformation, the meteorological data obtained for these three well locations were decomposed and the denoised coefficients were used with SVR for groundwater level prediction. The results obtained from the proposed WP–SVR model was compared with the basic SMO-SVR model for each well location. From Tables 7–9, it is clear that among the Wavelet packet coefficients the daubechis 4 wavelet with decomposition level 4 is giving better results. The SVR model with 4 input variables is performing better. The Wavelet packet denoising of input variables is found to improve the forecasting accuracy of SVR model. The developed WP–SVR model is relatively more efficient than the basic SVR prediction in groundwater level forecasting. The Wavelet packet decomposition better addresses the meteorological variables with non-stationary and non-linear features.

This study analysis and results reveal the following conclusions.

⇒ The proposed WP–SVR model outperforms the basic SMO-SVR model for a month lead time forecast.

⇒ The SVR modeling is mainly dependent on tuning of hyperparameters, selection of kernel function, and formulation of suitable optimization algorithm.

⇒ The improvement of results in WP–SVR model is due to dividing the data-set into multi-frequency bands and denoising the unnecessary peaks using discrete WPT.

⇒ The efficiency of any regression analysis mainly depends on the quality of the data that are being fed into the model and henceforth, proper pre-processing of the data using techniques like wavelets, Wavelet packets, or discretization using any other methods helps in improving the model performance of regression analysis. The work done in this study is a best example for the aforementioned conclusion.

The work can be extended for higher lead time forecasting and also models can be developed for monitoring seasonal groundwater level fluctuations by dividing the data according to the monsoon and non-monsoon periods.

9. Limitations of the study

- The wells selected for this study are monitored by Department of Mines and Geology, Government of Karnataka essentially to measure groundwater levels. No pumping is done. Pumping from wells will have a significant impact on the groundwater level fluctuations. Since this factor is unaccounted in the study, the model suitability is limited and doesn't hold good to real-time situations where there is pumping from wells.

- Other influential factors that are supposed to be included in modeling such as evaporation and evapotranspiration data could have been beneficial if available during modeling.

Acknowledgments
The authors would like to thank the Department of Mines and Geology, Government of Karnataka; Indian Meteorological Department, Pune, for providing the necessary data required for research and the Department of Applied Mechanics & Hydraulics, National Institute of Technology Karnataka for the necessary infrastructural support. The authors would like to thank two anonymous reviewers for their valuable suggestions and comments.

Funding
The authors received no direct funding for this research.

Author details
Sujay Raghavendra. N[1]
E-mail: sujayraghavendran@ymail.com

ORCID ID: http://orcid.org/0000-0002-0482-1936
Paresh Chandra Deka[1]
E-mail: pareshdeka@yahoo.com
ORCID ID: http://orcid.org/0000-0003-1771-2496

[1] Department of Applied Mechanics and Hydraulics, National Institute of Technology Karnataka, Surathkal, Mangalore 575025, India.

References

Abghari, H., Ahmadi, H., Besharat, S., & Rezaverdinejad, V. (2012). Prediction of daily pan evaporation using wavelet neural networks. *Water Resources Management, 26,* 3639–3652. doi:10.1007/s11269-012-0096-z

Adamowski, J., & Chan, H. F. (2011). A wavelet neural network conjunction model for groundwater level forecasting. *Journal of Hydrology, 407,* 28–40. doi:10.1016/j.jhydrol.2011.06.013

Batu, V. (1998). *Aquifer hydraulics: A comprehensive guide to hydrogeologic data analysis* (1st ed., p. 752). Wiley. Retrieved from http://as.wiley.com/WileyCDA/WileyTitle/productCd-0471185027.html

Behzad, M., Asghari, K., & Coppola, Jr., E. A. (2010). Comparative study of SVMs and ANNs in aquifer water level prediction. *Journal of Computing in Civil Engineering, 24,* 408–413. Retrieved from http://dx.doi.org/10.1061/(ASCE)CP.1943-5487.0000043

Campisi-Pinto, S., Adamowski, J., & Oron, G. (2012). Forecasting urban water demand via wavelet-denoising and neural network models. Case study: City of Syracuse, Italy. *Water Resources Management, 26,* 3539–3558. doi:10.1007/s11269-012-0089-y

Cherkassky, V., & Ma, Y. (2004). Practical selection of SVM parameters and noise estimation for SVM regression. *Neural Networks, 17,* 113–126. doi:10.1016/S0893-6080(03)00169-2

Chitsazan, M., Rahmani, G., & Neyamadpour, A. (2013). Groundwater level simulation using artificial neural network: a case study from Aghili plain, urban area of Gotvand, south-west Iran. *JGeope, 3,* 35–46. Retrieved from http://jgeope.ut.ac.ir/article_31930_4306.html

Coifman, R. R., & Wickerhauser, M. V. (1992). Entropy-based algorithms for best basis selection. *IEEE Transactions on Information Theory, 38,* 713–718. doi:10.1109/18.119732

Cortes, C., & Vapnik, V. (1995). Support-vector networks. *Machine Learning, 20,* 273–297. doi:10.1007/BF00994018

Daliakopoulos, I. N., Coulibaly, P., & Tsanis, I. K. (2005). Groundwater level forecasting using artificial neural networks. *Journal of Hydrology, 309,* 229–240. doi:10.1016/j.jhydrol.2004.12.001

Fan, X., & Zuo, M. J. (2006). Gearbox fault detection using Hilbert and wavelet packet transform. *Mechanical Systems and Signal Processing, 20,* 966–982. doi:10.1016/j.ymssp.2005.08.032

Fletcher, R. (2000). *Practical methods of optimization* (Vol. 53, p. 456). New York, NY: Wiley. doi:10.2307/2008742

Graps, A. (1995). Introduction to wavelets. *IEEE Computational Science & Engineering, 2,* 50–61. doi:10.1109/99.388960

Hall, M., Frank, E., Holmes, G., Pfahringer, B., Reutemann, P., Witten, I. H., & Witten, I. H. (2009). The WEKA data mining software. *ACM SIGKDD Explorations Newsletter, 11,* 10–18. doi:10.1145/1656274.1656278

Healy, R. W., & Cook, P. G. (2002). Using groundwater levels to estimate recharge. *Hydrogeology Journal, 10,* 91–109. doi:10.1007/s10040-001-0178-0

Hiscock, K. M., Rivett, M. O., & Davison, R. M. (2002). Sustainable groundwater development. *Geological Society, London, Special Publications, 193,* 1–14. doi:10.1144/GSL.SP.2002.193.01.01

Hu, G., Zhu, F., & Ren, Z. (2008). Power quality disturbance identification using wavelet packet energy entropy and weighted support vector machines. *Expert Systems with Applications, 35,* 143–149. doi:10.1016/j.eswa.2007.06.005

Jalalkamali, A., Sedghi, H., & Manshouri, M. (2011). Monthly groundwater level prediction using ANN and neuro-fuzzy models: A case study on Kerman plain, Iran. *Journal of Hydroinformatics, 13,* 867. doi:10.2166/hydro.2010.034

Kisi, O., & Shiri, J. (2011). Precipitation forecasting using wavelet-genetic programming and wavelet-neuro-fuzzy

conjunction models. *Water Resources Management, 25,* 3135–3152. doi:10.1007/s11269-011-9849-3

Learned, R. E., & Willsky, A. S. (1995). A wavelet packet approach to transient signal classification. *Applied and Computational Harmonic Analysis, 2,* 265–278. doi:10.1006/acha.1995.1019

Lei, J., Meyer, Y., & Ryan, R. D. (1994). Wavelets: Algorithms & applications. *Mathematics of Computation, 63,* 822. doi:http://dx.doi.org/10.2307/2153305

Liu, J., Chang, J., & Zhang, W. (2009). Groundwater level dynamic prediction based on chaos optimization and support vector machine. In *2009 Third International Conference on genetic and evolutionary computing* (pp. 39–43). Guilin: IEEE. doi:10.1109/WGEC.2009.25

Manimala, K., Selvi, K., & Ahila, R. (2012). Optimization techniques for improving power quality data mining using wavelet packet based support vector machine. *Neurocomputing, 77,* 36–47. doi:10.1016/j.neucom.2011.08.010

Misiti, M., & Misiti, Y. (1996). *Wavelet toolbox.* MathWorks. Retrieved from http://feihu.eng.ua.edu/NSF_TUES/w7_1a.pdf

Moosavi, V., Vafakhah, M., Shirmohammadi, B., & Behnia, N. (2013). A wavelet-ANFIS hybrid model for groundwater level forecasting for different prediction periods. *Water Resources Management, 27,* 1301–1321. doi:10.1007/s11269-012-0239-2

Nayak, P. C., Rao, Y. R. S., & Sudheer, K. P. (2006). Groundwater level forecasting in a shallow aquifer using artificial neural network approach. *Water Resources Management, 20,* 77–90. doi:10.1007/s11269-006-4007-z

Nourani, V., Kisi, Ö., & Komasi, M. (2011). Two hybrid artificial intelligence approaches for modeling rainfall–runoff process. *Journal of Hydrology, 402,* 41–59. doi:10.1016/j.jhydrol.2011.03.002

Nourani, V., Mogaddam, A. A., & Nadiri, A. O. (2008). An ANN-based model for spatiotemporal groundwater level forecasting. *Hydrological Processes, 22,* 5054–5066. doi:10.1002/hyp.7129

Ping, J., Qiang, Y., & Xixia, M. (2013). A combination model of chaos, wavelet and support vector machine predicting groundwater levels and its evaluation using three comprehensive quantifying techniques. *Information Technology Journal, 12,* 3158–3163. doi:10.3923/itj.2013.3158.3163

Platt, J. C. (1998). Fast training of support vector machines using sequential minimal optimization. In *Advances in Kernel methods - support vector learning* (pp. 185–208). Cambridge, MA: MIT Press. ISBN: 0-262-19416-3.

Raghavendra, N. S., & Deka, P. C. (2014a). Forecasting monthly groundwater table fluctuations in coastal aquifers using support vector regression. In S. Anadinni (Ed.), *International Multi Conference on innovations in engineering and technology (IMCIET-2014)* (pp. 61–69). Bangalore: Elsevier Science and Technology. Retrieved from http://media.wix.com/ugd/e1dfa2_a52f0f07953f4fe38603bd104530a71e.pdf

Raghavendra, N. S., & Deka, P. C. (2014b). Support vector machine applications in the field of hydrology: A review. *Applied Soft Computing, 19,* 372–386. doi:10.1016/j.asoc.2014.02.002

Shevade, S. K., Keerthi, S. S., Bhattacharyya, C., & Murthy, K. K. (2000). Improvements to the SMO algorithm for SVM regression. *IEEE Transactions on Neural Networks, 11,* 1100–1193. doi:10.1109/72.870050

Shi, B., & Zhu, C. (2009). Groundwater level prediction using ARMA-ANN model. In *2009 International Conference on test and measurement* (Vol. 2, pp. 295–298). Hong Kong: IEEE. doi:10.1109/ICTM.2009.5413048

Shinde, A., & Hou, Z. (2004). A wavelet packet based sifting process and its application for structural health monitoring. In *American Control Conference*

(Vol. 5, pp. 4219–4224). Boston, MA: IEEE. Retrieved from http://ieeexplore.ieee.org/xpls/abs_all. jsp?arnumber=1383970&tag=1

Sreekanth, P. D., Geethanjali, N., Sreedevi, P. D., Ahmed, S., Kumar, N. R., & Jayanthi, P. D. K. (2009). Forecasting groundwater level using artificial neural networks. *Current Science, 96*, 933–939. Retrieved from http://www. currentscience.ac.in/php/show_article.php?volume=096&i ssue=07&titleid=id_096_07_0933_0939_0&page=0933

Sreenivasulu, D., Deka, P. C., & Nagaraj, G. (2012). Investigation of the effects of meteorological parameters on groundwater level using ANN. *Artificial Intelligent Systems and Machine Learning, 4*, 39–44.

Subramanian, S., & Henry, J. (2010). Wavelet packet transform and support vector machine based discrimination of power transformer inrush current from internal fault currents. *Modern Applied Science, 4*, 67–82. Retrieved from http://connection.ebscohost.com/c/articles/51360808/ wavelet-packet-transform-support-vector-machine-based-discrimination-power-transformer-inrush-current-from-internal-fault-currents

Sudheer, C., Shrivastava, N. A., Panigrahi, B. K., & Mathur, S. (2011). Groundwater level forecasting using SVM-QPSO. In Hutchison, D., Kanade, P. N., Kittler, J., & Kleinberg, J. M. (Eds.), *Swarm, evolutionary, and memetic computing* (Vol. 7076, pp. 731–741). Berlin, Heidelberg: Springer Berlin Heidelberg. doi:10.1007/978-3-642-27172-4

Suryanarayana, C., Sudheer, C., Mahammood, V., & Panigrahi, B. K. (2014). An integrated wavelet-support

vector machine for groundwater level prediction in Visakhapatnam, India. *Neurocomputing, 145*, 324–335. doi:10.1016/j.neucom.2014.05.026

Tapak, L., Rahmani, A. R., & Moghimbeigi, A. (2014). Prediction the groundwater level of Hamadan-Bahar plain, west of Iran using support vector machines. *Journal of Research in Health Sciences, 14*, 81–86. Retrieved from http://www. ncbi.nlm.nih.gov/pubmed/24402856

Tong, W., Song, X., Lin, J., & Zhao, Z. (2006). Detection and classification of power quality disturbances based on wavelet packet decomposition and support vector machines. In *2006 8th International Conference on signal processing* (Vol. 4). Beijing: IEEE. doi:10.1109/ ICOSP.2006.346074

Vapnik, V. N. (1999). An overview of statistical learning theory. *IEEE Transactions on Neural Networks, 10*, 988–999. doi:10.1109/72.788640

Wanakule, N., & Aly, A. (2005). Artificial neural networks for forecasting groundwater levels. In L. Soibelman & P.-M. Feniosky (Eds.), *Computing in Civil Engineering* (pp. 1–10). Reston, VA: American Society of Civil Engineers. doi:10.1061/40794(179)128

Wickerhauser, M. V. (1994). *Adapted wavelet analysis: From theory to software* (p. 504). Natick, MA: A. K. Peters.

Yoon, H., Jun, S.-C., Hyun, Y., Bae, G.-O., & Lee, K.-K. (2011). A comparative study of artificial neural networks and support vector machines for predicting groundwater levels in a coastal aquifer. *Journal of Hydrology, 396*, 128–138. doi:10.1016/j.jhydrol.2010.11.002

9

Isotherms describing physical adsorption of Cr(VI) from aqueous solution using various agricultural wastes as adsorbents

Pushpendra Kumar Sharma[1]*, Sohail Ayub[2] and Chandra Nath Tripathi[3]

*Corresponding author: Pushpendra Kumar Sharma, Department of Civil Engineering, Hindustan College of Science and Technology, Farah, Mathura, Uttar Pradesh, India
E-mail: p.sharmaji10@gmail.com
Reviewing editor: Shashi Dubey, Hindustan College of Engineering, India

Abstract: Various agricultural wastes such as peels of pea (*Pisum sativum*) pod, tea (*Camellia sinensis*), and ginger (*Zingiber officinale*) and banana (*Musa lacatan*) waste were used to adsorb Cr(VI) from the aqueous solutions. A comparative adsorption efficiency study for all these adsorbents was done in laboratory for various pH, adsorbent doses, initial chromium concentrations, contact time, adsorbent sizes, temperature, and mixing speeds up to the optimization. The equilibrium sorption data were fitted into Langmuir, Freundlich, and Temkin isotherms and also the various thermodynamic parameters were determined. The value of R^2 was determined for Freundlich, Langmuir, and Temkin as 0.964, 0.963, and 0.858 (pea pod peels waste (PPP)), 0.969, 0.986, and 0.841(tea & ginger waste (T&G)), 0.985, 0.982, and 0.886 (banana peel waste (BW)). The maximum monolayer coverage (Q_0) from Langmuir isotherm model for pea pod, tea & ginger and banana peels waste were found to be 4.33 mg/g, 7.29 mg/g, and 10 mg/g, respectively, with separation factors (R_L) 0.0331, 0.0343, and 0.0756 which are well within favorable sorption. From Freundlich isotherm model, the sorption intensity (n) for the same adsorbents was also less than unity showing normal sorption. The heat of sorption (B) was also determined from Temkin isotherm model as 0.215, 0.271, and 0.271, respectively, vividly proving a favorable physical sorption. The Gibbs free energy was found maximum for BW as 6.0679 joule/mole. Out of the above said combination, BW was found the best low-cost adsorbent with high potential for the removal of hexavalent chromium from aqueous solutions.

Subjects: Pollution; Waste & Recycling; Water Engineering

Keywords: agricultural wastes; adsorbents; aqueous solution; isotherms; thermodynamic parameters; environmental sustainability engineering

1. Introduction
There has been a worldwide public awareness for the heavy metal contamination and toxicity in aquatic environment. Many heavy metals like lead (Pb), mercury (Hg), cadmium (Cd), arsenic (As),

ABOUT THE AUTHOR
Pushpendra Kumar Sharma is a reader/assistant professor in Civil and Environmental Engineering Department of Hindustan College of Science and Technology, Farah, Mathura. His research interests include water treatment, wastewater treatment, physiochemical treatment processes, environmental water resources engineering, etc., air pollution and control.

PUBLIC INTEREST STATEMENT
The study reveals that agricultural wastes such as pea pod peels, used tea and ginger, and banana peels can be reused as potential adsorbents for the removal of hexavalent chromium from tannery and metal finishing wastewaters. It is beneficial for protection water quality in rivers and ground water.

chromium (Cr), zinc (Zn), and copper (Cu), etc. have widespread usage in industries and enter the environment wherever they are produced, used, or discarded. All these metals become seriously toxic as ions or compound being soluble in water and readily absorbable by living organisms. In micro amount, these are essential nutrients in human bodies but in access cause severe physiological and health disorder. The chromium is of main concern here. Trivalent chromium as an essential trace element in the human diet enhances sugar metabolism (Katz, 1991) but hexavalent chromium is very detrimental for all living beings. Chromium intake may cause epigestric pain, nausea, vomiting, diarrhea, and hemorrhage (Browing, 1969). The chief chromium ores in nature are chromites (Cr_2O_3) or chrome ore ($FeOCr_2O_3$) mostly used in the electroplating of metal to prevent corrosion, plastic coating of surfaces to prevent from water and oil adverse effects, tanning leather, finishing metal, pigmenting and wood preservative, etc. resulting hexavalent chromium bearing wastewater discharge. Hence it is now a challenge for environmental engineers to dispose off such chromium contaminated wastewaters safely. Precipitation, co-precipitation, concentration, coagulation, filtration, etc. reduce chromium from wastewater (Shen & Wang, 1995) but produce solid residues with toxic compounds disposed by landfilling involving high costs and ground water contamination which is nothing but conversion of pollutants from liquid to solid phase, which is not a proper solution to the environmental problem. The presence of organic ligands or acidic conditions in the environment might have increased Cr(III) mobility and MnO_2, in the soil to oxidize Cr(III) to more toxic and mobile hexavalent Cr(VI) (Heary & Ray, 1987). And that is why, the practice of landfilling and land application of chromium containing sludge is being discouraged nowadays. Chemical precipitation, reverse osmosis, and some other modern technologies become costlier for micro concentrations. The process of adsorption proved to be the potential alternative for such situations (Huang & Morehart, 1991).

In developing countries like India, the physical adsorption has been proved an effective, potential, and economic method for the removal of chromium from wastewaters offering flexibility in design and producing high-quality treated effluents of desired standards for safe disposal and moreover the adsorbents can also be regenerated by suitable desorption for reuse (Ajmal, Rao, Ahmad, & Ahmad, 2000; Ajmal, Rao, & Siddiqui, 1995; Ayub, Ali, & Khan, 1998, 1999, 2001, 2002, 2003; Huang & Wu, 1975; Raji & Anirudhan, 1998; Sharma, Ayub, & Tripathi, 2013).

Many researchers have recently explored the possibility of various agro and horticultural wastes and byproducts as potential adsorbents for the removal hexavalent chromium from aqueous solutions (Ayub, Ali, & Khan, 2006; Ayub, Sharma, & Tripathi, 2014; Ayub et al., 1998, 1999, 2001, 2002; Baskaran, Venkatraman, Hema, & Arivoli, 2010; Chand, Aggarwal, & Kumar, 1994; Cimino, Paserini, & Toscano, 2000; Deo & Ali, 1992; Drake, Lin, Rayson, & Jackson, 1996; Huang & Wu, 1975; Periasamy, Srinivasan, & Murugan, 1991; Sharma, Ayub, & Tripathi, 2015; Sharma, Dubey, & Rehman, 2010; Sharma et al., 2013; Shukhla & Sakhardane, 1991; Siddiqui & Paroor, 1994; Vaishya & Prasad, 1991; Veena Devi et al., 2012; Weber, 1996). Continuing the same, here in this study we have substituted commercial expensive activated carbon by unconventional, low-cost and locally available agricultural wastes in plenty as adsorbents.

Chromium is an odorless and tasteless metal which is naturally found in rocks, plants, soil and volcanic dust, humans, and animals. The most common forms occurring in natural waters are trivalent and hexavalent chromium.

Cr(III) is an essential human dietary element and occurs naturally in many vegetables, fruits, meats, grains, and yeast. Cr(VI) occurs naturally in the environment from the erosion of natural chromium deposits and also can be produced by industrial processes. There are demonstrated instances of chromium being released to the environment by leakage, poor storage, or inadequate industrial waste disposal practices.

As far as the chemistry of chromium is concerned it is highly active transition metal existing in a number of oxidation states and exhibiting wide range stability. Thermodynamically, the reduced form of chromium, Cr(III), is most stable in both acidic and basic solutions as per evidence by the reduction potential (E^0) equations (Source: Shupack, 1991) as follows:

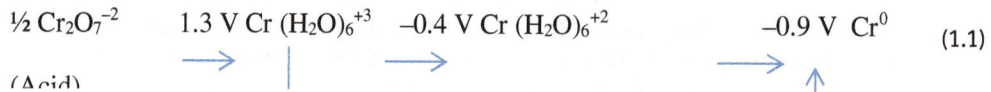

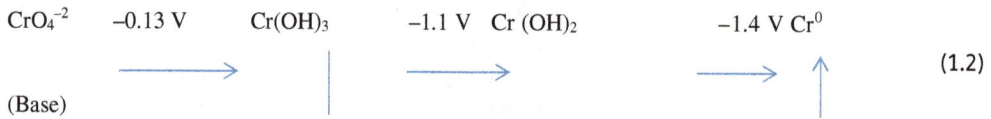

$$\tfrac{1}{2}\,Cr_2O_7^{-2} \quad\xrightarrow{\ 1.3\ V\ } Cr\,(H_2O)_6^{+3} \quad\xrightarrow{\ -0.4\ V\ } Cr\,(H_2O)_6^{+2} \qquad\xrightarrow{\ -0.9\ V\ }\ Cr^0 \qquad (1.1)$$

(Acid)

$$CrO_4^{-2} \quad\xrightarrow{\ -0.13\ V\ } Cr(OH)_3 \quad\xrightarrow{\ -1.1\ V\ } Cr\,(OH)_2 \quad\xrightarrow{\ -1.4\ V\ } Cr^0 \qquad (1.2)$$

(Base)

$+E^0$ values indicates that the reaction will proceed spontaneously (ΔG^0) to the right i.e. reduction under the standard conditions, whereas $-E^0$ values indicate that the oxidized species is favored and this is why Cr(III) is the most stable because it would require significant energy to reduce Cr(III) to Cr^0 or to oxidize Cr(VI).

Reduction, oxidation, and pH conditions severely affect the stability of various species of chromium. The stable domain for various species of chromium in aqueous environment specifically depends on oxidation potential (E^0) and pH (Landigran & Halllowell, 1975). For industrial wastewater, the predominant species are bichromate $HCrO_4^-$, dichromate $Cr_2O_7^{-2}$, and Cr^{+3} under redox conditions at pH < 3 with temperature ranging 20–30°C. One more thing interesting is that the divalent chromium ions Cr^{+2}, are mostly found in extremely reducing environment. The total chromium present governs the concentration distribution between $HCrO_4^-$ and $Cr_2O_7^{-2}$. The fraction $Cr_2O_7^{-2}$ remains only significant at high concentrations of total chromium (VI) (Stummm & Morgon, 1970).

Chrome tanning wastewater is typically treated in two stages, one is the conversion of hexavalent chromium to trivalent which freely bonds to hydroxide in stage-2 and finally a nontoxic precipitate: chromium hydroxide $Cr(OH)_3$. The most possible methods for reducing hexavalent chromium to trivalent chromium may be using chemical reducing agents such as sulfur dioxide (SO_2), sodium bisulfate ($NaHSO_3$), or sodium metabisulfite ($Na_2S_2O_5$). The following reduction reaction may take place using sulfur dioxide and maintaining pH between 2 and 3, in stage-1.

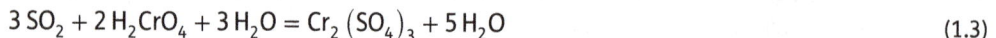

$$3\,SO_2 + 2\,H_2CrO_4 + 3\,H_2O = Cr_2\,(SO_4)_3 + 5\,H_2O \qquad (1.3)$$

After the completion of stage-1 reaction as above, lime i.e. calcium hydroxide $Ca(OH)_2$ is used to maintain pH ≥ 8.0 for the precipitation of chromium hydroxide as it can be easily adsorbed, separated, or disposed by any of the usual methods and it is less toxic also. The following possible reaction of precipitation may take place in stage-2.

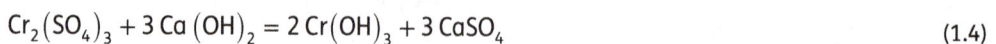

$$Cr_2\,(SO_4)_3 + 3\,Ca\,(OH)_2 = 2\,Cr(OH)_3 + 3\,CaSO_4 \qquad (1.4)$$

As far as the mechanism of adsorption is concerned, due to imbalance in surface forces, adsorbate molecules in the solution form a surface layer on adsorbent when in contact with solid surface during physical adsorption resulted from molecular condensation in the capillaries of the solid, whereas in chemical adsorption the molecular layer of adsorbate on the surface is formed through forces of residual valence of the surface molecule. Generally, high molecular weighted substances are easily adsorbed and this is why heavy metals are removed from the aqueous solutions of contaminated wastewater by the adsorption process. There is a quick establishment of an equilibrium interfacial concentration after slow diffusion into particles. The rate is inversely proportional to the square of the solute. The rate of adsorption varies directly proportional to the square root of contact period with the adsorbent. pH of the medium also affects the rate due to change in surface charges. The adsorption capacities also vary with the change of combinations of adsorbent and adsorbate. There is also a severe effect of complexity of wastewaters on adsorption capacities because of the interference between the various contaminants in the aqueous solutions.

The study compares the feasibility and potentials of agricultural waste and byproduct materials such as pea pod peels (PPP), tea and ginger waste (T&G), and banana waste (BW) to adsorb hexavalent chromium Cr(VI) from industrial wastewaters. pH, adsorbent dosage, initial chromium concentration, contact time, adsorbent grain size, temperature, and agitation speeds were optimized for sorption. The samples of the adsorbents were characterized before and after adsorption in the batch performance. Electron microscopic technique was used to characterize the surface of the adsorbent. Thermodynamic nature of the process was also studied.

2. Materials and methods

2.1. Sorbents and sample preparation
In the present study, various agricultural wastes such as peels of pea (*Pisum sativum*) pod (PPP), tea (*Camellia sinensis*) and ginger (*Zingiber officinale*) waste (T&G) and (*Musa lacatan*) BW were used to adsorb Chromium (VI) from the aqueous solution using batch adsorption processes.

2.1.1. Adsorbents
One week sun-dried BW from fruit seller and Pea pod peels from a farmhouse were first dried in oven for three days at a temperature of 90°C for 5 h daily and the T&G from a tea stall was well washed many times with water and finally with distilled water then spread for sun drying for a week, and then oven dried for three days at a temperature of 90°C for 5 h daily. After grinding, the waste powders were sieved through a 225 mesh (Indian Standard Sieve) to get an average and uniform sized particle, we washed these waste powders several times with distilled water to get rid of lighter materials and other impurities. The adsorbents were dipped in 0.1 N NaOH for a period of 9 h and washed several times with distilled water to remove the lignin and then dried again. The adsorbents were again rinsed separately with double-distilled water two times and dipped into 0.1 NH_2SO_4 for the period of 9 h again to remove traces of alkalinity. The acid-treated adsorbents were washed thoroughly with double-distilled water. Then the dry powder so prepared was exposed to sun and stored in a desiccator. Before using these powders as adsorbents a size range was also achieved by sieving through a sieve set 1.18 mm, 600 μm, 425 μm, 300 μm, 150 μm, 75 μm and pan i.e. less than 75 μm (Indian Standard Sieve) so as to optimize the sizes of these adsorbents.

2.1.2. Experimental
To understand the adsorption behavior, a number of batch studies were conducted according to the Standard Method to investigate the effect of pH, adsorbent dosage, initial chromium concentration, contact time, adsorbent grain size, temperature, and agitation speed. In the present study, synthetic wastewater of various concentrations of Cr(VI) was prepared from the "Mono-Element Standard Solution for AAS" 1,000 mg/l Chromium (Cr) in 0.5 mol/l Nitric acid (HNO_3) with a certified value 1001.2 mg/l ± 9.22 as 95% level of confidence from LOBA CHEMIE PVT. Ltd. and kept separately in glass-stoppered conical flasks. Then, varying doses of adsorbent were added to the synthetic wastewater in a conical flask of 250 ml. To achieve the adequate time of contact between adsorbent and the metal ions, the system was equilibrated by shaking the contents of the flasks at room temperature on a mechanical shaker for varying times (Indian Scientific Instruments factory, Ambala Cant, India) at Environmental Engineering Laboratory, Civil Engineering Department, Hindustan College of Science and Technology, Farah, Mathura, Uttar Pradesh (India).

The suspension was filtered through Whatman (No. 1) filter paper and the filtrate was analyzed to evaluate the concentration of Cr(VI) metal in the treated wastewater using Atomic Absorption Spectrometer (Make and Model-Perkin Elmer, AAS, Pin AAcle 900 K) installed at Civil Engineering Department, Environmental section, Z. H. College of Engineering and Technology, Aligarh Muslim University, Aligarh (India).

Adsorption studies were made for various times. Ultimate saturation time was also determined for each dose. All the studies were made at the room temperature varying from 25 to 28°C in summer. In case of low temperature study the adsorption conical flasks were put in refrigerator up to

10°C and that for the higher temperature studies the temperature was maintained by external heaters in closed chamber up to 50°C.

2.2. Thermodynamic parameters

Thermodynamic constants of equilibrium sorption processes can be determined using Gibbs free energy equation,

$$\Delta G = -RT \ln K \tag{2.1}$$

where ΔG = change in free energy (joule/mole), T = the process temperature in Kelvin, R = universal gas constant = 8.314.

The linear plot between "ΔG vs. T" is drawn for the equilibrium temperature and then simulated using mathematical modeling Sharma, Khan, and Ayub (2012a, 2012b) to the linear equation:

$$\Delta G = \Delta H^* - T\Delta S \tag{2.2}$$

where ΔH = changes in enthalpy and ΔS = changes in entropy of activation (Ayub et al., 2006).

Plotting the graph between ΔG and T as shown in Figure 9(a)–(c) for PPP, T&G and BW, respectively, we get the values of ΔH and ΔS for all these adsorbents used in the present study, the values so obtained are given in Table 1.

For all the adsorbents PPP, T&G, and BW Gibbs Free energy diagrams were plotted as shown in Figure 9(a)–(c), respectively, and then using best fit line and simulation thermodynamic constants such as the change in apparent enthalpy (ΔH), free energy (ΔG), and entropy (ΔS) of sorption were calculated using thermodynamic equations and values at 301 K have been tabulated in Table 1. The negative enthalpy ΔH values for all PPP, T&G and BW confirm the exothermic nature of the sorption process and suggested the poor binding. The positive value of free energy change ΔG indicates the random feasibility. Adsorption at a solid solution interface generally shows an increase in entropy ΔS showing faster interaction during the forward adsorption. Association, fixation of adsorbate on the interface between two phases result in loss of the degree of freedom thereby showing a negative entropy effect (Ayub et al., 2006).

2.3. Langmuir isotherms

Langmuir adsorption parameters were determined by transforming the Langmuir equation

$$q_e = \frac{(q_o K_L C_e)}{(1 + K_L C_e)} \tag{2.3}$$

In to linear form as

$$\frac{1}{q_e} = \frac{1}{q_o} + \left(\frac{1}{q_o K_L}\right)\left(\frac{1}{C_e}\right) \tag{2.4}$$

where C_e = the equilibrium concentration of adsorbate in mg/l; q_e = the amount of metal adsorbed per gram of the adsorbent at equilibrium (mg/g); q_o = maximum monolayer coverage capacity (mg/g) K_L = Langmuir Isotherm Constant (L/mg).

The values of q_{max} and K_L were computed from the slope and intercept of the Langmuir plot $\frac{1}{q_e}$ vs. $\frac{1}{C_e}$

The separation factor or equilibrium parameter R_L which is a dimensionless quantity and its value indicates the nature of adsorption means if $R_L > 1$ unfavorable, linear if $R_L = 1$, $0 < R_L < 1$ favorable and irreversible if $R_L = 0$.

Table 1. Summary of thermodynamic parameters and isotherms constants for pea pod peels waste (PPP), tea and ginger waste (T&G), and banana peels waste (BW)

At 301 K	Langmuir constants					Freundlich constants			Temkin constants			Thermodynamic parameters		
Adsorbent	(Max. monolayer coverage capacity, mg/g) q_{max}	$1/q_o$	(Langmuir constant) K_L	R_L (favorable if $0 < R_L < 1$)	R_2 (Best error distribution or perfect correlation)	k	$1/n$	Cc	b_T	A_T	B	ΔG (kJ/ mole)	ΔH (kJ/ mole)	ΔS (kJ/ mole)
PPP	4.329004	0.231	0.939	0.0331	0.955 (should be <1)	2.133	−0.093	0.911	11639.6	5637.523	−0.215	3.499981	−1.575	−2.537
T&G	7.29927	0.137	0.9059	0.0343	0.5555	2.558	−0.105	0.969	9234.37	3139.13	−0.271	4.792711	−1.915	−3.354
BW	10	0.1	0.3745	0.0756	0.5831	2.606	−0.108	0.985	9234.37	3093.01	−0.271	6.068	−3.580	−4.824

$$R_L = \frac{1}{[1 + [1 + K_L C_o]]} \tag{2.5}$$

where C_o = initial concentration of adsorbate = 30 mg; K_L = Langmuir constant related to the energy of adsorption (Dada, Olalekan, Olatunya, & Dada, 2012).

All the above Langmuir constants were determined for all the adsorbents of which plots have been shown in Figures 10(a) PPPL, 11(a) T&GL, and 12(a) BWL for PPP, T&G and BW, respectively, later. The values of thermodynamic constants so obtained for various adsorbents are given in Table 1.

2.4. Freundlich isotherms

The adsorption observations at a fixed initial concentration of Chromium in synthetic wastewater with varying adsorbent doses were fitted to linearized Freundlich adsorption isotherm as following.

$$\log \left(\frac{X}{m} \right) = \log K + \frac{1}{n} \log C_e \tag{2.6}$$

where (x/m) is the amount of Chromium (VI) adsorbed per unit mass of adsorbent expressed in mg/mg and C_e is the equilibrium concentration of aqueous solution. K is a constant, which is a measure of adsorption capacity, and $1/n$ is the measure of adsorption intensity (Dada et al., 2012).

The value of K and $1/n$ are given in Table 1 and plots for Freundlich Isotherms for the adsorbents PPP, T&G, and B/W are shown in Figures 10(b), 11(b) and 12(b) respectively.

2.5. Temkin isotherm

This isotherm has a factor which explicitly shows the interaction between adsorbent and adsorbate. Ignoring very low and high concentrations, this model assumes that heat of adsorption of molecules in the subsequent layers decreases linearly rather than logarithmic with coverage (Aharoni & Ungarish, 1977; Tempkin & Pyzhev, 1940). As shown in equation, its derivation is characterized by uniform distribution of binding energies up to a maximum value and was obtained by plotting the amount adsorbed q_e vs. ln C_e and the constants have been determined from slope and intercept. The model equation is as follows (Tempkin & Pyzhev, 1940)

$$q_e = \frac{RT}{b} \ln (A_T C_e) \tag{2.7}$$

$$q_e = \frac{RT}{b_T} \ln A_T + \left(\frac{RT}{b} \ln C_e \right) \tag{2.8}$$

using,

$$B = \frac{RT}{b_T} \qquad (9)$$

The equation becomes,

$$q_e = B \ln A_T + B \ln C_e \qquad (10)$$

A_T = Temkin isotherm equilibrium binding constant (L/g); b_T = Temkin isotherm constant; R = universal gas constant (8.314 J/mol/K); T = Temperature at 301 K and B = Constant related to heat of sorption (J/mol).

From the Temkin plot shown in Figures 10(c) PPPT, 11(c) T&GT, and 12(c) BWT for all the adsorbents PPP; T&G and BW, respectively, the estimated values are given in Table 1.

2.6. Scanning electron microscopy
Scanning Electron Microscopy (SEM) by Scanning Electron Microscope Model No. JSM 6510 LV Make; JEOL (Japan), at USIF, AMU, Aligarh, was also done for all the raw and used adsorbents to study the changes in adsorbent texture before (raw) and after (used) adsorption process as shown in Figures 13(a)–(f) for raw and used PPP, T&G, and BW, respectively, below. The samples were non-conducting so before SEM these were gold coated and SEM analysis results show that the adsorbents have potential surface area and cavities to adsorb chromium within for binding.

2.7. Desorption and hydrolysis test
For leaching properties of agro-based materials so as to ensure the quality of treated effluent desorption test was also conducted for the best adsorbent after comparison among PPP, T&G, and BW used as adsorbent to adsorb chromium i.e. BW. About 10 g of saturated adsorbent was kept in 300 ml capacity BOD bottle full of distilled water and shaken for two hours and then filtered. The filtrate was analyzed for the chromium content desorbed of which negligible amount shows that no desorption occurred.

3. Results and discussion
Using observations of the sorption studies to optimize the pH, adsorbent dosage, initial chromium concentration, contact time, adsorbent grain size, temperature, and agitation speed various adsorption isotherms have been plotted to determine the feasibility of sorption system along with its mechanisms through kinetics. Thermodynamic parameters clearly tell the story of the feasibility of the process. The results of batch studies for various adsorbents used in the study have been compared.

3.1. Comparison of pea pod peels waste, tea & ginger waste and banana waste
As shown in Figure 1, the best results were obtained in the pH range 2.5–3.5. The better adsorption capacity at low pH values would have been due to the large number of H$^+$ ions, neutralizing the hydroxyl group (OH$^-$) on adsorbed surface and thus reducing hindrance to the diffusion of dichromate ions. At higher pH, low adsorption may be due to excess OH$^-$ ions causing high hindrance to diffusion of positively charged chromium ions. It is well-known fact that the surface adsorbs anions favorably in low pH range due the presence of H$^+$ ions, whereas the surface is active for the adsorption of cations at higher pH values due to the accumulation of OH$^-$ ions (Huang & Stumm, 1973). Same results were found during the studies conducted on dihydric phenol removal on activated carbon by Mahesh, Rama, Praveen, and Usha (1999) and Sharma and Forster (1993) who used sphagnum moss peat for the removal of chromium.

As shown in Figure 2, the percentage chromium sorption is directly proportional to adsorbent dose. A dose of BW @ 10 g/l is sufficient to adsorb more than 70% Cr(VI) having 30 mg/l initial concentration of chromium within 1.0 h. On further increasing the adsorbent dose to 20 g/l more than 90% removal efficiency was observed. Similarly, 10 g/l adsorbent dose of T&G adsorbs more than

pH Vs Sorption

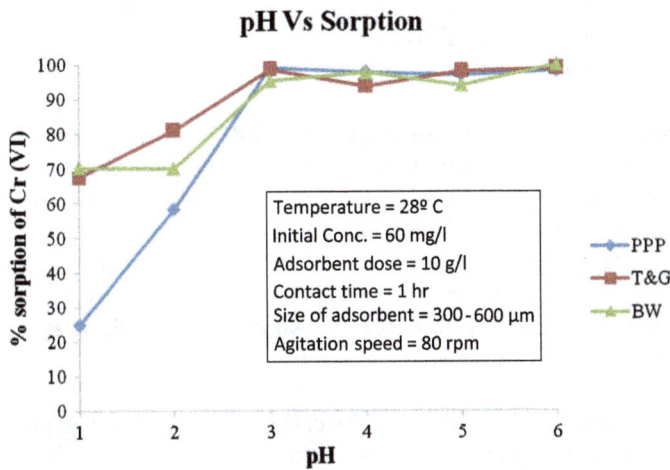

Figure 1. Effect of pH on Chromium Cr(VI) Sorption.

75% at pH 2.0; pea pod peels waste adsorbs 65% at pH 2.0. The same results were also reported by Ayub et al. (2001, 2002, 2014), Bansal and Sharma (1992), Kim and Joltek (1977), Mall (1992), Rao, Parwate, and Bhole (2002).

As shown in Figure 3; the removal efficiency was found dependent on the initial chromium concentration, adsorbent dose, contact time, particle size, and pH. Several researchers have reported that percent adsorption with increase in metal ion concentration considerably reduces (Panday, Prasad, & Singh, 1984). T&G is found to be best adsorbent, used BW better and pea pod peels waste equally good, with respect to the removal efficiency of the chromium metal in 10 g/l adsorbent dose and having 30 mg/l initial concentration of the solution for the average size of adsorbent ranging 600–300 μm @ 1 h contact time and pH 2.0. The removal efficiency ranges from 65 to 90% at 28°C in summer in the month of May 2015 (India).

As shown in Figure 4, in the beginning chromium Cr(VI) removal efficiency is observed to be very quick and later slows down. In first 45 min, T&G and BW adsorb more than 90% chromium, whereas PPP adsorbs 65% chromium but later by slow rate it reaches up to 80% at initial chromium concentration 30 mg/l, 10 g/l adsorbent dose, average size of adsorbent grains from 600 to 300 μm, pH 2.0. So it was observed that first hour of contact period is very important for quick adsorption which is very important for designing continuous adsorption systems in real problems.

Sorbent dose Vs % Cr (VI) Sorption

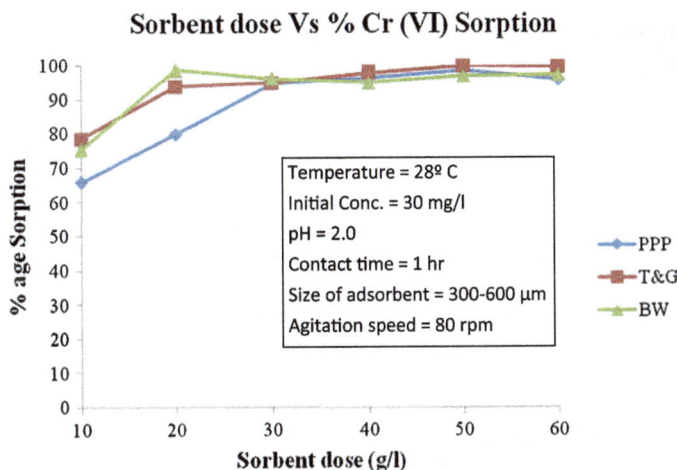

Figure 2. Effect of sorbent dose on Sorption.

From Figure 5 it was observed that neither too small nor too large grain size of adsorbent adsorbs the chromium but an average grain size ranging from 600 to 300 μm is most the efficient adsorbent for all the adsorbents used here in this study.

As shown in Figure 6, the optimum temperature range was from 25 to 30°C below and above this range the adsorption efficiency was observed decreasing. The initial start up agitation speed for mixing the adsorbent and adsorbate so as to achieve good contact in between these two was also studied, as a result it was observed that an average agitation speed of 80–100 rpm for first 15 min is sufficiently enough to optimum results, as shown in Figure 7. A small batch set up was also run for all the optimized affecting parameters for all the adsorbents i.e. PPP, T&W, and BW so as to get highest adsorption of chromium and the observed results have been shown in concluding Figure 8, the optimized affecting parameters have also been indicated there.

Figure 9(a)–(c) showing thermodynamic graphs where from the apparent enthalpy change (ΔH), free energy (ΔG), and entropy (ΔS) of sorption process have been calculated using thermodynamic equations and values at 28°C and shown in summary Table 1. The negative value of enthalpy change (ΔH) reveals exothermic nature of adsorbent. Negative values of ΔG reveal the process is feasible

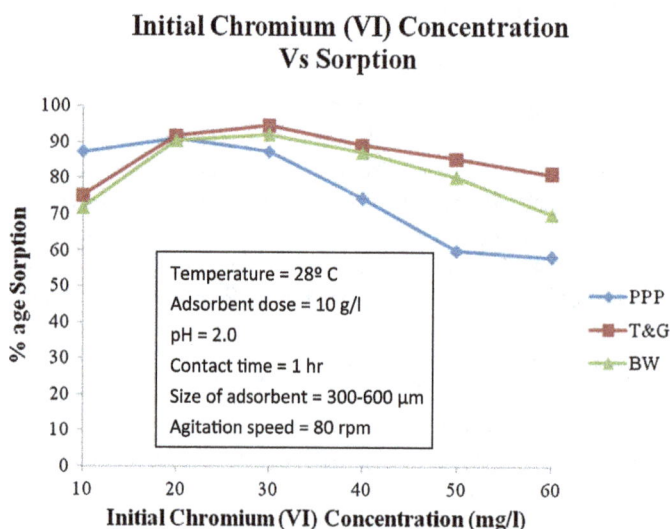

Figure 3. Effect of Initial Chromium(VI) Concentration on Sorption.

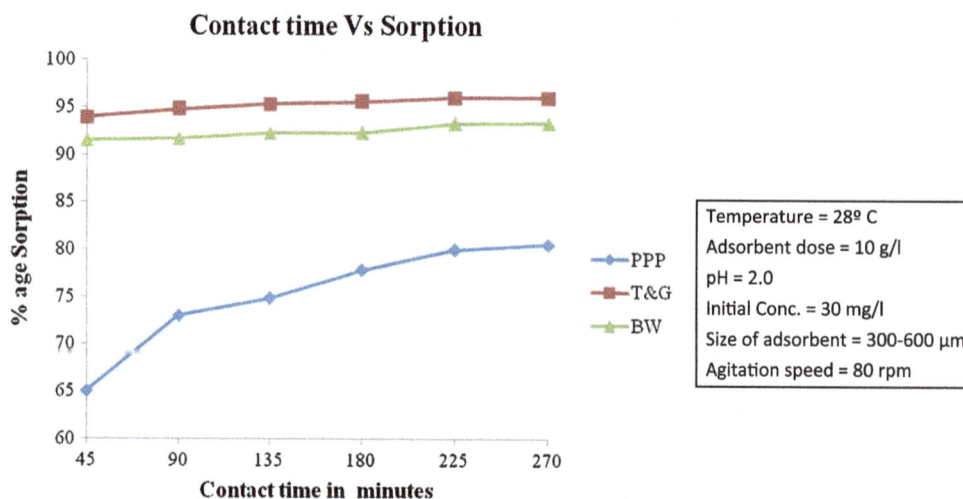

Figure 4. Effect of Contact time on Sorption.

Adsorbent Size Vs Sorption

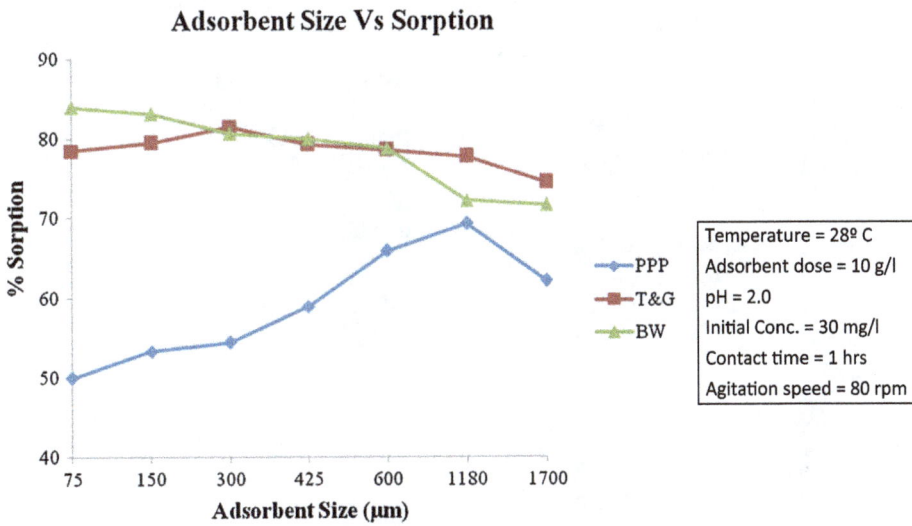

Figure 5. Effect of Adsorbent Size on Sorption.

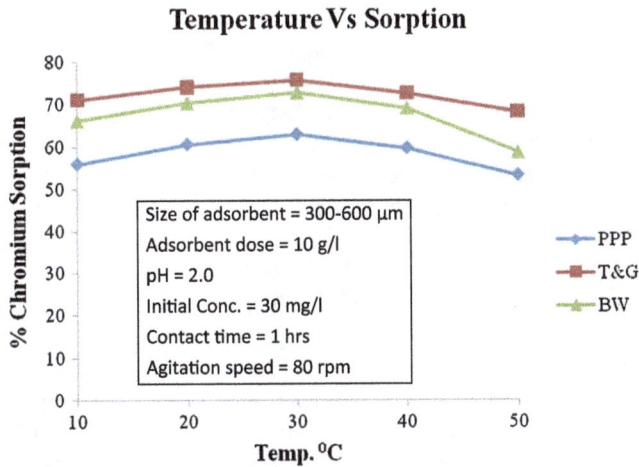

Temperature Vs Sorption

Figure 6. Effect of Temperature on Sorption.

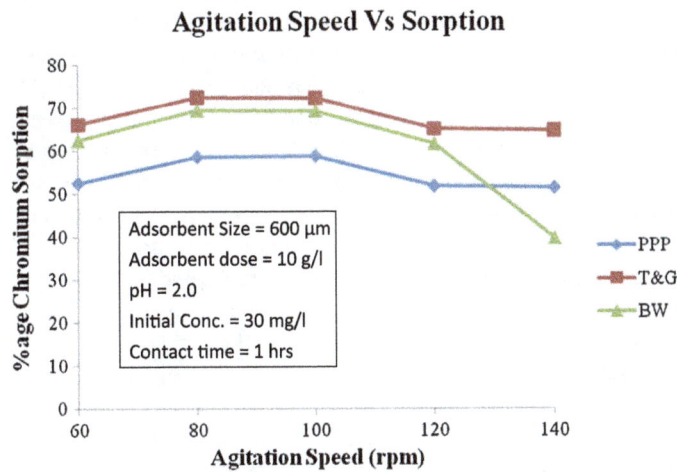

Agitation Speed Vs Sorption

Figure 7. Effect of Agitation Speed on Chromium Cr(VI) Sorption.

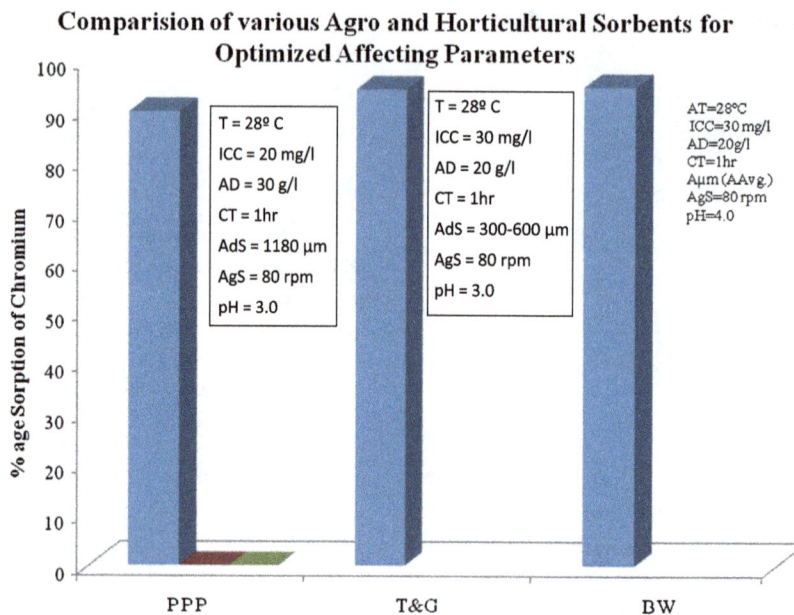

Comparision of various Agro and Horticultural Sorbents for Optimized Affecting Parameters

PPP:
T = 28º C
ICC = 20 mg/l
AD = 30 g/l
CT = 1hr
AdS = 1180 μm
AgS = 80 rpm
pH = 3.0

T&G:
T = 28º C
ICC = 30 mg/l
AD = 20 g/l
CT = 1hr
AdS = 300-600 μm
AgS = 80 rpm
pH = 3.0

BW:
AT=28°C
ICC=30 mg/l
AD=20g/l
CT=1hr
Aμm (AAvg.)
AgS=80 rpm
pH=4.0

Figure 8. Result of Optimized Affecting parameters for various sorbants.

and spontaneous. The negative values of free energy change for a system indicates the spontaneity of sorption process. Adsorption at a solid solution interface generally shows an increase in entropy. This shows a quicker interaction during the forward adsorption. Association, fixation, or immobilizations of sorbate on the interface between two phases result in loss of the degree of freedom thereby resulting in a negative entropy effect.

The positive entropy (ΔS) attributes to an increase in translational entropy due to randomness of displaced water molecules from the surface of the adsorbents. The negative values of entropy reported here show the decrease in translational entropy by regular attachment of water molecules to the surface of the adsorbent and supported by the adsorption process of hexavalent chromium on fly ash-wallastonite (Panday et al., 1984), chromium adsorption on blast furnace flue dust and COD on fly ash (Baisakh, Patnaik, & Patnaik, 1996) and adsorption on coal char (Baisakh & Patnaik, 2002).

The Langmuir, Freundlich, and Temkin isotherms have been plotted and shown in Figures 10(a) PPPL, 11(a) T&GL, 12 BWL, the Langmuir adsorption isotherms for PPP, Figures 10(b) PPPF, 11(b) T&GF, 12(b) BWF, the Freundlich isotherm curve for T&G and Figure 10(c) PPPT, 11(c) T&GT, 12(c) BWT, the Temkin isotherm curve for BW along with the summary of isotherms constants have been given in Table 1. The value of R^2 was determined for Freundlich, Langmuir, and Temkin as 0.964, 0.963, and 0.858 (PPP), 0.969, 0.986, and 0.841 (T&G), 0.985, 0.982, and 0.886 (BW). The maximum monolayer coverage (Q_0) from Langmuir isotherm model for PPP, T&G, and BW were found to be 4.33 mg/g, 7.29 mg/g, and 10 mg/g, respectively, with separation factors (R_L) 0.0331, 0.0343, and 0.0756 which are well within favorable sorption. From Freundlich isotherm model, the sorption intensity (n) for the same adsorbents was also less than unity showing normal sorption. The heat of sorption (B) was also determined from Temkin isotherm model as 0.215, 0.271, and 0.271, respectively, vividly proving a favorable physical sorption. The Gibbs free energy was found maximum for BW as 6.0679 joule/mole. Out of the above said combination BW was found the best low-cost adsorbent with high potential for the removal of hexavalent chromium from aqueous solutions.

Column study for a flow rate of 1.0 l/d containing initial chromium concentration of 50 mg/l at pH 1.5 was made for the best adsorbent i.e. BW until all adsorbent exhausted and the treated effluent was analyzed at different time intervals. Column capacities were greater than the batch capacities.

(a)

(b)

(c)

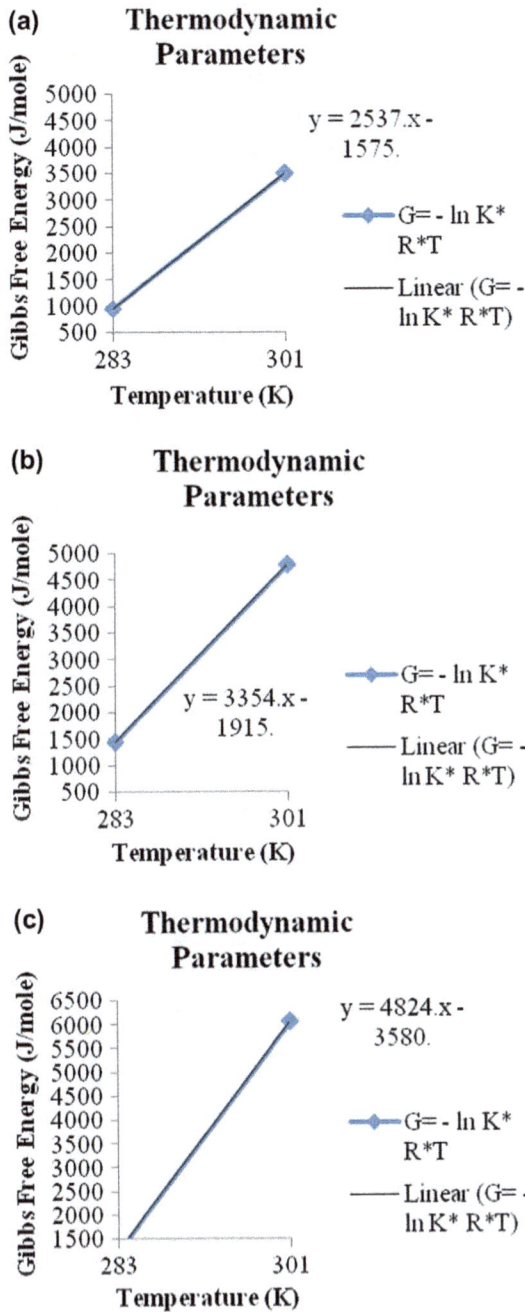

Figure 9. Thermodynamic parameters for (a) PPP, (b) T&G, and (c) BW.

As per desorption test for best adsorbent i.e. BW no leaching property of adsorbent was found which showed the required quality of treated effluent.

4. Conclusions

(1) The study reveals that agricultural wastes such as pea pod peels, used T&G, and BW can be reused as potential adsorbents for the removal of hexavalent chromium from tannery and metal finishing wastewaters. It is investigated that adsorption behavior depends on the nature of adsorbent and all the agricultural waste materials do not show similar behavior.

(a)

(b)

(c)

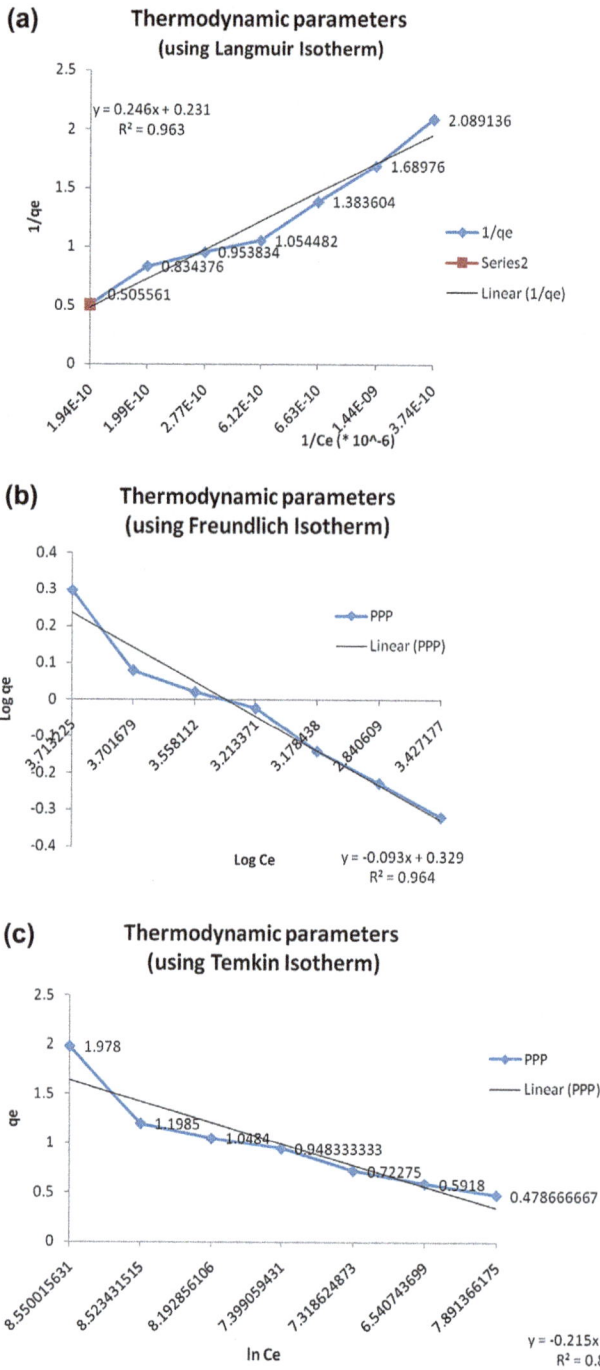

Figure 10. Thermodynamic parameters for (a) PPPL (using Langmuir isotherm), (b) PPPF (using Freundlich isotherm), and (c) PPPT (using Temkin isotherm).

(2) The chromium adsorption capacity of all the agricultural adsorbents is highly dependent on pH which is best suited in the range 2.0–3.0.

(3) Chromium uptake capacity follows first-order rate equation for all the adsorbents used in present study.

(4) The isotherm data obtained were best fitted to Freundlich adsorption isotherm.

(a)

Thermodynamic parameters
(using Langmuir Isotherm)

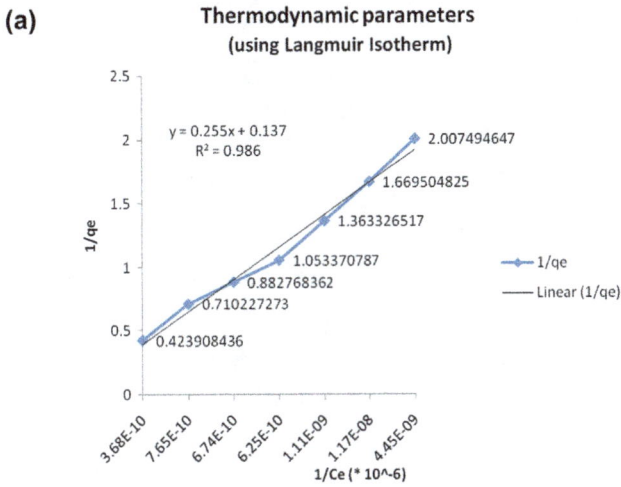

$y = 0.255x + 0.137$
$R^2 = 0.986$

2.007494647
1.669504825
1.363326517
1.053370787
0.882768362
0.710227273
0.423908436

1/qe
Linear (1/qe)

1/qe

1/Ce (* 10^-6)

3.68E-10 7.65E-10 6.74E-10 6.25E-10 1.11E-09 1.17E-08 4.45E-09

(b)

Thermodynamic parameters
(using Freundlich Isotherm)

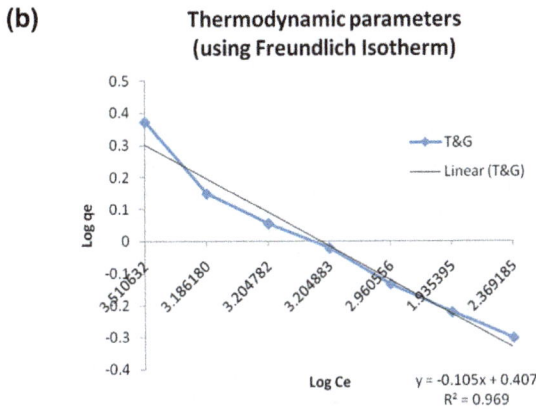

T&G
Linear (T&G)

Log qe

Log Ce

$y = -0.105x + 0.407$
$R^2 = 0.969$

3.610632 3.186180 3.204782 3.204883 2.960566 1.935395 2.369185

(c)

Thermodynamic parameters
(using Temkin Isotherm)

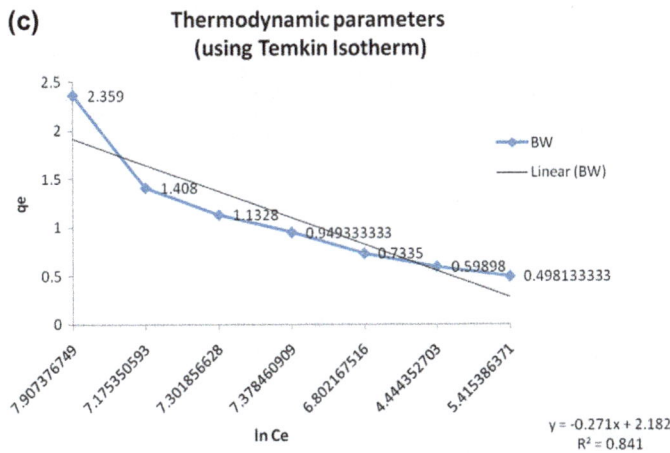

2.359

BW
Linear (BW)

1.408
1.1328
0.949333333
0.7335
0.59898
0.498133333

qe

7.907376749 7.17350593 7.301856628 7.378460909 6.802167516 4.44352703 5.415386371

In Ce

$y = -0.271x + 2.182$
$R^2 = 0.841$

Figure 11. Thermodynamic parameters for (a) T&GL (using Langmuir isotherm), (b) T&GF (using Freundlich isotherm), and (c) T&GT (using Temkin isotherm).

(5) The negative enthalpy (ΔH) confirms exothermic nature of adsorbent and poor binding. Positive values of Gibbs free energy (ΔG) indicate the non spontaneous adsorption. Negative values of ΔS indicate the less affinity of the adsorbent material.

(6) Leaching test after sorption studies indicate traces of chromium showing poor binding between adsorbate and adsorbents but in case of BW it was found satisfactory.

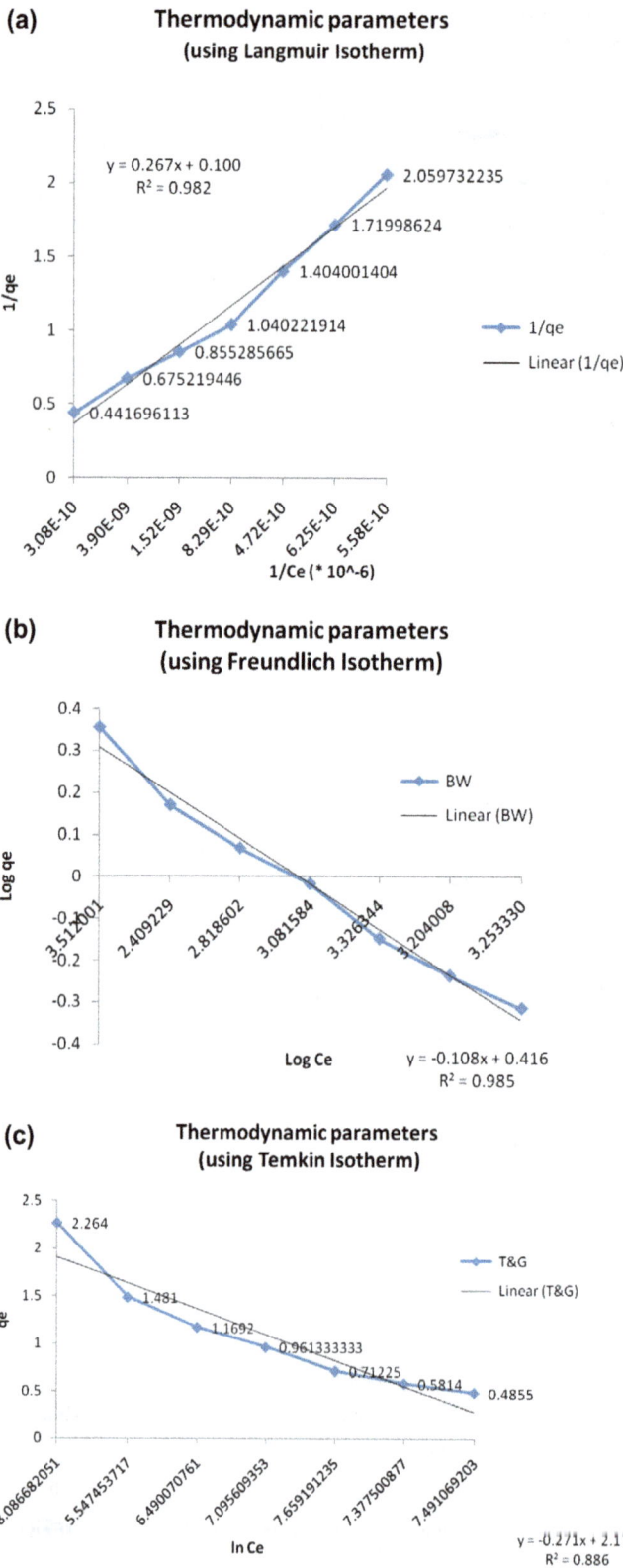

Figure 12. Thermodynamic parameters for (a) BWL (using Langmuir isotherm),
(b) BWF (using Freundlich isotherm), and (c) BWT (using Temkin isotherm).

(7) The agricultural waste sorbents do not hydrolyze and BOD of leachate remains unchanged which shows that the agro-based adsorbents act as stable materials under the process conditions.

(8) The exhaustive adsorption capacity of the sorbent is found satisfactory in batch process.

(a)

(b)

Figure 13. Changes in adsorbent texture before (raw) and after (used) adsorption process (a) PPP raw, (b) PPP used, (c) T&G raw, (d) T&G used, (e) BW raw, (f) BW used.

(c)

(d)

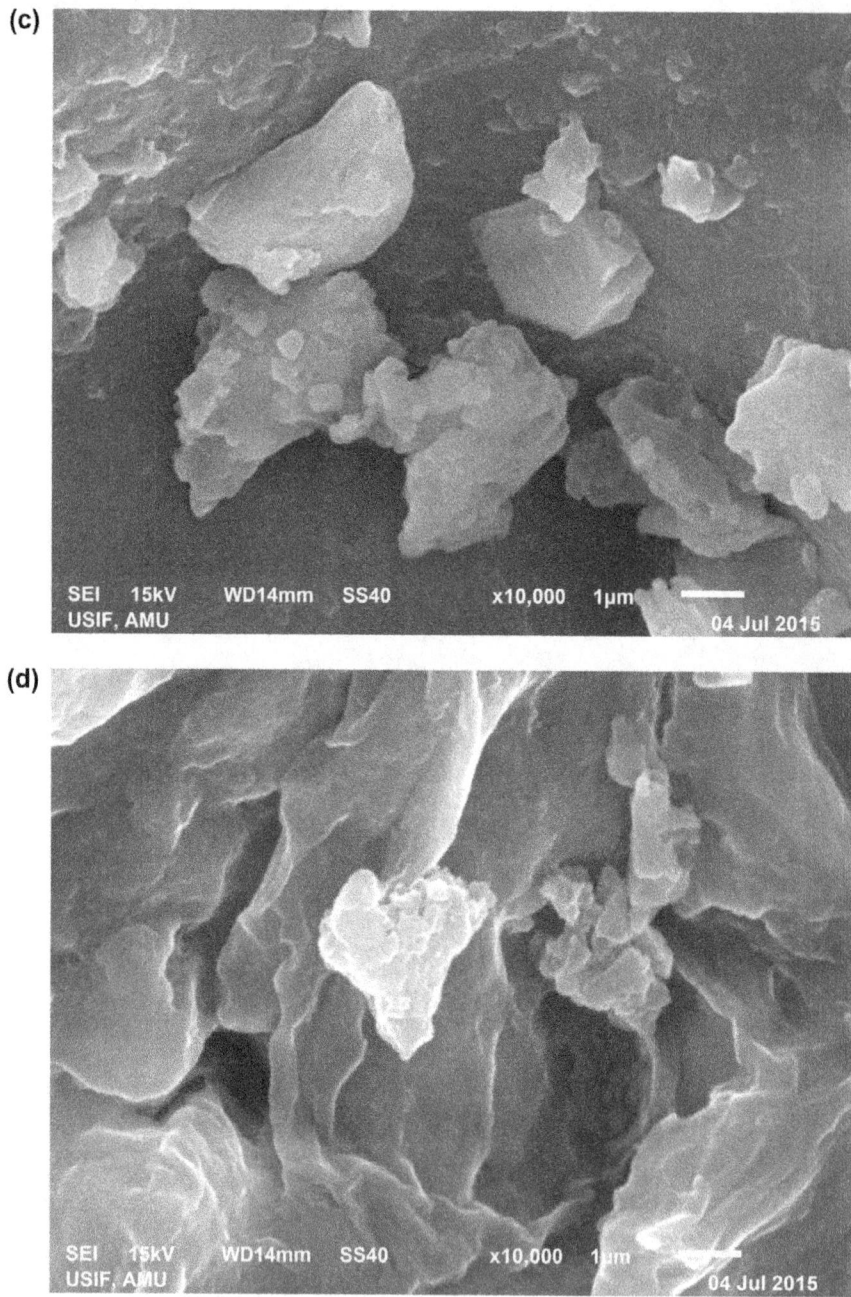

Figure 13. *(Continued).*

(e)

SEI 15kV WD14mm SS40 x10,000 μm
USIF, AMU 04 Jul 2015

(f)

SEI 15kV WD14mm SS40 x10,000 1μm
USIF, AMU 04 Jul 2015

Figure 13. *(Continued).*

(9) Being relatively cheaper and locally available, the agricultural waste adsorbents are superior than activated carbon due to the evolution of more many modern technologies waste disposal has also become easier in rural areas now days. All these agro and horticultural waste adsorbents require only alkali/acid treatment to increase their efficiency of sorption.

(10) SEM results show that the adsorbents have potential surface area and cavities to adsorb chromium within for binding.

Better column capacities may be due to continuously large concentration at the interface of the sorption zone as the sorbate solution passes through the column, whereas in the case of batch process the concentration gradient decreases.

Acknowledgements

The authors are thankful to Prof. Sartaj Tabassum, Coordinator, University Sofisticated Instrument Facility (USIF) Centre, Aligarh Muslim University, Aligarh and Mr Abul Maz, the technician in laboratory of University Sofisticated Instrument Facility (USIF), without their all kinds of support and techniques the above Scanning Electron Microscopy (SEM) study would not have been so successfully possible.

Funding

The authors received no direct funding for this research.

Author details

Pushpendra Kumar Sharma[1]
E-mail: p.sharmaji10@gmail.com
Sohail Ayub[2]
E-mail: sohailayub@rediffmail.com
Chandra Nath Tripathi[3]
E-mail: cntripathi01@gmail.com

[1] Department of Civil Engineering, Hindustan College of Science and Technology, Farah, Mathura, Uttar Pradesh, India.

[2] Department of Civil Engineering, Aligarh Muslim University, Aligarh, Uttar Pradesh, India.

[3] Environmental Engineering, Hindustan College of Science and Technology, Farah, Mathura, Uttar Pradesh, India.

References

Aharoni, C., & Ungarish, M. (1977). Kinetics of activated chemisorption. Part 2. Theoretical models. *Journal of the Chemical Society, Faraday Transactions 1: Physical Chemistry in Condensed Phases, 73*, 456–464. http://dx.doi.org/10.1039/f19777300456

Ajmal, M., Rao, R. A. K., & Siddiqui, B. A. (1995). Adsorption studies and the removal of dissolved metals using pyrolusite as adsorbent. *Environmental Monitoring and Assessment, 38*, 25–35. http://dx.doi.org/10.1007/BF00547124

Ajmal, M., Rao, R. A. K., Ahmad, R., & Ahmad, J. (2000). Adsorption studies on *Citrus reticulata* (fruit peel of orange): Removal and recovery of Ni(II) from electroplating wastewater. *Journal of Hazardous Materials, 79*, 117–131. http://dx.doi.org/10.1016/S0304-3894(00)00234-X

Ayub, S., Ali, S. I., & Khan, N. A. (1998). Treatment of wastewater by agricultural wastes. *Environmental Pollution Control Journal, 2*, 5–8.

Ayub, S., Ali, S. I., & Khan, N. A. (1999). Extraction of chromium from the wastewater by adsorption. *Environmental Pollution Control Journal, 2*, 27–31.

Ayub, S., Ali, S. I., & Khan, N. A. (2001). Efficiency evaluation of neem (*Azadirachta indica*) bark in the treatment of industrial wastewater. *Environmental Pollution Control Journal, 4*, 34–38.

Ayub, S., Ali, S. I., & Khan, N. A. (2002). Adsorption studies on the low cost adsorbent for the removal of Cr(VI) from the electroplating industries. *Environmental Pollution Control Journal, 5*, 10–20.

Ayub, S., Ali, S. I., & Khan, N. A. (2003). Chromium removal by adsorption on coconut shell.*Journal of Indian association for Environmental Management* (NEERI), *30*, 30–36.

Ayub, S., Ali, S. I., & Khan, N. A. (2006). Comparative study of different agro based adsorbents for the treatment of wastewater. *Current World Environment, 1*, 109–116.

Ayub, S., Sharma, P. K., & Tripathi, C. N. (2014, July–September). Removal of hexavalent chromium using agro and horticultural wastes as low cost sorbents from tannery wastewater: A review. *International Journal of Research in Civil Engineering, Architecture & Design, 2*, 21–35, ISSN Online: 2347-2855, Print: 2347-8284, DOA: 17082014.

Baisakh, P. C., & Patnaik, S. N. (2002). Removal of hexavalent chromium from aqueous solution by adsorption on coal char. *Indian Journal of Environmental Health, 4*, 189–196.

Baisakh, P. C., Patnaik, S. N., & Patnaik, L. N. (1996). Removal of COD from textile mill effluent using fly ash. *Indian Journal of Environmental Protection, 16*, 135–139.

Bansal, T. K., & Sharma, H. R. (1992). Chromium removal by adsorption on rice husk ash. *Indian Journal of Environmental Protection, 13*, 198–201. (Quoted by Rai, et al., 1995).

Baskaran, P. K., Venkatraman, B. R., Hema, M., & Arivoli, S. (2010). Adsorption studies of copper ion by low cost activated carbon. *Journal of Chemical and Pharmaceutical Research, 2*, 642–655.

Browing, E. (1969). *Toxicity of industrial metals.* London: Bullerworths (Quoted by Huang and Wu, 1975).

Chand, S., Aggarwal, V. K., & Kumar, P. (1994). Removal of hexavalent chromium from the wastewater by adsorption. *Indian Journal of Environmental Health, 36*, 151–158.

Cimino, G., Paserini, A., & Toscano, G. (2000). Removal of toxic cations and Cr(VI) from aqueous solution by hazelnut shell. *Water Research, 34*, 2955–2962. http://dx.doi.org/10.1016/S0043-1354(00)00048-8

Dada, A. O., Olalekan, A. P., Olatunya, A. M., & Dada, O.(2012). Langmuir, Freundlich, Temkin and Dubinin–Radushkevich isotherms studies of equilibrium sorption of Zn^{2+} unto phosphoric acid modified rice husk. *IOSR Journal of Applied Chemistry, 3*, 38–45. ISSN: 2278-5736.

Deo, N., & Ali, M. (1992). Use of a low cost material as an adsorbent in the removal of Cr(VI) from dilute aqueous solution. *Indian Journal of Environmental Pollution, 12*, 439–441.

Drake, L. R., & Lin, S., Rayson, G. D., & Jackson, P. J. (1996). Chemical modification and metal binding studies of datura innoxia. *Environmental Science & Technology, 30*, 110–114. http://dx.doi.org/10.1021/es950131d

Heary, S. K., & Ray, J. (1987). Processes affecting the remediation of chromium–contaminated sites. *Environmental Health Perspective, 92*, 24–40.

Huang, C. P., & Morehart, A. L. (1991). Proton competition in Cu(II) adsorption by fungal mycelia. *Water Research, 25*, 1365–1375. http://dx.doi.org/10.1016/0043-1354(91)90115-7

Huang, C. P., & Stumm, W. (1973). Specific adsorption of cations on hydrous γ-Al2O3. *Journal of Colloid and Interface Science, 43*, 409–420. http://dx.doi.org/10.1016/0021-9797(73)90387-1

Huang, C. P., & Wu, M. H. (1975). Chromium removal by carbon adsorption. *Journal of Water Pollution Control Federation, 47*, 2443–2446.

Katz, S. A. (1991). The analytical biochemistry of chromium. *Environmental Health Perspective , 92*, 13–16.

Kim, J., & Joltek, J. (1977). Chromium removal with activated carbon. *Progress in Water Technology, 9*, 143–155.

Landigran, R. B., & Halllowell, J. B. (June 1975). *Removal of chromium from plating rinse water using activated carbon* (EPA, 670/2-75-055). *Environmental protection technology series, EPA Grant No.S802113, Programme Element No.1DB0JG, Cincinnati, OH. National Environmental research centre, Office of R&D, US, EPA Retrieved from www.nepis.epa.gov/Exe/ZyPURL.cgi?Dockey=40001K3C.TXT

Mahesh, S., Rama, B. M., Praveen, K. N. H., & Usha, L. K. (1999). Adsorption kinetics of di hydric phenol hydroquinone on activated carbon. *Indian Journal of Environmental Health, 41*, 317–325.

Mall, I. D. (1992). *Studies on the treatment of pulp and paper mill effluent* (PhD theses). Banarus Hindu University, Varanasi (Quoted by Rai, et al., 1995).

Panday, K. K., Prasad, G., & Singh, V. N. (1984). Removal of Cr(VI) from aqueous solution by adsorption on fly ash wall astonite. *Technology and Biotechnology, 34A,* 367–374.

Periasamy, K., Srinivasan, K., & Murugan, P. K. (1991). Studies on chromium (VI) removal by activated groundnut husk carbon. *Indian Journal of Environmental Health, 31,* 433–439.

Raji, C., & Anirudhan, T. S. (1998). Sorptive behavior of chromium (VI) on saw dust carbon in aqueous media. *Ecology, Environment and Conservation, 31,* 2–33.

Rao, M., Parwate, A. V., & Bhole, A. G. (2002). Utilization of low cost adsorbents for the removal of heavy metals from wastewater. *Environmental Pollution Control Journal, 5,* 12–23.

Sharma, D. C., & Forster, C. F. (1993). Removal of hexavalent chromium using sphagnum moss peat. *Water Research, 27,* 1201–1208. http://dx.doi.org/10.1016/0043-1354(93)90012-7

Sharma, P. K., Ayub, S., & Tripathi, C. N. (2013, August). Agro and horticultural wastes as low cost adsorbents for removal of heavy metals from wastewater: A review. *International Refereed Journal of Engineering and Science, 2,* 18–27.

Sharma, P. K., Ayub, S., & Tripathi, C. N. (2015). Sorption studies of synthesized horticultural waste seeds for removal of hexavalent chromium from aqueous solutions. *International Journal of Environmental Science and Toxicology Research, 3,* 101–114.

Sharma, P. K., Dubey, S. K., & Rehman, Z. (2010, October 29–30). A study on arsenic removal from ground water by adsorption process. In *International Conference on Emerging Technologies for Sustainable Environment, (ETSE-2010)* (pp. 164–167). Aligarh: Department of Civil Engineering, A.M.U.

Sharma, P. K., Khan, N. A., & Ayub, S. (2012a, September). Mathematical modelling in anaerobic process. *International Journal of Civil, Structural, Environmental and Infrastructure Engineering Research and Development, 2,* 52–65. ISSN 2249-6866.

Sharma, P. K., Khan, N. A., & Ayub, S. (2012b). Modeling of cod reduction in a UASB reactor. *Global Journal of Engineering and Applied Sciences, 2,* 178–182. ISSN 2249-2631 (online): 2249 2623 (Print)–Rising Research Journal Publication.

Shen, K., & Wang, J. (1995). Hexavalent chromium removal in two stage bio reactor system. *Journal of environmental engineering* (ASCE), *121.*798–804.

Shukhla, S. R., & Sakhardane, V. D. (1991). Column studies on metal iron removal by dyed cellulosic materials. *Journal of Applied Polymer Science , 11,* 284–289.

Shupack, S. I. (1991). The chemistry of chromium and some resulting analytical problems. *Environmental Health Perspectives,92,* 17–24.

Siddiqui, Z. M., & Paroor, S. (1994). Removal of Cr(VI) by different adsorbents a comparative study. *International Journal of Environmental Protection, 14,* 273–278.

Stummm, W., & Morgon, J. J. (1970). *Aquatic chemistry.* Neywork, NY: Wiley.

Tempkin, M. I., & Pyzhev, V. (1940). Kinetics of ammonia synthesis on promoted iron catalyst. *Acta Physico-Chimica, 12,* 327–356.

Vaishya, R., & Prasad, S. C. (1991). Adsorption of copper on saw dust. *Indian Journal of Environmental Protection, 11,* 284–299.

Veena Devi, B., Jahagirdar, A. A., & Zulfiqar Ahmed, M. N. (2012, September–October). Adsorption of chromium on activated carbon prepared from coconut shell. *International Journal of Engineering Research and Applications, 2,* 364–370. ISSN: 2248-9622. Retrieved from www.ijera.com

Weber, C. W. (1996). In vitro binding capacity of wheat bran, rice bran and oat fiber for Ca, Mg, Cu and Zn alone and in different combinations. *Journal of Agricultural and Food Chemistry, 44,* 2067–2072.

10

Qualitative and quantitative assessment of outdoor daylight on vertical surfaces for six climate specific Indian cities

K.N. Patil[1*], S.C. Kaushik[1] and S.N. Garg[1]

*Corresponding author: K.N. Patil, Centre for Energy Studies, Indian Institute of Technology Delhi, Hauz Khas, New Delhi 110016, India
E-mail: kalmeshnp@gmail.com
Reviewing editor: Amir H. Alavi, Michigan State University, USA

Abstract: This article deals with experimental validation of solar illuminance model for vertical plane surfaces against measured solar global illuminance at New Delhi. The paper also deals with computation of solar illuminance (daylight), both quantitatively and qualitatively, on four cardinal vertical surfaces at six Indian stations. The stations identified for the study are: New Delhi, Jodhpur, Pune, Kolkata, Shillong and Leh. Quantitative analysis of daily average solar illuminance shows that most of the stations under study receive maximum illuminance on north, east and west surfaces, during summer months of April–July, whereas the south surface receives maximum illuminance during winter months of November–February. Higher range of daily average solar illuminance on four vertical surfaces is evaluated as follows: 45–55 klx on south surface, 20–35 klx on west surface (all the stations except Pune), 20–30 klx on east (Pune) and 14–15 klx on north surface. Qualitative analysis shows that for the global illuminance in the range of 20–40 klx, for four vertical surfaces, the percentage of annual day time is: 3–8% for north surface, 25–30% for south surface, 20–30% for west surface and 20–25% for east surface.

Subjects: **Clean Technologies, Environmental, Renewable Energy**

Keywords: **sky diffuse illuminance, daylight on vertical surface, cumulative frequency of solar illuminance**

ABOUT THE AUTHORS

K.N. Patil is a research scholar in Centre for Energy Studies, IIT Delhi. He is working on the research area, "Energy conservation in Building through Daylighting Ventilation and Space conditioning" and he has several research papers in national and international conferences and journals of repute.

S.C. Kaushik is professor at Centre for Energy studies, IIT Delhi. His research fields of activities include Thermal Science and Engineering, Energy Conservation and Heat Recovery, Solar Refrigeration and Air conditioning, and Solar Architecture. He is a pioneer researcher on Exergy Analysis and Finite time thermodynamics of Energy Systems at National and International level. He has been former Head of Centre for Energy Studies during 2007-2010.

S.N. Garg was scientific officer in Centre for Energy Studies, IIT Delhi. His fields of specialisation are: Solar Radiation Measurement and Computation, Energy Conservation in Buildings and Energy Efficient Windows.

PUBLIC INTEREST STATEMENT

Qualitative and quantitative assessment of daylight availability on vertical surfaces provides crucial inputs for the building designer so that daylight could be exploited to the full potential in the building. Daylight data for vertical surfaces is required for optimising the size and glazing material for the window, so as to conserve lighting energy in a building. Solar radiation data are available for most of the cities around the world but daylight (solar illuminance) data are hardly available. There is a need to generate daylight data from solar radiation data, through empirical or semi-empirical models, as daylight measurement is a very costly affair. This study provides inputs to the Strategists and policy-makers, at regional and national level of India, in benchmarking the energy conservation potential of lighting in the building sector.

1. Introduction

Lighting is the second largest consumer of energy in building sector. It accounts for 20% of total electric energy consumed in India (World Bank Report, 2008). Daylight is one of the most preferred options, and should be exploited to the fullest extent in order to conserve artificial lighting energy in buildings. Recent advent and popularity of advanced daylight systems such as light pipes, light guides, energy efficient windows and glazed wall, have created a demand for daylight data for vertical surfaces, so that daylight systems could be optimised for the fullest exploitation of the daylight.

Qualitative and quantitative assessment of daylight on vertical surfaces finds prime significance due to the fact that daylight could be exploited substantially through the vertical surfaces, mounted with window and glazed wall of a building. Daylight data are useful for the estimation of energy conservation potential of the daylight in lighting sector of the buildings for local, regional and even at national level. There is a need of outdoor daylight data, for different climatic conditions, which provides inputs for the design, simulation and optimisation of the glazing area, so that use of daylight could be maximised and hence, minimise the use of artificial light in a building.

International Commission on Illumination (CIE) model, Perez et al. models and Igawa model are some of the widely used empirical/semi-empirical models to generate daylight data for the stations, where measured daylight data are hardly available. Perez, Seals, and Michalsky (1993) developed and evaluated an all-weather model for instantaneous sky luminance, and evaluated six other models from recorded sky scans. Frame work of the model is based on CIE standard clear sky general formula. Li and Lam (2003) formulated a sky luminance model from the measurements at Hong Kong. Their model performed better than Perez et al. model. Li, Lam, Cheung, and Tang (2008) developed and validated a luminous efficacy model for sky diffuse illuminance, on four vertical surfaces under 15 CIE standard skies, using measured illuminance and irradiance data for Hong Kong. De Rosa, Ferraro, Kaliakatsos, and Marinelli (2008), Robledo and Soler (2003), developed luminous efficacy models for diffuse illuminance and global illuminance, respectively, for vertical surfaces. Their models were validated and found to give a better performance as compared to existing models. Ferraro, Mele, and Marinelli (2012) assessed the Perez et al. model, Igawa model and CIE model against one year measured sky luminance data for Osaka (Japan), and for Arcavacata (Italy). Ferraro found that Igawa model provides accurate results when used for the evaluation of absolute sky diffuse luminance, whereas Perez et al., all-weather models are better for the evaluation of relative sky diffuse luminance.

We find little literature concerned to annual daylight assessment on vertical plane surfaces for Indian stations representing all the climatic zones. The novelty in this research is the use of weather data for typical meteorological year (TMY) for generating solar illuminance data. TMY weather data reflects most prevailing weather data for the previous 20 years. And use of the empirical and semi-empirical models for solar illuminance, followed by the experimental validation of these models, by the authors.

2. Theoretical basis and methodology

2.1. Solar illuminance model for vertical surface

Total solar illuminance on a vertical surface is the sum of beam illuminance, sky diffuse illuminance and ground reflected illuminance, as illustrated in Figure 1. Solar illuminance model for vertical surface consists of the parametric model for sky diffuse illuminance, mathematical models for ground reflected illuminance and beam illuminance. Sky diffuse illuminance is evaluated using Perez et al. (1993) all-weather model while, ground reflected illuminance and beam illuminance are evaluated using the relevant equations, used for ground reflected irradiance and beam irradiance on a vertical surface. Solar illuminance on the horizontal surface required by these models is evaluated using Perez et al., luminous efficacy model.

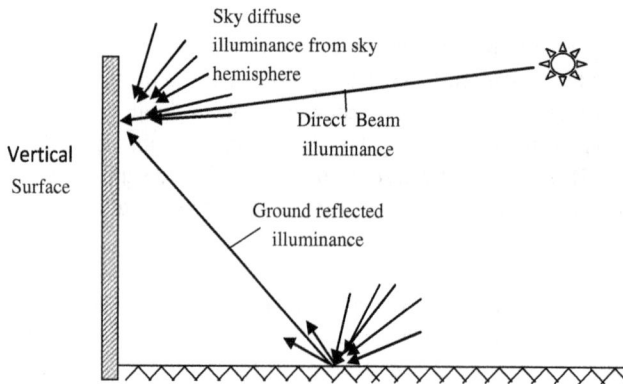

Figure 1. Solar illuminance on vertical surface.

2.1.1. Perez et al. luminous efficacy model

Luminous efficacy of solar irradiance is defined as the ratio of solar illuminance on an unobstructed horizontal surface to the corresponding solar irradiance (Kittler, Kocifaj, & Darula, 2012). Perez, Ineichen, Seals, Michalsky, and Stewart (1990) formulated luminous efficacy models from extensive measurements of both solar illuminance and solar irradiance on a horizontal surface. Various parameters like solar zenith angle, sky clearness, sky brightness and their effect on luminous efficacy are taken into account by the model coefficients, derived through empirical analysis. Solar geometrical angles required for the models have been referred from Duffie and Beckman (2006).

Solar global illuminance on horizontal surface (L_{gh}), is given as (Perez et al., 1990)

$$L_{gh} = I_{gh} \left[a' + b'w + c' \cos(\theta_z) + d' \log(\Delta) \right] \tag{1}$$

where I_{gh} is the solar global irradiance (W/m²); a', b', c' and d' are model coefficients, which are selected based on the value of sky clearness and sky ratio (Perez et al., 1990); w is the atmospheric precipitable water content (cm); θ_z is the solar zenith angle (radian); and Δ is the sky brightness.

Sky brightness, $$\Delta = \frac{I_{dh} \cdot m}{I_o} \tag{2}$$

where I_{dh} is the solar diffuse irradiance (W/m²); I_o is the solar extraterrestrial irradiance (W/m²); m is the optical air mass at local atmospheric pressure (Kasten & Young, 1989) and the same is given as

$$m = \frac{m_o \cdot P}{1013.25} \tag{3}$$

where m_o is the optical air mass at atmospheric pressure at average sea level (Kasten & Young, 1989) given as

$$m_o = \left[\sin \alpha_s + 0.50572 (\alpha_s + 6.07995)^{-1.6364} \right]^{-1} \tag{4}$$

where α is the solar altitude (radian); P is the local atmospheric pressure (mbar) (Lunde, 1980), is given as

$$P = 1013.25 \cdot \exp(-0.0001184 \cdot h) \tag{5}$$

where h is the height of the station above the average sea level, 216 m for New Delhi.

$$\text{Sky Clearness,} \quad \varepsilon = \frac{\dfrac{I_{dh}+I_{bn}}{I_{dh}}+1.041\cdot\theta_z^3}{1+1.041\cdot\theta_z^3} \tag{6}$$

where I_{bn} is the solar normal beam irradiance (W/m²).

$$\text{Sky Ratio,} \quad SR = \frac{I_{dh}}{I_{gh}} \tag{7}$$

Solar diffuse illuminance on horizontal surface (L_{dh}) is given as

$$L_{dh} = I_{dh}\left[a'' + b''w + c'' \cos(\theta_z) + d'' \log(\Delta)\right] \tag{8}$$

where a'', b'', c'' and d'' are the model coefficients which depend on sky ratio and sky clearness (Perez et al., 1990). Readers are suggested to refer Perez et al. (1990, 1993) and Patil, Garg, and Kaushik (2013) for details of model coefficients, available in tabular form.

2.1.2. Perez et al. all-weather model

Perez et al. (1993) developed an all-weather model for relative sky luminance of a sky segment. Model is a generalisation of CIE standard clear sky formula. Luminance distribution, due to different sky conditions (clear sky to overcast sky), is obtained by adjusting the five critical coefficients of the model. Figure 2 shows the geometrical angles related to the sky segment and the sun position in sky hemisphere. Relative luminance of a sky element (l_r), in terms of zenith angle of sky segment (θ), solar zenith angle (θ_z) and angle between sky segment and sun position (ξ), is given as

$$l_r(\theta,\xi) = \frac{f(\theta,\xi)}{f(0,\theta_z)} = \frac{\left[1+a_i\cdot\exp\left(\frac{b_i}{\cos\theta}\right)\right]\times\left[1+c_i\cdot\exp(d_i\cdot\xi)+e_i\cdot\cos^2\xi\right]}{\left[1+a_i\cdot\exp(b_i)\right]\times\left[1+c_i\cdot\exp(d_i\cdot\theta_z)+e_i\cdot\cos^2\theta_z\right]} \tag{9}$$

where a_i, b_i, c_i, d_i and e_i are the coefficients which are function of solar zenith angle, sky clearness and sky brightness. These are evaluated from model coefficients evaluated from empirical study which are represented by: $a_{ji}, b_{ji}, c_{ji}, d_{ji}$ and e_{ji} (j = 1 to 4, i(sky category) = 1 to 8) available in the tabular form, for eight categories of the sky conditions [2].

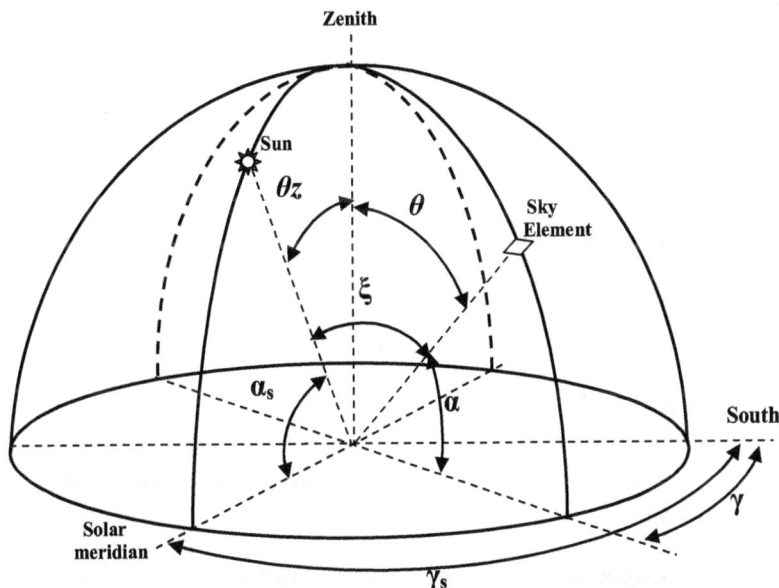

Figure 2. Position of sky element sun and zenith in sky hemisphere.

$$a_i = a_{1,i} + a_{2,i} \cdot \theta_z + \Delta(a_{3,i} + a_{4,i} \cdot \theta_z) \tag{10}$$

where b_i, c_i, d_i and e_i are evaluated similarly as that of Equation 10, with exception for c_1 and d_1 which are evaluated by equations 11 and 12

$$c_1 = \exp\left[\{\Delta(c_{1,1} + c_{2,1} \cdot \theta_z)\}^{c_{3,1}}\right] - 1 \tag{11}$$

$$d_1 = -\exp\left[\Delta(d_{1,1} + d_{2,1} \cdot \theta_z)\right] + d_{3,1} + \Delta d_{4,1} \tag{12}$$

From Figure 2 relation between ξ, θ and θ_z can be derived as

$$\cos\xi = \cos\theta_z \cdot \cos\theta + \sin\theta_z \cdot \sin\theta \cdot \cos(\gamma - \gamma_s) \tag{13}$$

$$\theta = \frac{\pi}{2} - \alpha$$

Relative sky luminance (Equation 9) is first expressed in terms of altitude (α) and azimuth angle (γ), and resulting equation is then substituted in Equation 14 to obtain the relative sky diffuse luminance (l_{rv}) on vertical plane surface.

$$l_{rv} = \int_0^\pi \int_0^{\pi/2} l_r(\alpha, \gamma) \cdot \cos^2\alpha \cdot \cos(\gamma - \gamma_{sur}) \, d\alpha \, d\gamma \tag{14}$$

where γ_{sur} is the surface azimuth angle of vertical plane surface.

Relative sky diffuse illuminance on horizontal surface (l_{rh}) can be evaluated by equation

$$l_{rh} = \int_0^{2\pi} \int_0^{\pi/2} l_r(\alpha, \gamma) \cdot \cos\alpha \cdot \sin\alpha \, d\alpha \, d\gamma \tag{15}$$

Analytical solution of Equations 14 and 15 is difficult; hence numerical method is used and the same is given by Equations 16 and 17.

For horizontal surface,

$$l_{rh} = \sum_{i=1}^{145} l_r(\alpha, \gamma) \cdot \cos\alpha \cdot \sin\alpha \cdot \Delta\alpha \cdot \Delta\gamma \tag{16}$$

For east and west vertical surface,

$$l_{rv} = \sum_{i=1}^{65} l_r(\alpha, \gamma) \cdot \cos^2\alpha \cdot \cos(\gamma - \gamma_{sur}) \, \Delta\alpha \, \Delta\gamma \tag{17}$$

For north and south vertical surface,

$$l_{rv} = \sum_{i=1}^{69} l_r(\alpha, \gamma) \cdot \cos^2\alpha \cdot \cos(\gamma - \gamma_{sur}) \, \Delta\alpha \, \Delta\gamma \tag{18}$$

where i is the number of sky elements under consideration. There are 145 sky segments visible to horizontal surface as shown in Figure 3.

Absolute sky diffuse illuminance is evaluated from relative sky diffuse luminance using normalisation ratio. The normalisation Ratio (NR) is defined as the ratio of absolute sky diffuse illuminance to the relative sky luminance on a horizontal surface.

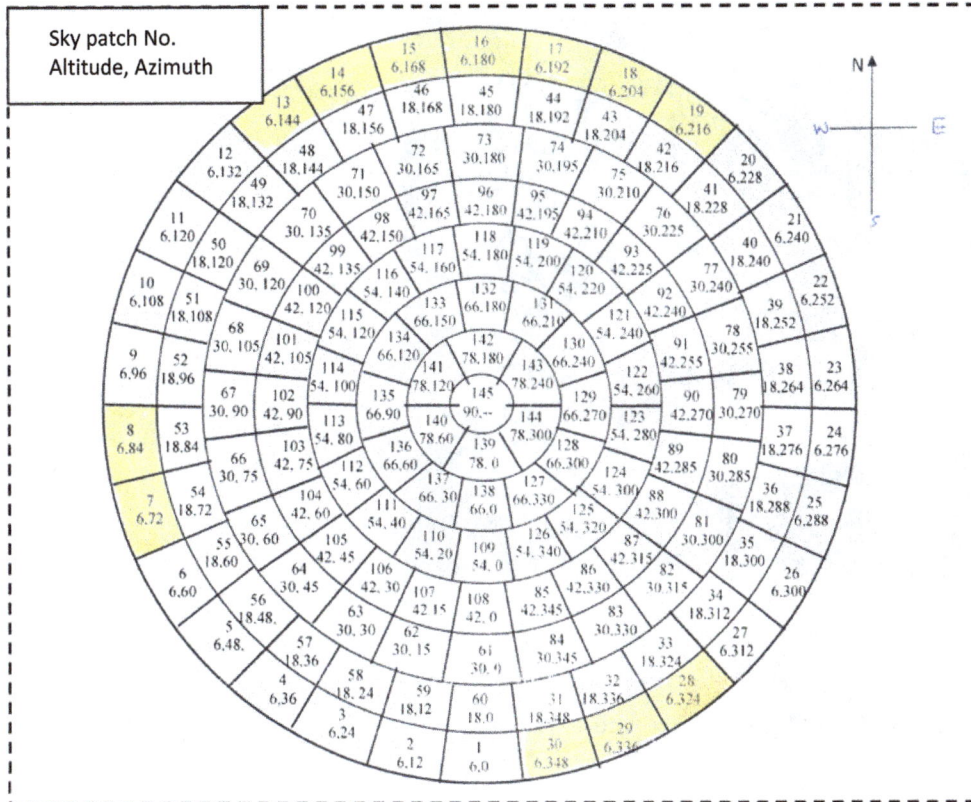

Figure 3. Evaluation points for sky luminance across sky hemisphere.

Normalisation Ratio (NR)

$$NR = \frac{L_{dh}}{I_{rh}}$$

where L_{dh} is either measured or evaluated using Perez luminous efficacy model given by Equation 8 Patil et al. (2013).

Absolute sky diffuse illuminance on vertical surface (L_v) is given as

$$L_v = NR \times I_{rv} \tag{19}$$

Beam illuminance on vertical surface (L_{bv}) is given as

$$L_{bv} = \frac{(L_{gh} - L_{dh}) \cdot \cos\theta_T}{\cos\theta_z} \tag{20}$$

where L_{gh} is measured or calculated using Perez et al., luminous efficacy model; θ_T is solar incident angle on vertical surface (radian).

Ground reflected illuminance is given as

$$L_{gr} = \frac{L_{gh} \times \rho}{2} \tag{21}$$

where ρ is ground reflectivity (assumed to be .2 for urban area).

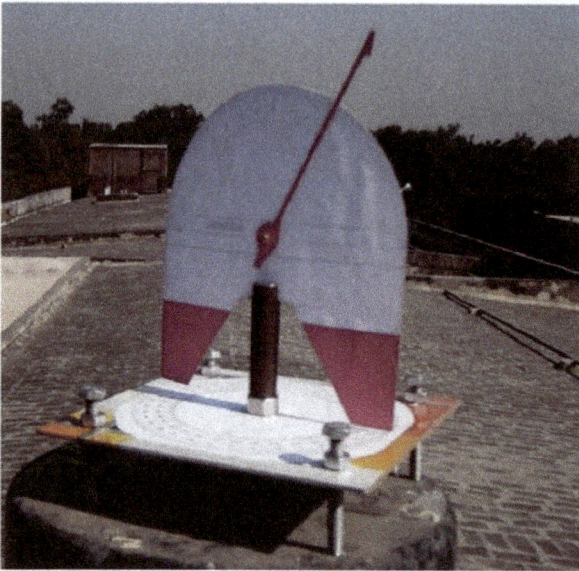

Figure 4. Protractor.

Total illuminance (L_{tv}) on vertical surface is given as

$$L_{tv} = L_v + L_{gr} + L_{bv} \tag{22}$$

Figure 3 shows the projection of sky hemisphere on the horizontal plane. Sky hemisphere is divided into 145 segments. Each sky segment is identified by its altitude (α) and azimuth (γ). Initially, the location of measurement is inspected for its exposure to sky hemisphere. Any sky segment, invisible for the instrument due to the obstruction posed by surrounding buildings, walls, vegetation, etc., is identified and marked by respective altitude and azimuth of sky segment, using the protractor (Figure 4). Obstructed sky segments are shaded as illustrated in Figure 3 and these sky segments are excluded from the evaluation of sky diffuse illuminance.

3. Experimental set-up

Initially, solar cardinal directions (north, south, east and west) at the location of measurement are marked using a pole shadow method. Further, the location is investigated for its exposure to sky hemisphere using the protractor as shown in Figure 4. Experimental set-up consists of a pyranometer (Kipp & Zonen make) to measure solar global irradiance and a photo sensor (LICOR make) to measure solar global illuminance on the vertical surface. Specifications of these two instruments are given in Table 1. Both the sensors are mounted on either side of the vertical surface of the wooden stand, as shown in Figure 5. Wooden stand can be turned about vertical axis manually so as to face the sensors towards cardinal directions for the measurement purpose.

Sl. No.	Instrument	Specifications
\multicolumn{3}{l}{**Table 1. Specifications of instruments**}		
1	Pyranometer	Make: KIPP and ZONEN, Holland Model: CM11, Sensitivity: 5.03×10^{-6} V/Wm^2, Linearity: ±.5%
2	Photometer	Make: Li-COR, USA, Accuracy: ±.4%, Range: 0–199 klx, Linearity: ±.05%

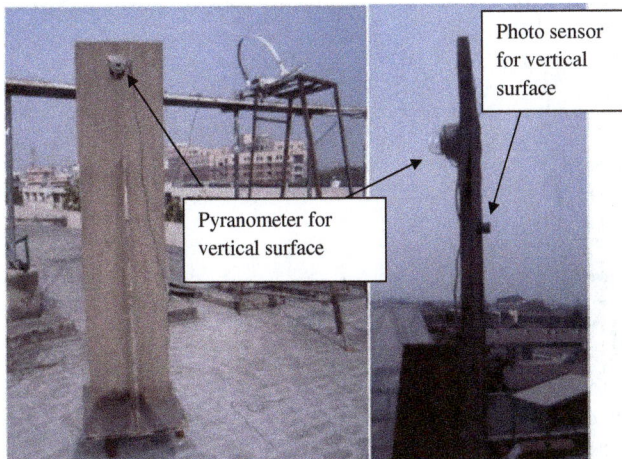

Figure 5. Measurement set-up for solar illuminance and solar irradiance on vertical surface.

4. Results and discussion

4.1. Indian climatic zones

Indian climates are highly diverse in nature and are classified, on the basis of ambient temperature and humidity(Bansal & Minke, 1995), into six climatic zones namely Hot and dry, Warm and humid, Composite, Moderate, Cold and cloudy, and Cold and sunny as shown in Figure 6. The main features of these climatic zones, as reported by Bansal and Minke (1995), are as follows:

4.1.1. Hot and dry climate

This climatic zone has flat with sandy ground, it receives high solar radiation in the range, 800–950 W/m^2. In summer, ambient temperature reaches to 40–45°C and in winter, maximum temperature reaches to 20–30°C. Relative humidity (RH) is usually in the lower range of 25–40% due to low vegetation and few surface water bodies.

4.1.2. Warm humid climate

This climatic zone constitutes mainly coastal part of India. Cloudy sky, high summer temperatures (30–35°C), high winter temperature (25–30°C) and high humidity (70–90% throughout the year) are the main features of this climate.

4.1.3. Composite climate

It consists of central part India, variable landscape and seasonal vegetation are the features of this climate. Solar intensity is very high in summer, low intensity in monsoon. Day time temperature reaches to 32–43°C during day and 27–32°C during night. In winter, ambient temperatures are in the range of 10–25°C during day time and 4–10°C during night time. RH is about 20–25% in dry period and 55–95% in wet period. Sky is overcast in monsoon, whereas it is clear in winter and hazy in summer.

4.1.4. Moderate climate

This climate is neither hot nor cool; it consists of hilly and high plateau regions. Abundant vegetation, uniform radiation, lower ambient temperatures are the outstanding features of this climate. In summer, highest temperature is in the range of 30–34°C. RH levels are in the range of 20–55% during winter and summer, whereas in monsoon it is 55–90%.

4.1.5. Cold and cloudy

This climate covers northern part of India; situated in high altitude with abundant vegetation even in summer season. Solar irradiance is low in winter with high percentages of diffuse component.

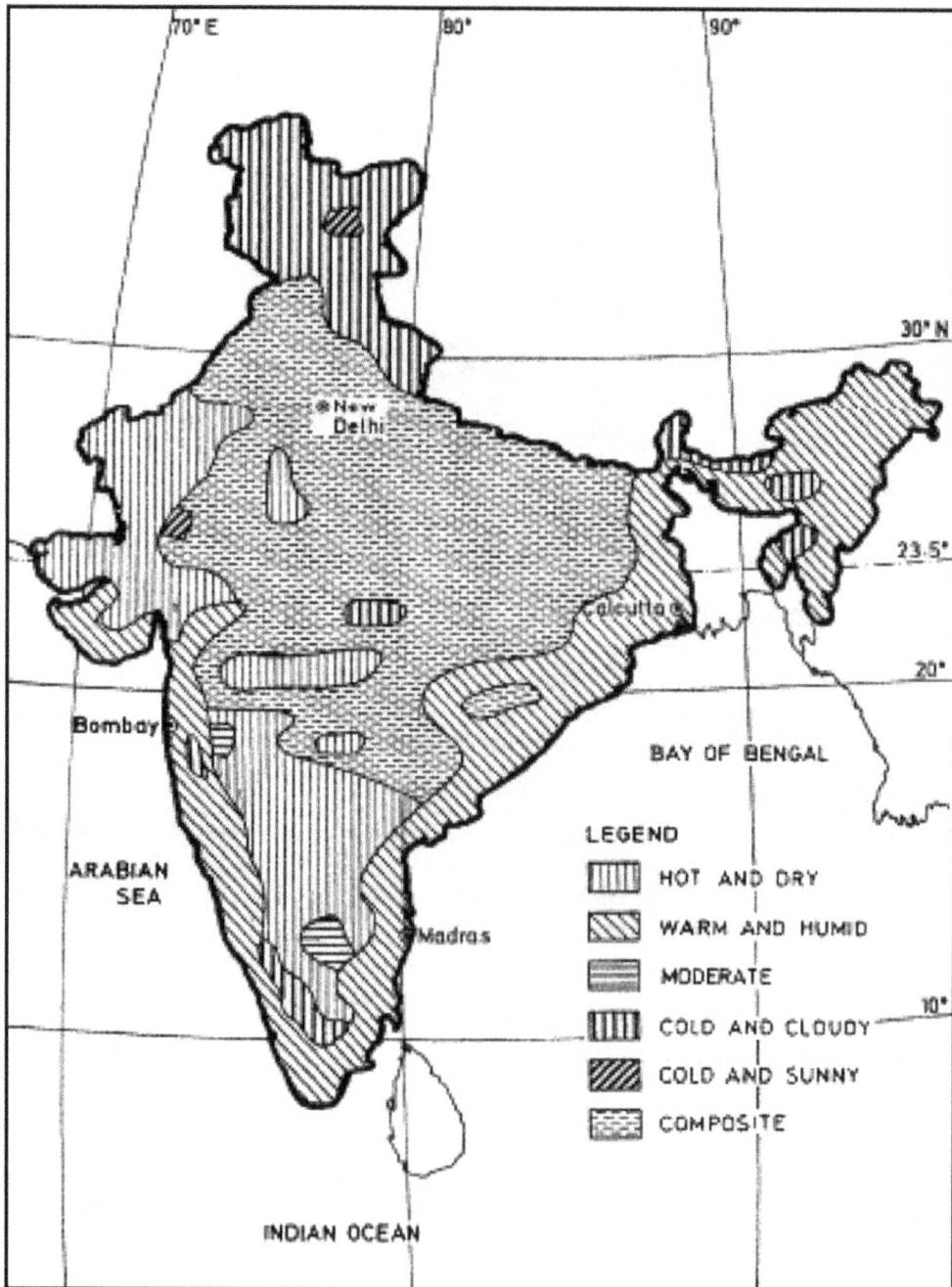

Figure 6. Bansal and Minke classification of Indian climates.

Maximum ambient temperature is in the range of 20–30°C in summer while in winter, it is extremely low (4–8°C) during day time. RH is in the high range of 70–80%. Sky is overcast for most part of the year except summer.

4.1.6. Cold and sunny

This climate mainly exists in the region, Leh and Ladhak of northern India. Little vegetation, cold desert, intense beam radiation are the main features of this zone. In summer, temperature reaches to 17–24°C during day and it is 4–11°C during night; RH is in lower range of 10–50%. Sky is fairly clear throughout the year.

Six prominent stations, one from each climatic zone, selected for the study are: New Delhi (28.63°N, 77.2°E, 216 m) for composite climate, Jodhpur (26.29°N, 73.03°E, 231 m) for Hot and dry climate, Pune (18.52°N, 73.85°E, 560 m) for moderate climate, Kolkata (22.57°N, 88.37°E, 9 m) for warm and humid climate, Shillong (25.57°N, 91.88°E, 1525 m) for cold and cloudy climate and Leh (34.17°N, 77.58°E, 3500 m) for cold and sunny climate.

4.2. Experimental validation of solar illuminance model

Solar global irradiance and solar global illuminance on vertical surfaces were measured for every half an hour from morning 7.30 am to evening 5.30 pm Measurements were carried out for 12 days, during March and April months of the year 2012, on roof top of the building, Centre for Energy Studies, IIT Delhi. Solar global illuminance model for vertical surfaces was assessed for its performance by statistical parameters such as mean bias error (MBE), root mean square error (RMSE), relative error (RE) and Coefficient of Determination (R^2) (Muneer, 2004).

Table 2 shows MBE and RMSE values of solar illuminance model for four vertical surfaces. It is observed that solar illuminance model underestimates the actual illuminance on north surface, whereas it overestimates the solar illuminance on other three surfaces. South surface shows lowest MBE (0.30%), followed by east surface (3.87%), west surface (7.89%) and north surface (−8.65%). RMSE values show decreasing trends from 19.01 to 2.68 for north, south, east and west surfaces accordingly.

The REs of the model derived results for four vertical surfaces are presented in Table 3. Solar illuminance model for vertical surfaces performed better for east surface with 70% of results for this surface fall within ±10% RE. For south surface, 70% of results fall within ±15% of RE and overall analysis shows 80–90% of results derived by the model for all the four surfaces fall in the range of ±25% of RE.

Figure 7 shows the relation between measured global illuminance and model derived global illuminance. The value of Coefficient of Determination (R^2) represents how well the predicted data correlate with measured data. It is observed that for all the four vertical surfaces, illuminance model was performed satisfactorily, as the model derived illuminance varies linearly with measured illuminance. R^2 value is more than .9 for east surface, west surface and south surface, and it is about .8 for north surface. The R^2 value higher than .9 indicates a good agreement between models derived

Table 2. MBE and RMSE of solar illuminance model

Surfaces	MBE (%)	RMSE (%)
North	−8.65	19.01
South	0.30	16.00
East	3.87	12.99
West	7.89	2.68

Table 3. Validation of solar illuminance model based on RE

Range of relative error (%)	Percentage of results for vertical surfaces			
	North	South	East	West
±05	18	28	38	28
±10	41	54	69	47
±15	56	71	79	67
±20	71	80	89	84
±25	82	88	92	91

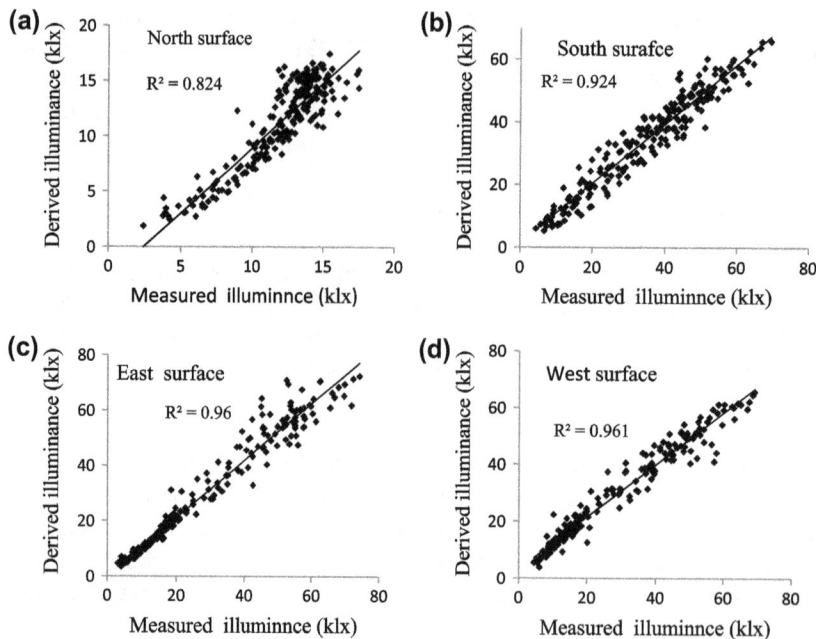

(a) North surface $R^2 = 0.824$ — Derived illuminance (klx) vs Measured illuminnce (klx)

(b) South surafce $R^2 = 0.924$ — Derived illuminance (klx) vs Measured illuminnce (klx)

(c) East surface $R^2 = 0.96$ — Derived illuminance (klx) vs Measured illuminnce (klx)

(d) West surface $R^2 = 0.961$ — Derived illuminance (klx) vs Measured illuminnce (klx)

Figure 7. Comparison of measured illuminance with evaluated illuminance for vertical surfaces.

illuminance and measured illuminance. North surface doesn't receive direct illuminance, and direct illuminance can be evaluated more precisely as compare to diffuse illuminance. Hence, R^2 value is slightly smaller for the north surface.

Figure 3 shows the projection of sky hemisphere on the horizontal plane. Sky hemisphere is divided into 145 segments according to international standard practice, and each sky segment is specified by its altitude (α) and azimuth (γ). Initially, the location of measurement is inspected for its exposure to sky hemisphere. The sky segments, invisible to the instrument due to the obstruction caused by the surrounding buildings, terrain and vegetation etc., are identified, and obstructing sky segments are marked by their respective altitude and azimuth angle, using the protractor (Figure 4). Obstructing sky segments are shown with shaded colour as shown in Figure 3, and shaded sky segments are excluded from the evaluation of sky diffuse illuminance.

4.3. Quantitative assessment of daylight on vertical surfaces

Hourly values of solar irradiance, solar geometric angles and other weather data for a TMY, required in the solar illuminance model, were generated through simulation in the TRANSYS software. Solar illuminance model is constituted by three models; first model is for sky diffuse illuminance, computed using Perez et al. all-weather model, second is for ground reflected illuminance and third is for beam illuminance. Solar illuminance on the horizontal surface, required for the evaluation of ground reflected and beam illuminance on vertical surface, was evaluated using Perez et al. model for the luminous efficacy of solar irradiance.

Daily average solar illuminance was computed from monthly average hourly global as well as diffuse illuminance on vertical surface for 12 months of a TMY. The study was carried out for all the six stations, and for four vertical surfaces, the results are presented in Figures 8–15.

4.3.1. Daily average solar illuminance on the north surface

Daily average solar global illuminance on the north surface for all the stations is shown in Figure 8. It is observed that for most of the stations, highest illuminance occurs during summer month of June, except for New Delhi station, where it is in the month of July. Peak values are found to be in the range of 14–15 klx for New Delhi, Jodhpur, Pune, and Kolkata, whereas for Shillong and Leh stations, it is in the range of 10–12 klx during July, August and September. A decreasing trend of illuminance

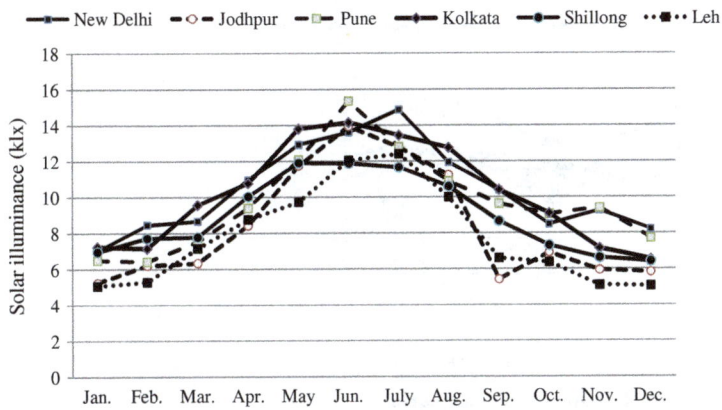
Figure 8. Daily average solar global illuminance on north surface.

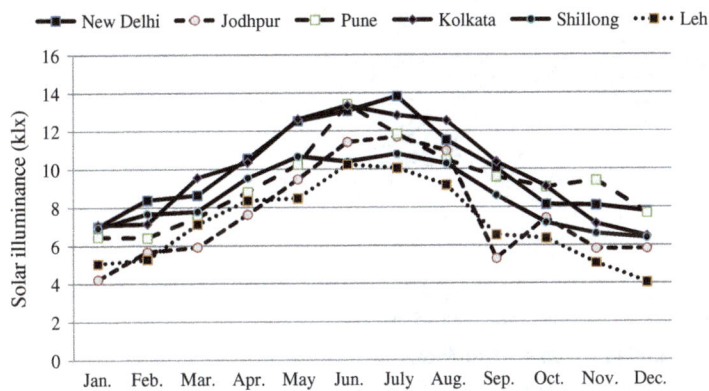
Figure 9. Daily average solar diffuse illuminance on north surface.

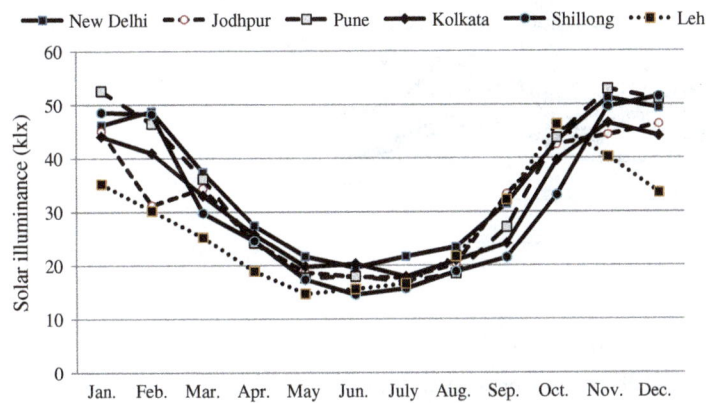
Figure 10. Daily average solar global illuminance on south surface.

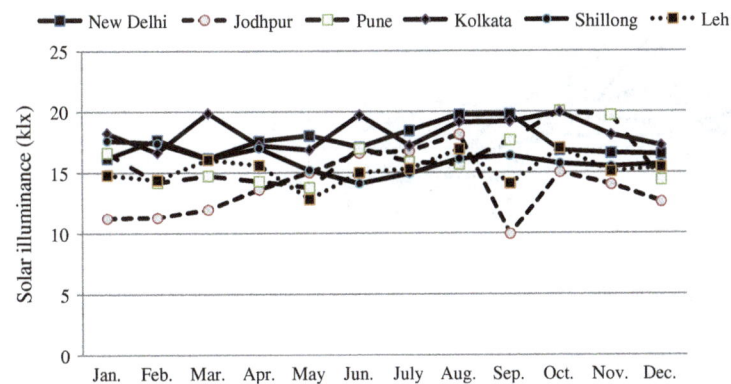
Figure 11. Daily average diffuse illuminance on south surface.

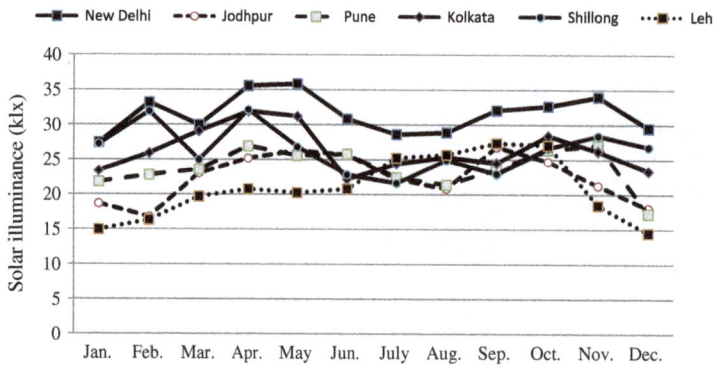

Figure 12. Daily average solar global illuminance on east surface.

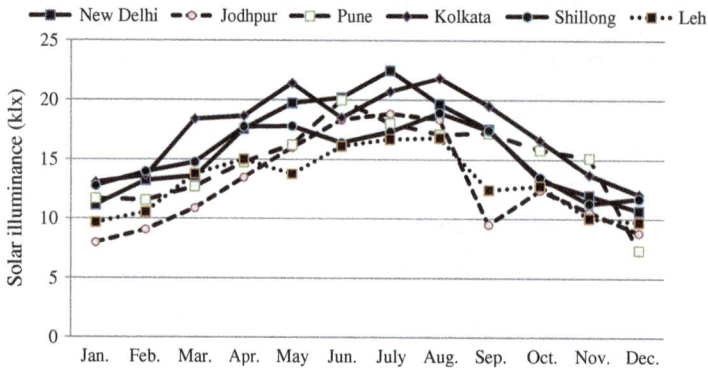

Figure 13. Daily average diffuse illuminance on east surface.

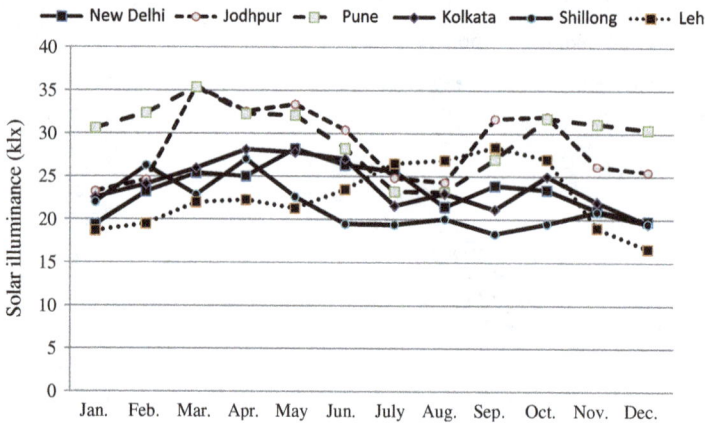

Figure 14. Daily average solar global illuminance on west surface.

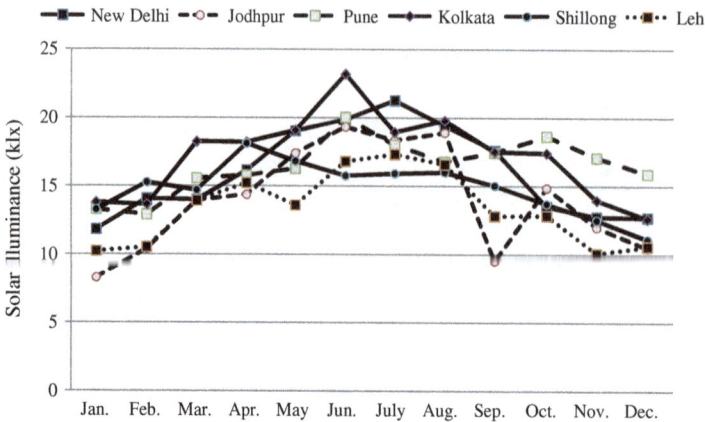

Figure 15. Daily average diffuse illuminance on west surface.

is seen in monsoon season. For most of the stations, daily average solar illuminance follows decreasing trend in monsoon and winter season, and minimum level is found in the range of 6–8 klx in the months of December and January. Daily average solar diffuse illuminance on north surface, as shown in Figure 9, follows almost same trends and levels of daily average global illuminance on this surface, as the north surface at all the stations receive only diffuse illuminance throughout the year.

4.3.2. Daily average illuminance on the south surface
South surface attracts higher attention for energy conservation through daylighting because; it receives highest solar irradiance as well as solar illuminance among all vertical surfaces. Daily average solar global illuminance on south surface for all the stations is shown in Figure 10. It is observed that for most of the stations, highest illuminance occurs during winter month of November and December. In winter, solar incident angle on the south surface is steeper, resulting in higher beam illuminance and hence higher global illuminance on the surface. Higher values of daily average illuminance for all the stations are found to be in the range of 45–55 klx. Peak values of daily average illuminance at each stations are as follows: New Delhi, 49–51 klx in the months of November and December; Jodhpur, 45–46 klx during December and January; Pune, 53 klx in November and January; Kolkata, 44–46 klx in November and January; Shillong, 50–52 klx in November and December; Leh, 40–46 klx in October and November. South surface receives lower daily average illuminance during the summer period due to the larger solar incidence angle that causes lesser beam illuminance, and hence lesser global illuminance on the south surface. Daily average global illuminance in summer season is as follows: 20–27 klx for New Delhi, 18–24 klx for Jodhpur and Pune stations, 20–26 klx for Kolkata station, 15–25 klx for Shillong station, whereas Leh station receives 15–20 klx. Global illuminance shows increasing trend during monsoon period for all the stations.

Figure 11 shows daily average diffuse illuminance on the south surface. It is almost uniform, and there is no drastic variation throughout the year for all the stations. Annual variations of diffuse illuminance for all the stations are: 16–20 klx for New Delhi and Kolkata, 11–18 klx for Jodhpur, 14–20 klx for Pune, Shillong, and Leh.

4.3.3. Daily average illuminance on the east surface
The daily average solar global illuminance on east surface for all the stations is shown in Figure 12. It is observed that for most of the stations, highest illuminance occurs during summer months of April and May. The peak daily average illuminance for all the stations is observed in the range of 20–35 klx (approx.). The peak daily average illuminance for each station is as follows: 31–36 klx for New Delhi, in the months of November and December; 26 klx for Jodhpur and Pune, in April and June; 22–32 klx (approx.) for Kolkata and Shillong, in April and June; 27 klx for Leh, in September and October. During monsoon, it is observed that all the stations except Shillong show decreasing trends of illuminance. Global illuminance during monsoon for each station is as follows: 30 klx for New Delhi; 22–23 klx for Jodhpur, Pune, and Shillong; 25–26 klx for Kolkata and Leh. Global illuminance in winter season (December to February) found to show lower levels except Shillong station, the range of values is: 27–30 klx for New Delhi station; 17–18 klx for Jodhpur and Pune stations; 23 klx for Kolkata. Shillong station receive high range of 27–32 klx, whereas Leh receives 14–15 klx.

Daily average solar diffuse illuminance (Figure 13) is found to attain maximum level at the end of summer season for the following three stations: New Delhi (22 klx), Jodhpur (19 klx) and Pune (20 klx), whereas it attains maximum during monsoon season for the following stations: Kolkata (22 klx), Shillong (19 klx), Leh (17 klx). During winter, Low range of diffuse illuminance (10–13 klx. approx.) is observed for all the stations.

4.3.4. Daily average illuminance on west surface
The daily average solar global illuminance on the west surface for all the stations is shown in Figure 14. For most of the stations, highest illuminance occurs during summer months of April and May. For each station, peak values are found to be in the range of 28–35 klx (approx.) and the details are as follows: 25–28 klx for New Delhi; 30–35 klx for Jodhpur and Pune; 27–28 klx (approx.) for

Kolkata and Shillong; 28 klx for Leh station. During the monsoon period, average illuminance received by this surface is as follows: 24 klx for New Delhi and Pune, 27 klx for Jodhpur, 19–22 klx for Shillong and Kolkata. Daily average global illuminance in winter season attains lower level for most of the stations, the range of values is: 19–23 klx for New Delhi; 23–26 klx for Jodhpur; 30–32 klx for Pune; 20–23 klx for Kolkata and Shillong, whereas Leh station receives 17–19 klx of illuminance. Results show that Jodhpur and Pune receive more global illuminance on the west surface as compared to those levels on the east surface. Whereas, New Delhi, Kolkata Shillong found to receive lower global illuminance on the west surface as compared to the corresponding levels on the east Surface.

Diffuse illuminance (Figure 15) on the west surface is found to attain maximum level at the end of summer season for the following three stations: New Delhi (22 klx), Jodhpur (19 klx) and Pune (20 klx). Whereas, it attains maximum during the monsoon season for the following three stations: Kolkata (22 klx), Shillong (19 klx) and Leh (17 klx). Diffuse illuminance falls to Lower levels (10–13 klx) during the winter season.

Overall quantitative assessment shows that most of the stations under study, receive solar global illuminance during summer months, as follows: 20–35 klx on east surface, 20–30 klx on west surface, 15–25 klx on south surface and 10–15 klx on north surface. During monsoon months, it is observed that south and east surfaces receive high range of global illuminance (20–30 klx) and west surface receives 20–25 klx of daily average global illuminance, whereas north surface receives low range of global illuminance (10–15 klx). During winter months, south surface receives highest daily average illuminance to the tune of 45–50 klx, east and west surfaces receive 20–30 klx, whereas north surfaces receive low range of 5–10 klx of daily average global illuminance.

4.4. Qualitative assessment of daylight on vertical surfaces

Magnitude of solar illuminance and duration of its availability are some of the most critical input parameters required for the design of a glazing in daylight conscious buildings. In this view, qualitative assessment of daylight on vertical surfaces for six Indian stations has been carried out through cumulative frequency of solar illuminance for a TMY. Qualitative assessment of daylight involves evaluation of the duration of occurrence (percentage of annual day time hours) for the availability of certain range of solar illuminance. The qualitative assessment of daylight for four vertical surfaces at six stations of India is presented in subsequent section.

4.4.1. Cumulative frequency of solar illuminance on the north surface

Cumulative frequencies of solar global illuminance and solar diffuse illuminance, on the north surface for all the stations, are shown in Figure 16(a) and (b) respectively. From Figure 16(a), it is observed that hourly global illuminance varies in the range of 0–40 klx, in a TMY for all the stations.

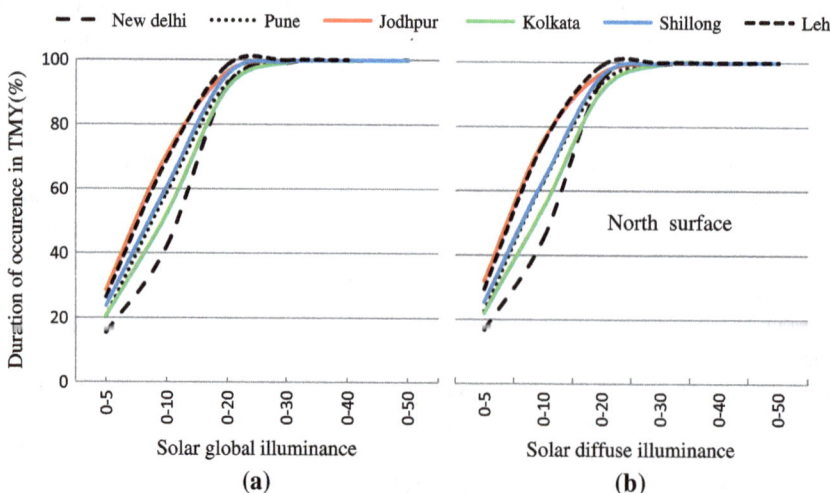

Figure 16. Cumulative frequency of solar illuminance on north surface.

For 68–77% of day time in a year, most of the stations receive solar global illuminance in the range of 5–20 klx. For the solar illuminance of 5–10 klx, annual occurrence time at each station is: 40% for Jodhpur, Pune, Shillong and Leh; 30% for Kolkata, whereas it is 27% for New Delhi. For solar illuminance in the range 10–20 klx, the available time in TMY is: 50% for New Delhi, 40% for Kolkata, 35% for Pune and Shillong, 30% for Leh and 25% for Jodhpur.

4.4.2. Cumulative frequency of solar illuminance on the south surface

Figure 17(a) and (b) shows cumulative frequencies of solar global illuminance and solar diffuse illuminance respectively, on the south surface at all the stations. It is interesting to note that most of the stations, except cold zone stations, receive global illuminance in the range of 5–20 klx for 25% of annual day time, whereas cold zone stations (Leh and Shillong) receive it for 30% of annual day time. For the global illuminance of 20–40 klx, the duration of occurrence is in the range of 25–30% of annual day time for all the stations. The stations with more duration of occurrence, in increasing order are: Jodhpur, Leh, New Delhi, Shillong, Pune and Kolkata. Most of the stations except cold zone station, receive global illuminance in the range 40–60 klx for 18–20% of annual day time, whereas cold zone stations receive it for 15% of annual day time.

Related to solar diffuse illuminance on the south surface, it varies in the range of 0–50 klx for the whole year. For the solar diffuse illuminance of 10–30 klx, the occurrence time is in the range of 55–65% of annual daytime. Stations with more duration of occurrence, in increasing order, within the range of 55–65%, are: Leh, Shillong, Kolkata, Pune, New Delhi and Jodhpur.

4.4.3. Cumulative frequency of solar illuminance on the east surface

Cumulative frequencies of solar global illuminance and solar diffuse illuminance, on the east surface for all the stations, are shown in Figure 18(a) and (b), respectively. For the global illuminance in the range 5–20 klx, annual duration of occurrence at all the stations is: 35% for New Delhi and Kolkata, 50% for Jodhpur and Leh, 45% for Pune and it is 40% for Shillong. Similarly, for the illuminance in the range of 10–30 klx, duration of occurrence is: 35% for New Delhi and Kolkata, 33% for Jodhpur and Shillong and 40% for Leh and Pune. For solar global illuminance in the range 30–60 klx, the annual occurrence time is about 18–28%. Related to solar diffuse illuminance on east surface, it varies in the range of 0–40 klx for the whole year. It is observed that for the diffuse illuminance of 5–20 klx, occurrence time is in the range of 40–60% of annual day time. Stations receiving illuminance in this range, in increasing order are: Kolkata, New Delhi, Shillong, Pune, Jodhpur and Leh.

4.4.4. Cumulative frequency of solar illuminance on the west surface

Cumulative frequencies of solar global illuminance and solar diffuse illuminance, on the west surface for all the stations, are shown in Figure 19(a) and (b), respectively. For the global illuminance in

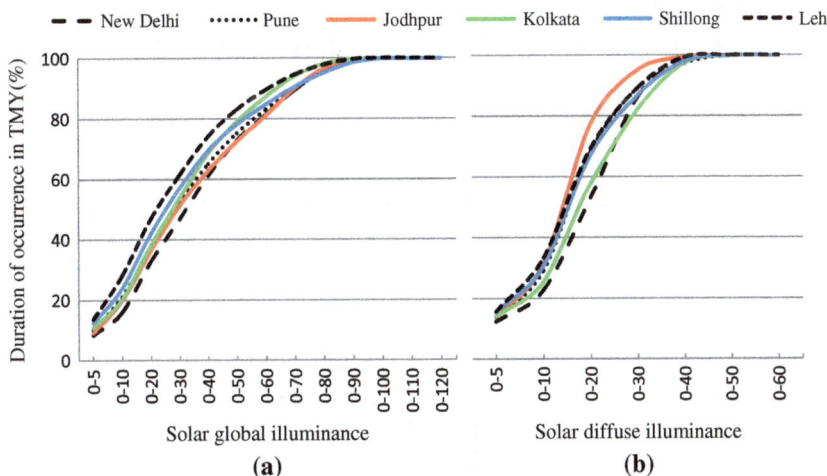

Figure 17. Cumulative frequency of solar illuminance on south surface.

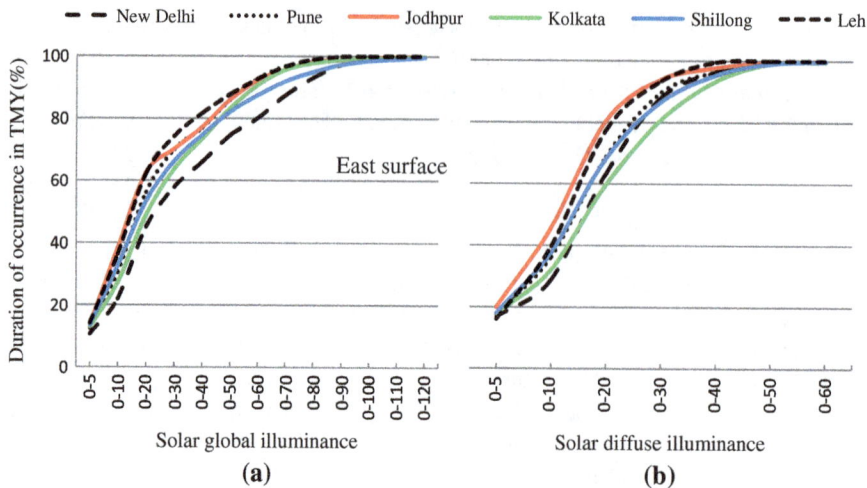

Figure 18. Cumulative frequency of solar illuminance on east surface.

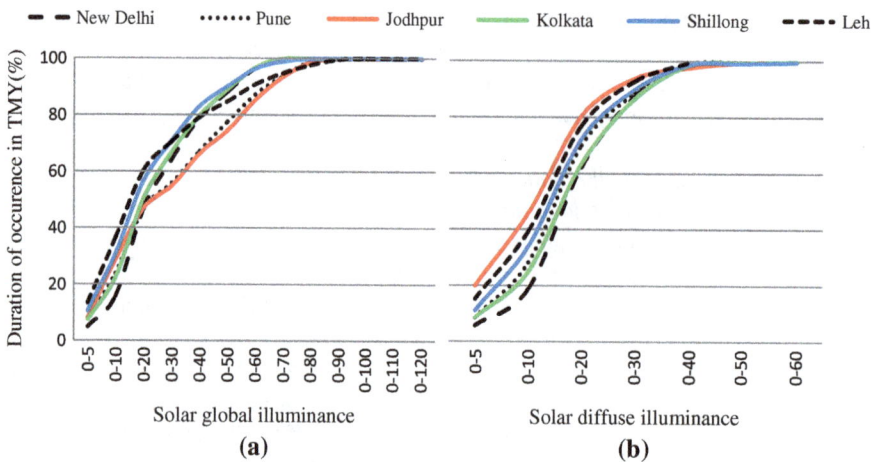

Figure 19. Cumulative frequency of solar illuminance on west surface.

the range of 5–20 klx, duration of occurrence in terms of percentage of TMY is: 43% for New Delhi and Kolkata, 39% for Jodhpur and Pune, 45% for Pune and Shillong and 47% for Leh. For the global illuminance in the range of 10–30 klx, duration of occurrence for the each station is: around 48% for New Delhi and Kolkata; 26% for Jodhpur; 40% for Shillong and Leh; and 33% for Pune. Similarly, for the global illuminance of 30–60 klx, duration of occurrence is in the range of 20–30%. The stations with more duration of occurrence, in increasing order are: Leh, Shillong, Jodhpur, Kolkata, Pune and New Delhi.

Related to solar diffuse illuminance on the west surface, it is found to vary in the range of 0–40 klx for the whole year. Figure 19(b) shows that solar diffuse illuminance, in the range of 5–20 klx, occurs for 55–60% of annual daytime for all the stations. Stations with higher occurrence time are: Kolkata, New Delhi, Shillong, Jodhpur, Pune and Leh. Whereas, for the range of 20–40 klx of illuminance, annual duration of occurrence is: 25% for cold zone stations, 29% for Pune, 35% for New Delhi and Kolkata, whereas it is 17% for Jodhpur.

Overall qualitative assessment shows that for the solar global illuminance of 5–20 klx, annual duration of occurrence is as follows: 70–75% for north surface; 25–30% for south surface; 35–50% for east surface; and 40–45% for west surface. For solar global illuminance of 20–40 klx, details of the annual duration of occurrence are as follows: 3–8% for north surface, 25–30% for south surface, 20–25% for east surface and 20–30% for west surface.

5. Conclusions

Qualitative and quantitative assessment of outdoor daylight is extremely important for the design of daylight systems, so that the diverse benefits of daylight for indoor visual comfort could be reaped and appreciated. Empirical model for solar global illuminance on four cardinal vertical surfaces at New Delhi has been experimentally validated. Model derived global illuminance on vertical surfaces is very close to measured illuminance with MBE of −8.65% for the north surface, 0.30% for the south surface, 3.87% for the east surface and 7.89% for the west surface. The model is further used for computation of daylight for vertical surfaces at different climate specific stations of India. Hourly values of daylight data are analysed qualitatively and daily average values of daylight are assessed quantitatively for all the stations. Approximate peak global illuminance for all the stations is found to be: 14–15 klx for north surface, 45–50 klx for south surface, 25–35 klx for both the east and west surface, whereas lower range of global illuminance is found to be: 5–7 klx for north surface, 15–20 klx for south surface, 15–25 klx for east and west surface. Qualitative analysis of hourly solar illuminance for the north surface shows that most of the stations, receive 5–20 klx of solar global illuminance, for 70–75% of annual daytime.

For global illuminance of 30–60 klx for south surface, annual daytime for all the stations is 35% for New Delhi and Kolkata, 30% for Jodhpur and Pune and 28% for Shillong and Leh. Similarly for global illuminance of 10–30 klx on east surface, the availability is 40% for Pune and Leh, 37% for New Delhi and Kolkata and 34% for Jodhpur and Shillong.

There is an ample scope for research on daylight data generation for vertical surfaces at the station for which solar irradiance data are available but daylight data are hardly available. Most of the stations of developing countries don't have daylight data. There is a need to generate daylight contour map for a country so that daylight potential and its energy conservation potential, CO_2 mitigation, could be quantified. Further, the study can be extended to develop analytical models and software tools for the evaluation of indoor illuminance from various daylighting systems in order to assess the economical and environmental benefits of the daylight in buildings.

Nomenclature

α	altitude of sky element (radian)
α_s	solar altitude angle (radian)
γ	azimuth of sky element (radian)
γ_s	solar azimuth angle (radian)
γ_{sur}	surface azimuth angle (radian)
ρ	ground reflectivity(dimensionless)
θ_T	solar incident angle on vertical surface (radian)
θ_z	solar zenith angle (radian)
I_{dh}	solar diffuse radiation on a horizontal surface (W/m²)
I_{bn}	solar beam radiation at normal incidence (W/m²)
I_{gh}	solar global radiation on a horizontal surface (W/m²)
I_o	extraterrestrial solar radiation at normal incidence (W/m²)
L_{bv}	beam illuminance on vertical surface (klx)
L_{dh}	diffuse illuminance on horizontal surface (klx)
L_{dm}	daily average illuminance (klx)
L_{gh}	global illuminance on horizontal surface (klx)

L_{gr} ground reflected illuminance on vertical surface (klx)

L_{tv} total illuminance on vertical surface (klx)

Acknowledgements

The author, K.N. Patil, gratefully acknowledges Centre for Energy Studies, IIT Delhi and SDM College of Engineering & Technology Dharwad (Karnataka State, India) for sponsorship under the quality improvement programme of the Government of India.

Funding

The authors received no direct funding for this research.

Author details

K.N. Patil[1]

E-mail: kalmeshnp@gmail.com

S.C. Kaushik[1]

E-mail: kaushik@ces.iitd.ac.in

S.N. Garg[1]

E-mail: sngarg@yahoo.com

[1] Centre for Energy Studies, Indian Institute of Technology Delhi, Hauz Khas, New Delhi 110016, India.

References

Bansal, N. K., & Minke, G. (1995). *Climatic zones and rural housing in India.* Julich: Forschungszentrum.

De Rosa, A., Ferraro, V., Kaliakatsos, D., & Marinelli, V. (2008). Calculating diffuse illuminance on vertical surfaces in different sky conditions. *Energy, 33,* 1703–1710. http://dx.doi.org/10.1016/j.energy.2008.05.009

Duffie, J. A., & Beckman, W. A. (2006). *Solar engineering of thermal processes.* New York, NY: Wiley.

Ferraro, V., Mele, M., & Marinelli, V. (2012). Analysis of sky luminance experimental data and comparison with calculation methods. *Energy, 37,* 287–298. http://dx.doi.org/10.1016/j.energy.2011.11.031

Kasten, F., & Young, A. T. (1989). Revised optical air mass tables and approximation formula. *Applied Optics, 28,* 4735–4738. http://dx.doi.org/10.1364/AO.28.004735

Kittler, R., Kocifaj, M., & Darula, S. (2012). *Daylight science and daylighting technology.* New York, NY: Springer. http://dx.doi.org/10.1007/978-1-4419-8816-4

Li, D. H. W., & Lam, J. C. (2003). An analysis of all-sky zenith luminance data for Hong Kong. *Building and Environment, 38,* 739–744. http://dx.doi.org/10.1016/S0360-1323(02)00232-9

Li, D. H. W., Lam, T. N. T., Cheung, K. L., & Tang, H. L. (2008). An analysis of luminous efficacies under the CIE standard skies. *Renewable Energy, 33,* 2357–2365. http://dx.doi.org/10.1016/j.renene.2008.02.004

Lunde, P. J. (1980). *Solar thermal engineering.* New York, NY: Wiley.

Muneer, T. (2004). *Solar radiation and daylight models.* Elsevier: Butterworth-Heinemann.

Patil, K. N., Garg, S. N., & Kaushik, S. C. (2013). *Journal of Renewable Sustainable Energy, 5,* 063120. Retrieved from http://dx.doi.org/10.1063/1.4841195 http://dx.doi.org/10.1063/1.4841195

Perez, R., Ineichen, P., Seals, R., Michalsky, J., & Stewart, R. (1990). Modeling daylight availability and irradiance components from direct and global irradiance. *Solar Energy, 44,* 271–289. http://dx.doi.org/10.1016/0038-092X(90)90055-H

Perez, R., Seals, R., & Michalsky, J. (1993). All-weather model for sky luminance distribution—Preliminary configuration and validation. *Solar Energy, 50,* 235–245. http://dx.doi.org/10.1016/0038-092X(93)90017-I

Robledo, L., & Soler, A. (2003). Estimation of global illuminance on inclined surfaces for clear skies. *Energy Conversion and Management, 44,* 2455–2469. http://dx.doi.org/10.1016/S0196-8904(03)00003-7

World Bank Report. (2008). *Residential consumption of electricity in India. Documentation of data and methodology* (Background paper-India). Strategies for low carbon growth. Retrieved from http://www.moef.nic.in/downloads/public-information/Residentialpowerconsumption.pdf

11

Spatial and temporal drought analysis in the Krishna river basin of Maharashtra, India

Dattatraya R. Mahajan[1]* and Basavanand M. Dodamani[1]

*Corresponding author: Dattatraya R. Mahajan, Department of Applied Mechanics and Hydraulics, National Institute of Technology Karnataka, Surathkal, Mangalore, India

E-mail: drcdmahajan196@gmail.com

Reviewing editor: Giorgio Mannina, Universita degli Studi di Palermo, Italy

Abstract: Droughts can be distinguished by three vital characteristics—spatial coverage, intensity, and duration. The objective of the present study was to study the temporal and spatial variation of drought incidences by using the Standardized Precipitation Index (SPI) and Percent of Normal Precipitation (PNP) at multiple time scales; computed using monthly precipitation data (1960–2012) in the study area of 59 rain gauge stations. Drought climatology based on these drought indices (PNP and SPI) has been studied for finding out their suitability for drought monitoring over Krishna basin in Maharashtra. Study results indicate that at dry and wet conditions, the SPI performs better than PNP in monitoring drought at multiple time scales. However, SPI-1 fails to recognize drought conditions in pre-monsoon and post-monsoon months. The spatially interpolated droughts maps are prepared which displays the variation of drought severity across the study area. The ranking of stations has been done as per drought severity to identify severely drought prone areas.

Subjects: Georisk & Hazards; Water Engineering; Water Science

Keywords: Standardized Precipitation Index (SPI); Percent of Normal Precipitation (PNP); spatial variation; temporal variation; drought climatology

ABOUT THE AUTHOR

Basavanand M. Dodamani, currently an associate professor in the Department of Applied Mechanics and Hydraulics, National Institute of Technology Karnataka, Surathkal, has over 21 years of experience in teaching undergraduate and postgraduate students. He earned his BE Civil degree from B. V. B College of Engineering & Technology, Hubli, Karnataka, MTech (WREM), and PhD from the prestigious National Institute of Technology, Karnataka. His current research interests are Hydraulics, Fluid Mechanics, Water Resources Engineering, Surface Water Hydrology, Ground Water Hydrology, Irrigation and Drainage Engineering, and Design of Hydraulics Structure. He has published numerous journal articles, national and international conference papers. He is a member of several profession bodies like Indian Society for Hydraulics, Indian Society for Technical Education and ACCE, etc.

PUBLIC INTEREST STATEMENT

Drought is one of the major water-related natural hazards. The investigation of spatial and temporal variation of rainfall is of great importance in water resources planning and management as it is related with food security and management of scarce water resource, which becomes critical in case of drought events. The aim of the present study was to study drought climatology over the Krishna basin in Maharashtra, at both local and regional level using two drought indices, namely, Percent of Normal Precipitation (PNP) and Standardized Precipitation Index (SPI). It was found that the SPI performs better than PNP in monitoring drought at multiple time scales. The drought analysis was also effective in managing Kharif, Rabi, and summer crops. The spatially interpolated drought maps, ranking of rain gauge stations will help the water managers/district administrators for taking appropriate measures in drought relief and prioritization of drought mitigation works.

1. Introduction

Drought is acknowledged as a significant natural disaster which leads to food, fodder, and water shortages along with destruction of vital ecological system. Drought is a natural hazard having negative effect on society and environment which is intensified by increasing water demand (Mishra & Singh, 2010). Drought is a phenomenon associated with scarcity of water due to delay in rainy season and/or reduction in "Normal" rainfall. Droughts can be experienced anywhere such as areas having little or high rainfall. It brings misery to large sections of the population and habitat. The drought characteristics vary significantly amid regions such as the Western Europe, North American Great Plains, southern Africa, Australia, and northwestern India (Wilhite & Buchanan-Smith, 2005). Drought is an environmental disaster which has concern with hydrology, environment, ecology, meteorology, geology, and agricultural scientists. The surface as well as ground water resources are affected by drought and can lead to reduced-productivity as well as crop failure, reduction in water supply, declined water quality, reduced power generation, disturbed riparian habitations, suspended entertaining events, and economic as well as social events (Riebsame, Changnon, & Karl, 1991). Demand of water has been increased many folds to cater a growing population in developing countries and for maintaining living standards as well as for recreation in the developed countries. Rise in conflicts for water sharing has been increased amongst urban population (drinking and domestic use), expansion of agriculture in rural sector (livelihood and food security), generation of power (hydropower and thermal power plants), and industrial sectors (processing and cleaning). The conflicts for water sharing between countries, states, regions, and districts had made people enemy of each other. The situation gets worsened in drought period when limited water resources get depleted faster than rejuvenation. Also water scarcity is also added by factors, such as water contamination, variation of rainfall pattern due to climate change, etc.

Indian economy is largely based on agriculture, as approximately 70% of the total population depends on it for their livelihood. Almost 75–90% of the yearly rainfall of India occurs in four months of the rainy season due to the southwest monsoon. Owing to both spatial and temporal abnormalities in monsoon precipitation, droughts frequently occurs in most part of the country. The net sown area in the country is about 140 Mha, out of which the 68% of the area is susceptible to drought and 50% area is classified as severely drought prone. (Kamble, Ghosh, Rajeevan, & Samui, 2010). Knowledge of the probability of occurrence of droughts along with duration, intensity, and spatial extent is critical in the planning as well as management of scarce water resources. To compare spatial and temporal variation of historical droughts with current conditions, drought indices are used and thus provide decision-makers a tool to measure drought events and reduce drought impacts.

The interpretation of raw data in analyzing drought conditions over an area is made simple by a single value of drought Index (Pai, Latha, Guhathakurta, & Hatwar, 2011). The Percent of Normal Precipitation (PNP) and Standardized Precipitation Index (SPI) are widely used worldwide to analyze the spatial and temporal variation of Meteorological drought occurred due to reduction in precipitation over a period of time in a region. To quantify precipitation variation at multiple time scales, the SPI was developed by McKee, Doesken, and Kliest (1993). As per their definition, a drought event at any time scale occurs when the SPI is negative and reaches −1. A drought event terminates when the SPI becomes positive. Drought classification based on SPI is provided in Table 1. After conceptualization, the SPI is used by many researchers to quantify drought in their studies. The main advantage of the SPI is that dry and wet periods can be monitored over a wide range of time scales starting from 1 to 72 months (Edwards & McKee, 1997).

In Asia, the SPI is getting recognition in the study of the magnitude, intensity, and spatial variation of droughts. Pai et al. (2011) examined droughts of various intensities at district as well as national level over India using PNP and SPI. They observed that district-wide drought climatology over India based on PNP is found to be highly influenced by aridity of the region, while that based on SPI is not biased. They also found that for district-wide drought monitoring, SPI is more suitable whereas for the nation-wide drought monitoring, both PNP and SPI are suitable during the southwest monsoon season. Mishra and Desai (2005) analyzed temporal variation of drought using SPI values for

SPI values	Class
Table 1. Weather classification based on SPI	
SPI values	**Class**
>2	Extremely wet
1.5 to 1.99	Very wet
1.0 to 1.49	Moderately wet
−0.99 to 0.99	Near normal
−1 to −1.49	Moderately dry
−1.5 to −1.99	Severely dry
<−2	Extremely dry

multiple time scales in the Kansabati river basin, India by developing quantitative relationships between frequency, drought severity and area. They found out that the local drought in the basin can be assessed by drought severity—area—frequency curves. Xie, Ringler, Zhu, and Waqas (2013) investigated the spatio-temporal variability of drought frequency by calculating SPI for 3-, 6-, and 12-month time scales using gridded precipitation data in Pakistan during 1960–2007. Their PCA of the calculated SPI fields revealed that the drought incidence in Pakistanis is characterized by a large spatial extent of affected area. A 16 years cycle having an intensive drought period lasting for some years is noticed by them. Mishra and Nagarajan (2011) examined drought characteristics using SPI in the Tel river basin in Kalahandi district of Odisha, India. Their study demonstrated that the study area is affected by severe and extreme droughts from time to time.

Drought climatology for Europe was studied by Hughes and Saunders (2002) by SPI at multiple time scales for the period of 1901–1999. Their study has shown that SPI is a simple and effective tool in analyzing European drought. Mihajlović (2006) analyzed the meteorological drought for the period of 2003–2004 at 32 stations in the Pannonian part of Croatia by means of the SPI at multiple time scales of 1, 3, 6, and 12 months. It was also shown that the drought progression from start to its end can be monitored by SPI at multiple time scales.

Many researchers had studied droughts in India using PNP. Ramdas (1950) defined drought when weekly rainfall is 50% of the Normal or less. Using PNP, Appa Rao (1991) categorized the drought prone and chronically drought affected areas. The various statistical features of country-wide drought incidences were examined by Chowdhury, Dandekar, and Raut (1989). A decreasing trend in the drought affected area in India was observed by Sen and Sinha Ray (1997). A detailed study consisting of the variability of drought incidence over districts of Maharashtra was done by Gore and Sinha Ray (2002). The probability of occurrence of drought at subdivision-wide and national-wide scale was done by Sinha Ray and Shewale (2001). Guhathakurta (2003) studied the probability of district-wide droughts in India.

In India, crop seasons are categorized as Kharif season, Rabi season, and hot weather/summer season. The Kharif season commences from June 15 to October 14. Paddy and groundnut are examples of Kharif crops. The Kharif season coincides with the rainfall season due to the southwest monsoon which occurs in June–September. About 80% of annual rainfall occurs during southwest monsoon, around 15% in northeast post-monsoon (October to December) and about 5% pre-monsoon (January to May). The Rabi season commences from October 15 to February 28/29 and coincides with post-monsoon season. Wheat and Gram are examples of Rabi crops. One or two intermittent rains are sufficient for Rabi crops. They survive on the soil moisture retained at the end of Kharif season and augmented by the ground water. Also the low temperature in the winter season causes very less evaporation losses.

It is very difficult to develop a single definition of a drought which is having diverse geographical and temporal distribution and also affecting the wide variety of sectors (Heim, 2002). The aim of this

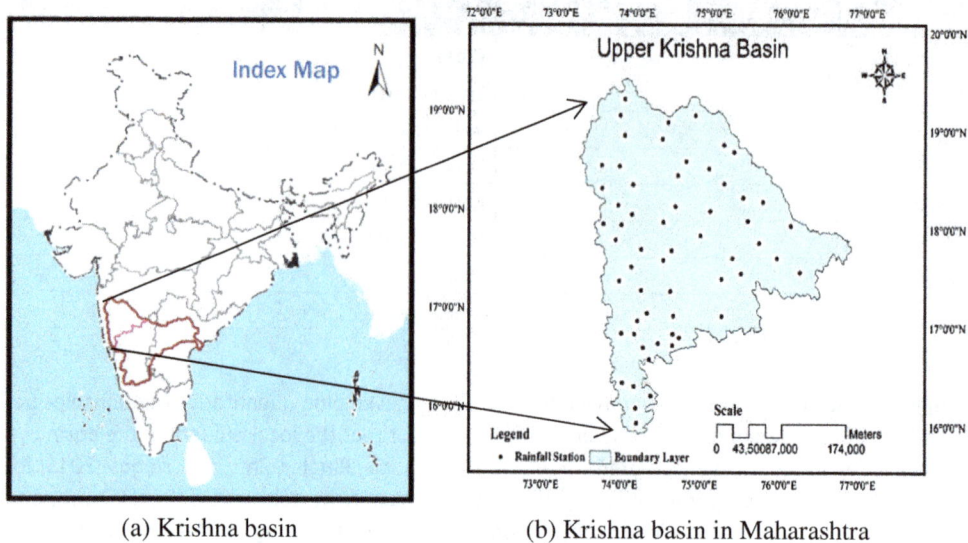

(a) Krishna basin   (b) Krishna basin in Maharashtra

Figure 1. (a) Krishna basin. (b) Krishna basin in Maharashtra.

study was to investigate the Meteorological drought, its spatial and temporal variation by the SPI and PNP at multiple time scales over the Krishna basin in Maharashtra. Also, an attempt is made to find out which of these two indices is better for drought monitoring over the study area. This paper evaluates the SPI on 1-, 3-, 6-, 9-, 12-, 18-, 24-, 36-, and 48-month time scales computed using long time series (1960–2012) of monthly precipitation data at 59 stations in the study area. PNP is analyzed using Annual precipitation and Water-Year precipitation of the same time series.

2. Study area and data used

A part of the Krishna basin which comes under administrative boundary of the state of Maharashtra, is selected to analyze the spatio-temporal variation of Meteorological drought. The Krishna river basin comes under semi-arid southern region. It ranks fourth considering annual discharge, and fifth largest basin in terms of surface area. The basin lies between 73°17′ to 81°9′ east longitudes and 13°10′ to 19°22′ north latitudes. It extends 701 km in terms of length and 672 km in width. In the Krishna river basin, predominant land is used for agriculture where the annual Normal rainfall (1969–2004) is 859 mm. Krishna basin occupies eight percent of Indian Territory with surface area of about 258,948 km². The basin area is shared by three states of erstwhile Andhra Pradesh (75,948 km²_ 30%), Karnataka (111,650 km²_ 44%), and Maharashtra (69,425 km²_ 26%). Out of total river length of about 1,400 km, Maharashtra shares 306 km while the remaining river flows through Karnataka (483 km) and at last Andhra Pradesh (306 km) (The Krishna basin report, India-WRIS, 2014). The study area under research is shown in Figure 1.

The monthly rainfall data is compiled from daily rainfall data of 59 rainfall stations in the Krishna basin in Maharashtra from 1960 to 2012 is used for the analysis. The mean rainfall varies from 5610 mm at Mahabaleshwar situated in Western Ghats of Maharashtra where the river originates to 498 mm at Dhond in the plateau.

3. Methodology

3.1. Steps in calculating SPI

The SPI was developed by McKee et al. (1993) and McKee Doesken, and Kleist (1995) to compute precipitation deficits on multiple time scales. The brief procedure to calculate SPI is as follows.

(1) The probability density function (PDF) is determined to describe the long-term time series of precipitation observations.

(2) The cumulative probability of an observed precipitation amount is computed.

(3) The inverse normal (Gaussian) function, with mean zero and variance one, is then applied to the cumulative probability resulting in the SPI.

The detailed procedure is referred from McKee et al. (1993-1995).

3.2. Steps in calculating PNP

PNP is simple form of determining amount of precipitation at any region or location. It is very effective when used for rainfall season for climatologically similar region. It is ratio of actual precipitation to Normal precipitation determined in percentage. The Normal precipitation is generally considered to be a 30-year mean. PNP can be calculated for a day, week, month, annual year, water year, and rainfall season. When calculated for annual year monthly precipitation totals from January to December is considered, while for water year monthly precipitation totals from June to May is considered. Table 2 denotes the weather classification based on PNP.

$$\text{PNP } \% = P/P_{\text{Normal}} \times 100$$

where P_{Normal} = Long Period Average (LPA). It is the mean rainfall value of particular time scale over a specified period.

Definition of a drought is region specific. The U.S. Weather Bureau identified drought when the rainfall of 21 days or above is less than 30% of the Normal value (Henry, 1906). In this study, we have considered the Meteorological drought which is defined by the India Meteorological Department as the seasonal rainfall deficiency exceeding 25% of the long-term average value of the rainfall. A local drought is considered to be "moderate" if the rainfall deficiency at that place is in between 26 and 50% of the Normal value and "severe" if it is more than 50% of the Normal value.

4. Results and discussion

In temporal analysis, temporal variation of droughts by SPI and PNP drought indices, weather classification by SPI at multiple time scales, and month wise variation of weather at SPI-1 time scale is done.

4.1. Temporal variation of droughts over the study area

PNP analysis is done under two categories namely water year and annual year. In water year, monthly precipitation totals from June to May considered, while for annual year, monthly precipitation totals from January to December is considered. SPI analysis is done using SPI-12 time scale for the month of May and December. SPI-12 of May resembles to water year (June-May) and SPI-12 of December resembles to annual year (January–December) of PNP index.

For local drought analysis under PNP index, district wise classification is used and for regional drought analysis, nation wise classification is used as per Table 2. Three severe droughts viz. 1972, 2003, and 2012 which are common as per PNP and SPI were compared by using Pearson Correlation co-eff. Local droughts as per rain gauge stations in these three drought years were compared using PNP and SPI index.

Table 2. Weather classification based on PNP		
Season rainfall classification	**Local drought**	**Regional drought**
No drought	>75% of LPA	0–10% of area is under local drought
Mild meteorological drought	–	10–20% of area is under local drought
Moderate meteorological drought	50–74% of LPA	21–40% of area is under local drought
Severe Meteorological drought	<50% of LPA	>40% of the area is under local drought

From Table 3 it is observed that (i) Pearson Correlation co-eff. between SPI-12 and PNP is nearly equal to 1. (ii) SPI has three classes (moderate, severe, and extreme) of drought as compared to two (moderate and severe) classes of PNP. (iii) Number of moderate drought incidences as per PNP is more than that of SPI-12. (iv) Sum of severe and extreme droughts as per SPI is more than severe drought as per PNP. (v) Number of rainfall stations under severe drought class as per annual year is more than water year.

4.2. Weather classification at multiple time scales of SPI

The weather classification by SPI is classified as per Table 1 and the results are tabulated in Table 4. Since the number of months available in each time scales differ from each other, all values are converted into percentage for easy comparison and displayed in Figure 2. It is observed from Figure 2 that there is decreasing trend in Wet class while increasing trend for Dry class from SPI-1 to SPI-12. SPI-1 shows most wet conditions. Dry conditions are more prominently shown by SPI-48. SPI-3 shows more percentage of normal rainfall than any other time scales.

4.3. Month wise variation of weather at SPI-1 time scale

Month wise weather variation is done by SPI-1 time scale, which resembles to average monthly rainfall over the study period and gives fair picture of monthly rainfall pattern.

From Table 5 it is observed that even though rainfall is very scanty over January–March, large positive SPI values are generated. Drought conditions are not observed in January–April and November–December. May has only one incidence of moderately dry condition. June–October has nearly equal numbers of Wet and Dry conditions.

Table 3. Local drought classification as per PNP and SPI-12

Drought classification	1972 SPI December	PNP 1972 annual	2003 SPI December	PNP 2003 annual	2012 SPI December	PNP 2012 annual	1973 SPI May	1973 PNP water year	2004 SPI May	2004 PNP water year
Moderate	7	21	11	32	17	28	6	21	14	35
Severe	14	34	16	21	15	12	12	33	26	17
Extreme	34	–	24	–	3	–	36	–	10	–
Total	55	55	51	53	35	40	54	54	50	52
Pearson co-eff.	0.92		0.89		0.94		0.91		0.87	

Table 4. Weather classification based on multiple time scales of SPI

SPI series	2.00 or more	1.50 to 1.99	1.00 to 1.49	−0.99 to 0.99	−1.00 to −1.49	−1.50 to −1.99	−2.00 or less	Total (months)
	Extremely wet	Very wet	Moderately wet	Near normal	Moderately dry	Severely dry	Extremely dry	
SPI-1	22	52	131	387	26	12	7	636
SPI-3	15	27	62	469	35	17	10	634
SPI-6	15	26	55	446	52	23	14	631
SPI-9	17	24	51	441	54	25	17	628
SPI-12	18	23	49	439	55	23	17	625
SPI-18	17	25	48	433	55	23	17	619
SPI-24	17	26	47	427	56	26	14	613
SPI-36	18	23	48	416	55	29	12	601
SPI-48	16	25	50	398	60	27	13	589

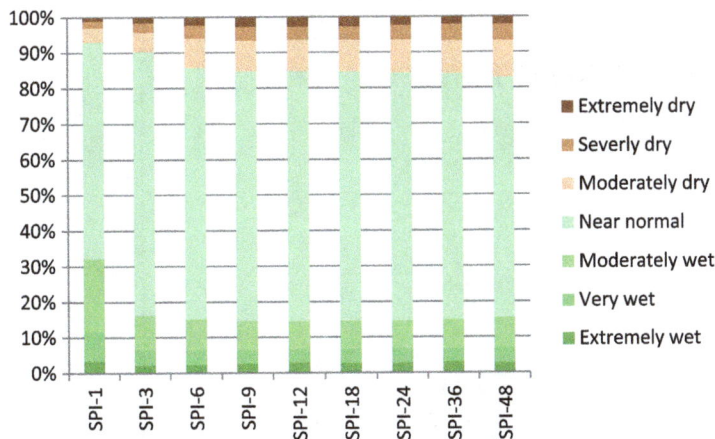

Figure 2. Weather classification on percentage basis.

SPI values	Class	January	February	March	April	May	June	July	August	September	October	November	December
Table 5. Month wise variation of weather on multiple time scales of SPI-1													
>2.00	Extremely wet	1	8	2	1	2	1	1	1	1	1	1	1
1.50 to 1.99	Very wet	15	14	4	3	2	2	2	2	2	2	2	2
1.00 to 1.49	Moder-ately wet	32	26	20	5	4	5	5	6	5	5	4	11
−0.99 to 0.99	Near Normal	5	5	27	43	44	37	36	35	36	36	44	39
−1.00 to −1.49	Moder-ately dry	0	0	0	0	1	4	5	5	4	7	0	0
−1.50 to −1.99	Severely dry	0	0	0	0	0	2	2	3	3	2	0	0
<−2.00	Extremely dry	0	0	0	0	0	1	2	1	1	1	0	0

4.4. Spatial variation of drought severity over the study area

The study area is bifurcated using Theissen Polygon tool in Arc GIS 10.2, into 59 polygons corresponding to 59 rainfall stations. Each of the polygons represents the influential area of the rainfall station as per its location in the study area measured in sq. km. and is expressed as fraction of the study area. The rainfall stations closely spaced will be assigned less area and vice versa (Figure 3). Even though some stations may designate drought conditions, a regional drought is acknowledged only when some major portion of the total study area is under drought.

Regional weather is classified as per Table 1 for SPI and Table 2 for PNP to all 59 rain gauge stations in the study area from 1960 to 2012. Area wise regional weather classification of every year is performed and the results are tabulated in Table 6.

From Table 6 it is observed that number of severe drought incidences in PNP analysis increases as compared to SPI. Number of Moderate drought incidences as per PNP water year is more than SPI-May. Number of Mild drought incidences as per SPI-May is more than PNP water year while it is nearly same for SPI-December and PNP-Yearly. The area wise weather classification for PNP-Water year is performed for all Severe and Moderate drought years. Sr. No. 1–7 are Severe droughts while Sr. No. 8–19 are Moderate droughts. The results are tabulated in Table 7 and displayed in Figure 4.

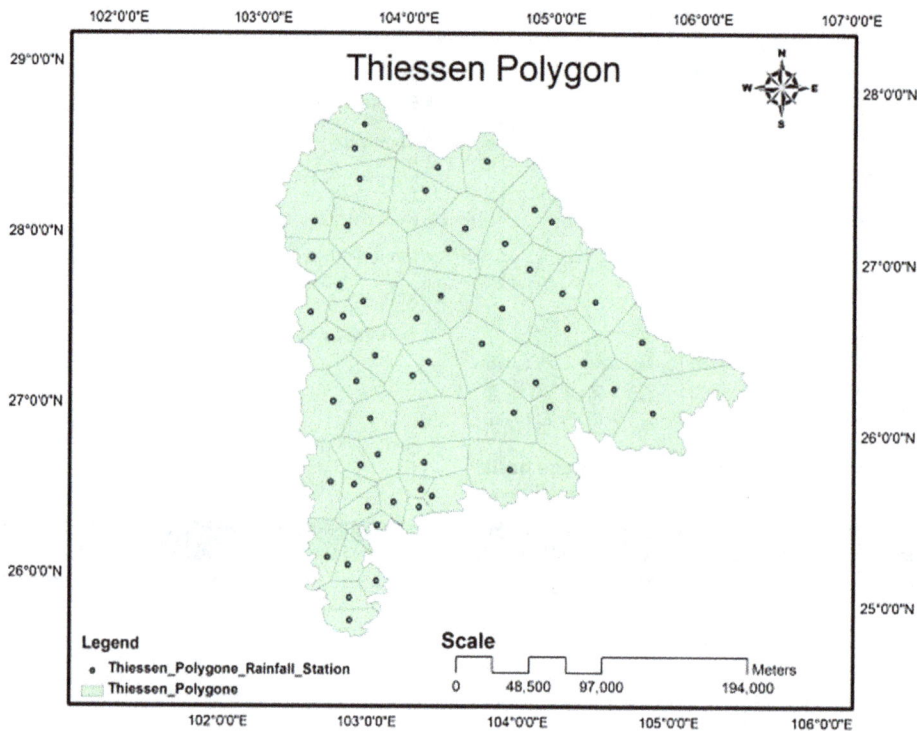

Figure 3. Bifurcation of the study area as per area of influence of rain gauge stations.

Table 6. Area wise regional drought classification as per PNP and SPI				
Drought classification	**PNP-water year**	**SPI-May**	**PNP-yearly**	**SPI-December**
Severe (>40%)	7	4	9	3
Moderate (21–40%)	12	7	12	13
Mild (10–20%)	10	14	11	10
Non Drought (<10%)	23	27	21	27
Total study years	52	52	53	53

From Table 7 it is observed that drought year 1972–1973 is the severest of all with 51% of study area under severe drought and 39.4% of moderate drought followed by 2003–2004 with 23.4% of study area under severe drought and 65.8% of moderate drought.

4.5. Drought analysis for managing the Kharif, Rabi, and summer/hot season crops

Table 8 shows the classification of rainfall stations by drought severity as per SPI-12 (December) and PNP (Yearly), similarly Table 9 shows the classification of rainfall stations by drought severity as per SPI-12 (May) and PNP (Water-Year) for various drought years. From Tables 8 and 9 it is observed that in comparison with SPI-12 analysis, PNP analysis shows more number of stations under drought. It is observed that even though the average PNP is greater than 100% for the year 1994–1995; the area is witnessed by Moderate drought as per regional drought classification shown in Table 7.

Drought analysis by SPI-12 (December) and PNP-(Water-Year) designates the water stress condition for January–December and June–May, respectively. By comparing average SPI-12 and PNP from Tables 8 and 9, it is observed that the drought severity is increased for all Water Years at par with antecedent Annual year except 1985–1986, 1986–1987,1992–1993, 2003–2004, and 2001–2002 where the drought severity is reduced. This is due to non-seasonal rainfall from January to May. The increase in drought severity for water year hampers the Summer/Hot season crops which are water stressed due to failure of monsoon and vice versa. Thus SPI-12 (May) and PNP (Water Year) in

Sr. No.	Drought year	Excess	Normal	Deficient	Moderate drought	Severe drought
		>110%	91–110%	75–90%	50–74%	<50%
1	1971–1972	1.5	20.1	34.4	37	7
2	1972–1973	0	3.8	6.1	39.1	51
3	1982–1983	5.3	19.8	27.5	45.1	2.3
4	1985–1986	1	4.4	50.1	39.4	5.1
5	2000–2001	8.9	26.5	23.1	41	0.5
6	2002–2003	1.3	9.8	33.1	47.2	8.6
7	2003–2004	0	4.6	6.2	65.8	23.4
8	1966–1967	16.6	24.3	36.4	22	0.7
9	1968–1969	4.6	34	33.1	23.3	5
10	1978–1979	17.1	26.4	32.1	22.6	1.8
11	1984–1985	7.9	30.3	35.5	26.3	0
12	1986–1987	4.7	20.8	44.8	28.8	0.9
13	1987–1988	26.1	21.6	13.5	32.3	6.5
14	1992–1993	3.4	37.2	34.2	25.2	0
15	1994–1995	53.3	14.2	5.4	20.1	7
16	1995–1996	10.6	21.4	37.5	29.6	0.9
17	2001–2002	9.5	22.4	32.1	36	0
18	2008–2009	11.9	34	30.1	24	0
19	2011–2012	14.6	27	23	30	5.4

Table 7. Area wise weather classification for PNP-water year for severe and moderate drought years

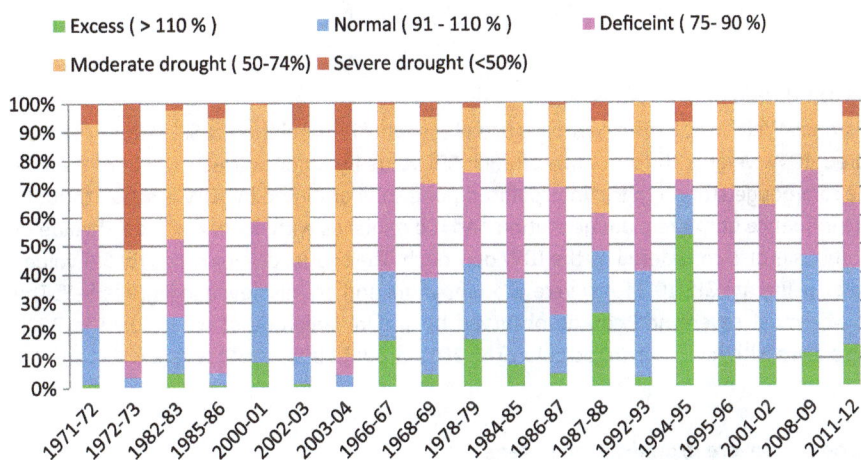

Figure 4. Weather classification for PNP-water year for severe and moderate drought years.

comparison with SPI-12 (December) and PNP (Yearly), helps to analyze the state of Summer/Hot season crops. Drought analysis by SPI-12(December) and PNP-Yearly designates the Water Balance due to southwest monsoon and northeast post-monsoon rains which terminates by the end of December. The rainfall stations under Moderate droughts (PNP: 50-74% and SPI-12: −1.00 to −1.49) for all drought years is witnessed by failure of Rabi crops with reduced productivity of Kharif crops. The rainfall stations under Severe droughts (PNP: <50%) and Severe/Extreme dry conditions (SPI-12: <−1.49) for all drought years is witnessed by failure of both Kharif as well as Rabi crops.

Table 8. Drought classification of rainfall stations as per SPI-12 (December) and PNP (yearly)

Year	SPI-12 (December)					PNP (yearly)			
	Extremely dry	Severely dry	Moderately dry	Total	Average SPI-12	Severe drought	Moderate drought	Total	Average PNP (%)
	−2.00 or less	−1.50 to −1.99	−1.00 to −1.49			<50%	50–74%		
1971	1	1	11	13	−0.50	1	15	16	83.85
1972	34	14	7	55	−2.00	34	21	55	49.48
1982	0	3	12	15	−0.49	0	20	20	84.27
1985	1	5	13	19	−0.78	3	20	23	76.19
2000	0	5	12	17	−0.52	1	18	19	84.79
2002	2	6	13	21	−0.79	3	25	28	76.96
2003	25	15	11	51	−1.82	21	32	53	52.77
2012	3	15	17	35	−1.12	12	28	40	67.19
1966	0	2	4	6	0.04	1	9	10	99.68
1968	2	4	6	12	−0.46	2	16	18	85.8
1978	0	1	2	3	0.12	0	5	21	102.38
1984	0	2	4	6	−0.30	0	11	11	89.21
1986	1	5	13	19	−0.76	2	28	30	75.95
1987	7	1	10	18	−0.27	2	15	17	96.86
1992	0	3	9	12	−0.54	0	24	24	81.83
1994	0	3	13	16	0.22	1	19	20	103.55
1995	0	0	8	8	−0.17	0	8	30	94.07
2001	0	2	16	18	−0.55	0	21	21	83.67
2008	0	0	6	6	−0.15	0	11	26	93.31
2011	0	4	9	13	−0.41	3	17	20	84.29

4.6. Spatial interpolation of SPI values

The Inverse Distance Weighting (IDW) method is used for spatial interpolation of SPI Values over the entire study area. IDW weights the influence of each SPI value by a normalized inverse of the distance from the rain gauge station to the interpolated point. It is assumed by IDW method that SPI value has more influence at the rain gauge station and it diminishes with distance. The SPI value of neighboring stations is also considered by the IDW approach. The spatial display maps for SPI values were prepared by using Arc GIS 10.2 for severe drought years and are displayed in Figures 5–7. The 1972 drought was most severe and can be observed by spatial coverage of red color. The 2003 drought is little severe with reduction in red color. The 2012 drought is very less severe with very few red spots.

4.7. Identification of severe drought prone areas

As per McKee et al. (1993), SPI greater than −2 is classified as Severely Dry. The summation of drought months having SPI value >−2 for all time scales is performed. The ratio of sum for each station to the highest value from all stations is multiplied by 10. This method will assign rank from 1 to 10. Rank 1 will be the least severe drought prone area while rank 10 will be the highest severe drought prone area as per Table 10. From Table 10 it is observed that Gadhingalaj station is least affected by droughts whereas Paud station is most affected by droughts.

5. Conclusion

Drought climatology over Krishna basin in Maharashtra is examined at both local and regional level using PNP and SPI. The PNP and SPI analysis was helpful to identify local and regional droughts.

Table 9. Drought classification of rainfall stations as per SPI-12 (May) and PNP (water-year)								
Year	SPI-12 (May)					PNP (water-year)		
	Extremely dry	Severely dry	Moderately dry	Total	Average SPI-12	Severe drought	Moderate drought	Total	Average PNP
	−2.00 or less	−1.50 to −1.99	−1.00 to −1.49			<50%	50–74%		(%)
1971–1972	3	2	13	18	−0.72	3	22	25	78.11
1972–1973	36	12	6	54	−2.03	33	21	54	49.29
1982–1983	1	8	17	26	−0.80	1	31	32	76.58
1985–1986	2	9	11	22	−0.91	4	23	27	73.09
2000–2001	1	3	16	20	−0.56	1	23	24	83.87
2002–2003	4	6	17	27	−0.92	4	28	32	74.08
2003–2004	10	26	14	50	−1.56	17	35	52	58.08
1966–1967	0	3	8	11	−0.32	1	16	17	88.89
1968–1969	3	4	9	16	−0.56	3	15	18	83.62
1978–1979	0	3	7	10	−0.26	2	12	14	90.96
1984–1985	0	1	7	8	−0.30	0	13	13	90.14
1986–1987	0	3	8	11	−0.56	1	19	20	81.68
1987–1988	7	5	8	20	−0.37	4	16	20	94.28
1992–1993	0	1	8	9	−0.45	0	16	16	84.78
1994–1995	0	5	11	16	0.21	4	15	19	103.65
1995–1996	0	4	5	9	−0.43	1	14	15	86.96
2001–2002	0	2	13	15	−0.47	0	20	20	86.44
2008–2009	0	0	10	10	−0.26	0	13	13	90.82
2011–2012	1	5	10	16	−0.49	3	19	22	82.77

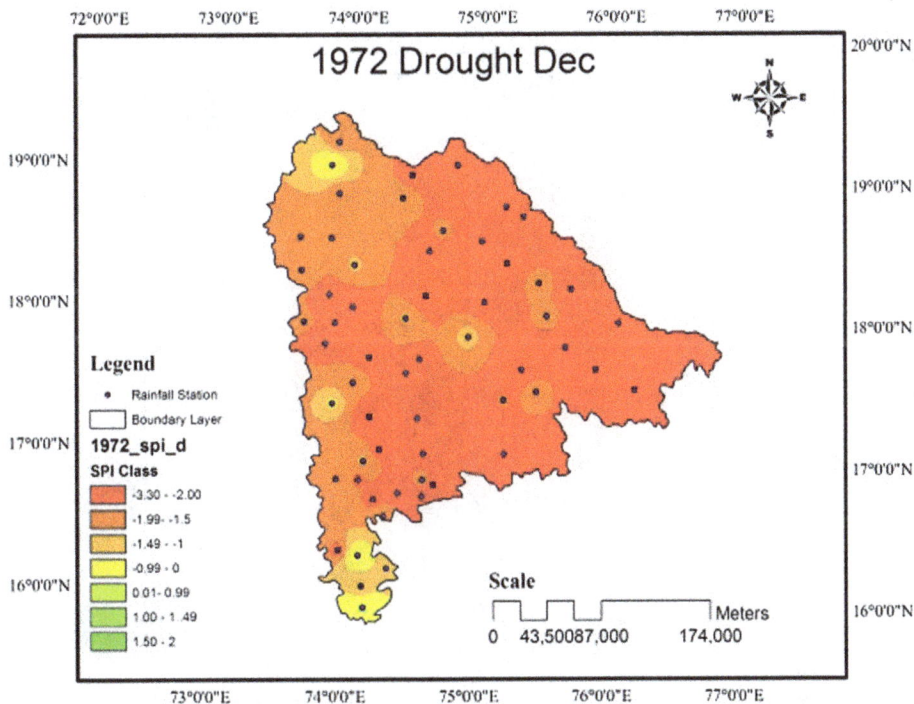

Figure 5. Spatial variation of 1972 drought.

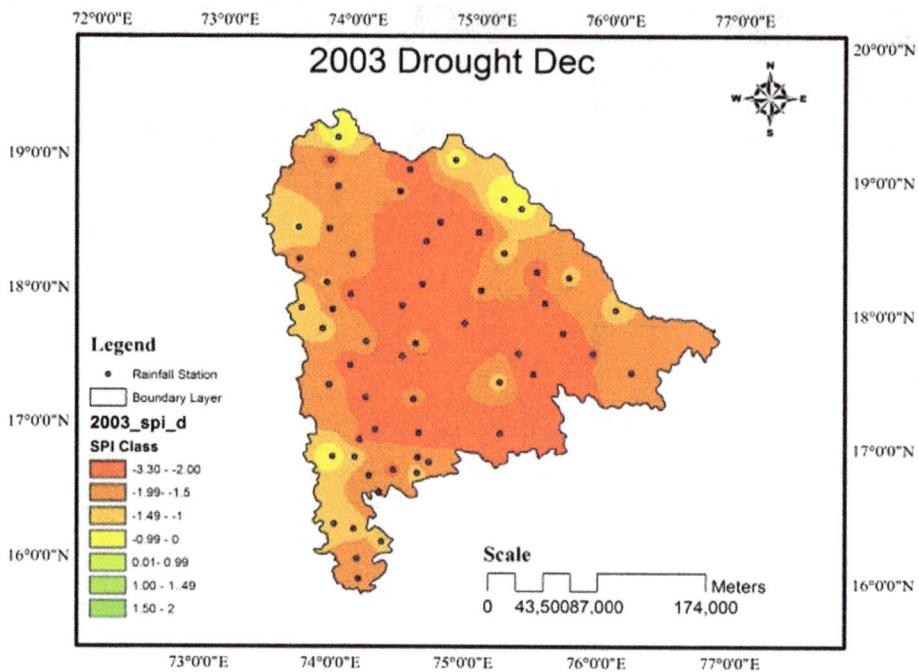

Figure 6. Spatial variation of 2003 drought.

However, there were significant differences in analysis of local and regional droughts. Since SPI is a normalized Index, the variability of SPI is proportional to that of the precipitation variance. SPI performs better than PNP in monitoring drought at multiple time scales. Drought severity of climatologically different locations or regions can be compared by SPI. From Table 4 it can be concluded that SPI-1 fails to recognize drought conditions in pre-monsoon and post-monsoon months. SPI-1 shows exacerbated picture of very wet conditions in January–March, even though rainfall is almost nil during these months. SPI-1 time scale evaluates wet or dry conditions only from June to October i.e.

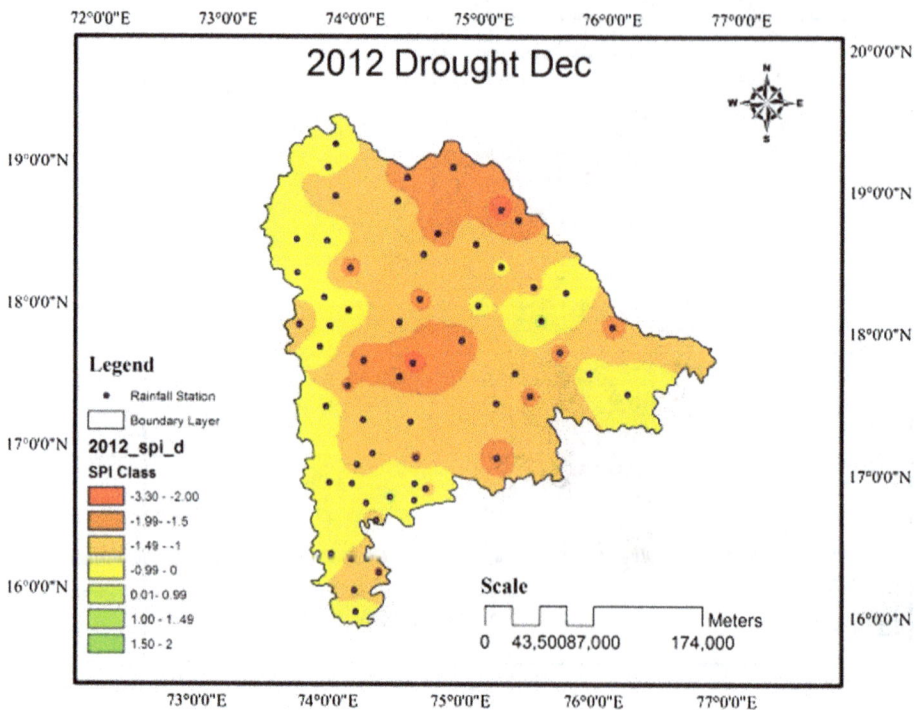

Figure 7. Spatial variation of 2012 drought.

Table 10. Ranking of rain gauge stations as per Local droughts at multiple time scales										
Station name	SPI-1	SPI-3	SPI-6	SPI-9	SPI-12	SPI-18	SPI-24	SPI-36	SPI-48	Total	Rank
Mahabaleshwar	6	12	13	12	14	15	11	0	0	83	3
Radhanagari	8	6	9	11	15	15	13	1	0	78	3
Chandgad	10	9	14	14	11	28	27	39	39	191	7
Velhe	7	5	4	4	1	5	8	12	14	60	2
Ajra	9	11	12	13	12	8	10	14	24	113	4
Shahuwadi	7	10	12	15	13	15	14	12	13	111	4
Patan	8	10	14	14	14	14	12	12	24	122	4
Paud	12	18	23	25	22	33	38	47	58	276	10
Panhala	10	7	9	9	11	7	2	0	0	55	2
Medha	5	9	11	14	15	14	12	1	10	91	3
Gargoti	10	10	12	13	15	18	21	31	29	159	6
Bhor	8	9	17	22	22	22	20	12	5	137	5
Kolhapur	8	10	16	14	10	14	8	17	21	118	4
Shirala	9	11	16	17	21	18	23	31	44	190	7
Gadhinglaj	7	8	7	7	0	0	0	1	2	32	1
Satara	6	13	13	14	20	20	18	10	14	128	5
Wai	4	10	14	17	19	23	20	15	24	146	5
Kagal	9	6	12	14	17	12	1	13	15	99	4
Ambegaon	3	10	11	14	9	8	7	11	2	75	3
Pune	6	8	10	8	7	5	0	0	0	44	2
Koregaon	5	8	12	17	13	14	4	1	1	75	3
Tuljapur	6	11	15	16	14	13	16	14	12	117	4
Karad	5	10	17	27	30	24	21	17	11	162	6
Hatkanangale	9	12	14	20	19	15	7	0	0	96	3
Islampur	10	11	13	19	20	21	19	17	22	152	6
Solapur	6	10	16	18	19	23	18	1	0	111	4
Jamkhed	4	8	18	17	21	16	12	11	10	117	4
Akalkot	8	12	15	14	12	15	18	10	1	105	4
Junnar	7	12	19	26	29	19	12	0	0	124	4
Khed	7	10	13	13	11	6	0	1	8	69	3
Barshi	3	9	18	24	31	30	28	23	23	189	7
Asti	5	12	22	30	28	23	24	12	7	163	6
Pandharpur	7	6	9	13	16	17	10	8	0	86	3
Sangola	6	9	10	12	9	6	3	0	0	55	2
Sangli	8	6	9	10	9	3	0	0	0	45	2
Shirol	8	7	13	13	11	16	13	22	14	117	4
Vita	5	8	15	16	12	17	10	13	24	120	4
Ahmednagar	4	8	18	17	21	16	12	11	10	117	4
Tasgaon	9	13	19	23	25	27	28	27	27	198	7
Madha	5	9	10	15	14	19	13	2	9	96	3
Miraj	9	8	10	12	12	14	2	4	4	75	3
Parenda	6	6	9	18	20	34	35	23	15	166	6
Saswad	4	7	13	16	19	9	2	0	0	70	3
Jath	11	10	13	14	11	30	34	34	46	203	7

(Continued)

Table 10. (Continued)

Station name	SPI-1	SPI-3	SPI-6	SPI-9	SPI-12	SPI-18	SPI-24	SPI-36	SPI-48	Total	Rank
Mohol	4	9	17	27	27	26	26	10	5	151	5
Karmala	5	9	19	28	31	22	13	13	1	141	5
Mangalvedha	6	12	14	17	16	25	35	19	28	172	6
Dahiwadi Man	5	9	12	15	11	17	11	10	11	101	4
Karjat	7	9	21	26	27	21	21	22	8	162	6
Malshiras	7	9	14	19	17	10	8	0	9	93	3
Indapur	4	8	16	17	9	2	0	0	0	56	2
Khandala	7	7	16	23	25	23	30	25	13	169	6
Parner	3	15	24	31	36	29	25	14	22	199	7
Vaduj	5	11	15	15	17	15	10	11	20	119	4
Shrigonda	7	17	20	22	17	21	18	19	10	151	5
Baramati	6	10	18	24	17	16	17	13	2	123	4
Phaltan	5	7	11	12	12	10	11	15	9	92	3
Shirur	10	13	19	18	12	10	9	20	29	140	5
Dhond	5	11	20	26	24	16	11	13	11	137	5

S-W monsoon period. From frequency distributions of weather conditions from Table 3, it can be concluded that SPI-1 values are most positively skewed while SPI-48 values are negatively skewed. SPI-9, SPI-12, and SPI-18 values have Normal distribution. Sum of severe and extreme droughts as per SPI is much more than severe drought as per PNP, while number of moderate droughts are more in PNP. Water year drought analysis will have more influence in managing summer crops, since we can assess the availability of soil moisture at the end of water year; similarly Yearly drought analysis will have more influence on managing Kharif and Rabi crops. Number of rainfall stations under severe drought class as per Annual rainfall is more than Water Year. The Theissen polygon method to assign influence area to each rain gauge station seems to be appropriate for regional drought analysis. The spatially interpolated drought maps will help water managers and district administrators for taking measures in drought relief and contingency planning. The ranking of rain gauge stations will help the water managers and administrators to allocate funds and deciding priority for drought mitigation works.

Acknowledgments
The authors are grateful to the Director (National Data Centre), Office of ADGM(R), India Meteorological Department, Shivajinagar, Pune 411005, for providing the daily rainfall data in the study area.

Funding
The authors received no direct funding for this research.

Author details
Dattatraya R. Mahajan[1]
E-mail: drcdmahajan196@gmail.com
ORCID ID: http://orcid.org/0000-0001-5450-9804
Basavanand M. Dodamani[1]
E-mails: bm.dodamani@gmail.com, mani@nitk.ac.in
[1] Department of Applied Mechanics and Hydraulics, National Institute of Technology Karnataka, Surathkal, Mangalore, India

References
Appa Rao, G. (1991). *Drought and southwest monsoon, training course on monsoon meteorology*. 3rd WMO Asian/African Monsoon Workshop, Pune.
Chowdhury, A., Dandekar, M. M., & Raut, P. S. (1989). Variability in drought incidence over India–A statistical approach. *Mausam, 40*, 207–214.
Edwards, D. C., & McKee, T. B. (1997, May 1–30). *Characteristics of 20th century drought in the United States at multiple scales* (Atmospheric Science Paper No. 634).
Gore, P. G., & Sinha Ray, K. C. (2002). Variability in drought incidence over districts of Maharashtra. *Mausam, 53*, 533–538.
Guhathakurta, P. (2003). Drought in districts of India during the recent all India normal monsoon years and its probability of occurrence. *Mausam, 54*, 542–545.
Heim, Jr., R. (2002). A review of twentieth-century drought indices used in the United States. *Bulletin of the American Meteorological Society, 83*, 1149–1165.
Henry, A. J. (1906). *Climatology of the United States* (Bulletin Q. U.S. Weather Bureau Bull. 361, pp. 51–58). Washington, DC.
Hughes, B. L., & Saunders, M. A. (2002). A drought climatology for Europe. *International Journal of Climatology, 22*, 1571–1592.
http://dx.doi.org/10.1002/(ISSN)1097-0088

Kamble, M. V., Ghosh, K., Rajeevan, M., & Samui, R. P. (2010). Drought monitoring over India through normalized difference vegetation index (NDVI). *Mausam, 61*, 537–546.

The Krishna basin report, India-WRIS. (2014). *A report on the Krishna river basin by Central Water Commission.* Hyderabad: Ministry of Water Resources, New Delhi-India and National Remote Sensing Centre, ISRO, Department of Space, Government of India.

McKee, T. B., Doesken, N. J., & Kliest, J. (1993, January 17–22). The relationship of drought frequency and duration to time scales. In *Proceedings of the 8th Conference on Applied Climatology* (pp. 179–184), Anaheim, CA. Boston, MA: American Meteorological Society.

McKee, T. B., Doesken, N. J., & Kleist, J. (1995). *Drought monitoring with multiple time scales.* Paper presented at 9th Conference on Applied Climatology. Dallas, TX: American Meteorological Society.

Mihajlović, D. (2006). Monitoring the 2003–2004 meteorological drought over Pannonian part of Croatia. *International Journal of Climatology, 26*, 2213–2225. http://dx.doi.org/10.1002/(ISSN)1097-0088

Mishra, A. K., & Desai, V. R. (2005). Spatial and temporal drought analysis in the Kansabati river basin, India. *International Journal of River Basin Management, 3*, 31–41. http://dx.doi.org/10.1080/15715124.2005.9635243

Mishra, A. K., & Singh, V. P. (2010). A review of drought concepts. *Journal of Hydrology, 391*, 202–216. http://dx.doi.org/10.1016/j.jhydrol.2010.07.012

Mishra, S. S., & Nagarajan, R. (2011). Spatio-temporal drought assessment in Tel river basin using standardized precipitation index (SPI) and GIS. *Geomatics, Natural Hazards and Risk, 2*, 79–93. http://dx.doi.org/10.1080/19475705.2010.533703

Pai, D. S., Latha, S., Guhathakurta, P., & Hatwar, H. R. (2011). District-wide drought climatology of the southwest monsoon season over India based on standardized precipitation index (SPI). *Natural Hazards-Springer Science.* doi:10.1007/s11069-011-9867-8

Ramdas, D. A. (1950). Rainfall and agriculture. *Industrial Journal of Meteriology and Geophysics, 1*, 262–274.

Riebsame, W. E., Changnon, S. A., & Karl, T. R. (1991). *Drought and natural resource management in the United States: Impacts and implications of the 1987–1989 drought* (p. 174). Boulder, CO: Westview Press.

Sen, A. K., & Sinha Ray, K. C. (1997). *Recent trends in drought affected areas in India.* In *Presented at International Symposium on Tropical Meteorology, INTROPMET-1997.* New Delhi: IIT.

Sinha Ray, K. C., & Shewale, M. P. (2001). Probability of occurrence of drought in various subdivisions of India. *Mausam, 52*, 541–546.

Wilhite, D. A., & Buchanan-Smith, M. (2005). Drought as hazard: Understanding the natural and social context. In D. A. Wilhite (Ed.), *Drought and water crisis. Science, technology, and management issues* (pp 3–29).London: CRC Press, Taylor and Francis. http://dx.doi.org/10.1201/CRCBKSPE

Xie, H., Ringler, C., Zhu, T., & Waqas, A. (2013). Droughts in Pakistan: A spatiotemporal variability analysis using the standardized precipitation index. *Water International, 38*, 620–631. doi:10.1080/02508060.2013.827889

12

Spatial distribution of various parameters in groundwater of Delhi, India

Parul Gupta[1] and Kiranmay Sarma[1]*

*Corresponding author: Kiranmay Sarma, University School of Environment Management, GGSIP University, Sector-16C, Dwarka 110078, New Delhi, India
E-mail: kiranmay@ipu.ac.in
Reviewing editor: Shashi Dubey, Hindustan College of Engineering, India

Abstract: The present study analyzed the spatial variability in groundwater quality and depth in National Capital Territory (NCT) of Delhi, India. The study classified the parameters into five distribution classes viz., low, moderately low, moderate, moderately high, and high. Spatial variability maps were generated using kriging tool in ArcGIS environment. Primary data collected seasonally during 2012–2014 were used for the generation of maps. Physico-chemical parameters were correlated with each other and groundwater depth. All the parameters were found to be negatively correlated with groundwater depth. Spatial distribution maps showed that maximum concentration of most parameters was found in the northern parts of the study area, while maximum depth was reported from the southern part. Maximum area of around 59% of total area of Delhi has low electrical conductivity, TDS, and hardness values. With groundwater depth improving toward north Delhi, groundwater quality is found to be improving toward south parts of Delhi.

Subjects: Earth Sciences; Environment & Agriculture; Environmental Studies & Management

Keywords: groundwater; Delhi; depth; water quality; Kriging; Environmental sustainability engineering

ABOUT THE AUTHORS

Parul Gupta is a PhD research scholar of University School of Environment Management, Guru Gobind Singh Indraprastha University, New Delhi. She is a recipient of selected Indraprastha Research Fellowship to carry out her research on groundwater resources of Delhi.

Kiranmay Sarma, PhD, is an associate professor in University School of Environment Management, Guru Gobind Singh Indraprastha University, New Delhi. He has got his MSc degree in Geoinformation Science and Earth Observation from International Institute of Geoinformation Science and Earth Observation (ITC), the Netherlands. He got awarded his PhD from North-Eastern Hill University, Shillong. He has expertise in Remote Sensing and GIS, natural resource management, wildlife conservation, forest inventory mapping, and disaster management. He has published more than 55 research papers in various reputed national and international journals, edited books, and conference proceedings. Besides, he has authored seven books/monographs in the sector of environment.

PUBLIC INTEREST STATEMENT

Groundwater is one of the cleanest freshwater resources available on earth and is considered as the most threatened resources. Since groundwater is non-uniformly distributed throughout world, its availability and quality are of the major concerns for urban and agricultural areas. The present work was conducted in National Capital Territory (NCT) of Delhi, one of the fast-growing capital cities of the world. Through the study, trend of groundwater depth and quality throughout the study area were found by collecting samples from certain sites. The study highlighted the depleting groundwater resources in southern parts of Delhi and deteriorating quality in northern parts instead of shallow water depth. Thus, the work can help authorities and planners to manage urban activities, especially urban and agricultural activities, to preserve groundwater resources and prevent its abuse.

1. Introduction

Groundwater is one of the most important sources of water for various human activities. With deteriorating surface water quality, there is an increasing reliance on groundwater resources in urban and rural areas. Thus, groundwater quality and quantity are continuously under threat due to abuse of resources. Problems of falling groundwater table and deteriorating groundwater quality have become a common problem throughout the world. According to State of India's Environment's (2015) report, India is one of the largest consumers of groundwater resources, distributed unevenly in the country. In India, there is lack of scientific data due to incomplete understanding in recharge and discharge processes of underground water and are mostly approximated. The number of groundwater withdrawal structures has increased rapidly leading to overexploitation and depletion of the resource (Singh & Singh, 2002). Das (2011) reported that the uneven alluvium region of Delhi is a potential groundwater reservoir of the city and diverse geology and topography of the area have significant control on movement of groundwater. With continuous and high influx of population, the capital of India is facing acute water shortage and groundwater is fulfilling about 50% of the water requirement (CGWB, 2006).

Estimation of groundwater quality and depth is vital for planning sustainable management of this vital natural resource, and thus requires handling of large amount of spatial and non-spatial data. Geographical Information System (GIS) is an important tool for dealing with such types of data (Chaudhary, Kumar, Roy, & Ruhal, 1996; Goyal, Chaudhary, Singh, Sethi, & Thakur, 2010; Sarma, Sarma, & Barik, 2012). Goovaerts (2000) stated that geostatistics was developed to deal with problems of spatially autocorrelated data (Moukana & Koike, 2008). Sarangi, cox, and Madramootoo (2005) attempted geostatistical techniques to know groundwater pollution and depth at unsampled locations. Kriging methods are considered for mapping spatial variations using spatial correlations between sampled points (Ella, Melvin, & Kanwar, 2001).

The present work is an attempt to find out the spatial variability in values of various physico-chemical parameters and depth of groundwater in NCT of Delhi using kriging interpolation tool of ArcGIS. The study was taken up as a reference from work by Dash, Sarangi, and Singh (2010) with primary data collected during 2012–2014.

2. *Study Area*

The National Capital Territory of Delhi lies between 28°24′15″ and 28°53′00″ N latitudes and 76°50′24″ and 77°20′30″ E longitudes with a total geographical area of 1,483 km². Delhi region is a part of the Indo-Gangetic Alluvial Plains, at an elevation ranging from 198 to 220 m above mean sea level (Dash et al., 2010). Physiographically, the region shows four major variations: (a) The Delhi ridge, which is a prolongation of Aravalli hills consisting of quartzite rocks, extending from the southern part of the territory to the western bank of Yamuna for a stretch of about 35 km. (b) The Chattarpur Alluvial basin, covering an area of about 48 km², is occupied by alluvium derived from the adjacent quartzite ridge. (c) Alluvial plains on the eastern and western sides of the ridge; and (d) Yamuna flood plain deposits (Kumar, Ramanathan, Rao, & Kumar, 2006). The primary source of groundwater in Delhi is from the southwest monsoon rainfall which occurs during the months of July–September (Datta & Tyagi, 1996).

For the present study, a total eight sites were selected based on the four land covers in Delhi viz., protected forest, trees outside forest, maintained park, and settlement area. Two sites under each land cover category were selected (Figure 1) for analysis.

3. Material and Methods

A total of 48 samples of groundwater were collected from the 8 selected study sites for physico-chemical analysis. Groundwater samples were collected in polyethylene bottles from tube wells and hand pumps after pumping the water for 10 min. Samples were collected during October 2012–June 2014, seasonally, for post-monsoon, winter, and summer. Groundwater depth was measured using Piezometers installed by Central Groundwater Board (CGWB). Groundwater temperature, pH, and

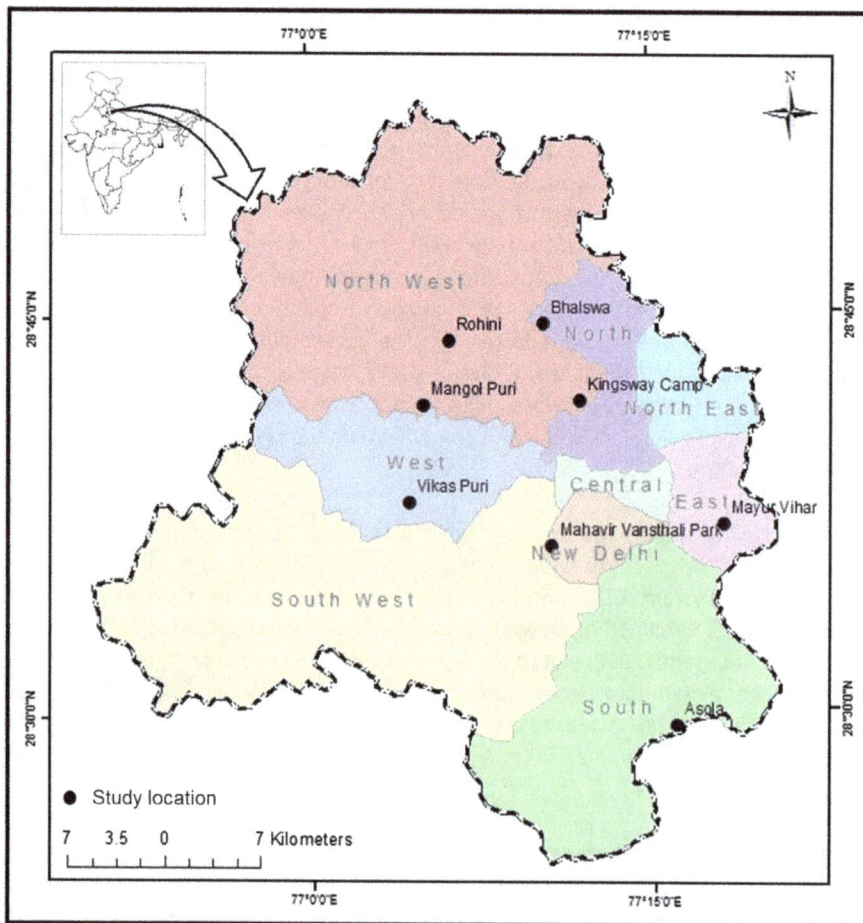

Figure 1. Map showing the study site locations in NCT of Delhi.

Electrical Conductivity (EC) were measured on-site using digital meters. Groundwater samples were collected in polyethylene containers and were analyzed for TDS, hardness (total, calcium, and magnesium), alkalinity, acidity, chloride, nitrate, sulfate, fluoride, sodium, and potassium according to the standard procedures given in APHA Standard Methods (1998). From all the physico-chemical parameters studied, nine parameters were selected for the present study using Principal Component Analysis (PCA). PCA and correlation of data were conducted by SPSS 19 software. Data acquired were used for generation of spatial variability maps by ArcGIS 10.2 software.

3.1. Kriging

Spatial interpolation techniques are used to find out values at points where data have not been estimated. Kriging is a geostatistical method for spatial interpolation. Kriging being a local and stochastic method of interpolation is better from deterministic methods in assessing prediction error with estimated variances (Chang, 2012). This method was used for generation of maps which is an advanced and unique interpolation technique that helps derive a predictive value for an unmeasured location. It is based on regionalized variable theory, which assumes that the spatial variation in the data being modeled is homogeneous across the surface. Kriging assumes that the distance or direction between sample points reflects a spatial correlation that can be used to explain variation in the surface. Kriging interpolation method suffers limitations in the case of outliers and non-stationarity in the data (Weise, 2001).

Two following tasks are necessary to make a prediction by kriging:

- Uncover the dependency rules
- Make the predictions

Thus, kriging goes through two steps to realize these two tasks:

(a) It creates variograms and covariance functions to estimate the statistical dependence values that depend on the model of autocorrelation.

(b) It predicts the unknown values.

Ordinary kriging which has been applied in the present study provides interpolated or kriged values from equations that minimize the variance of the estimation error. The coefficients of the linear combination known as weights depend on: (i) the distance between the sample point and the estimated point and (ii) the spatial structure of the variable. Ordinary kriging is based on the assumption that the mean of the process is constant and invariant within the spatial domain. This is expressed as:

$$z(x) = \mu + \epsilon(x)$$

where, μ is an unknown constant and generally considered the mean value of the regionalized variable; $z(x)$ is the value of regionalized variable at any location x with stochastic residual $\epsilon(x)$ with zero mean and unit variance (Dash et al., 2010).

4. Results and Discussion

Based on the PCA, nine physico-chemical parameters were selected for the present study (Belkhiri, Boudoukha, & Mouni, 2011; Choi et al., 2005), which are the contributing factors in the groundwater quality of NCT of Delhi. They are Electrical Conductivity (EC), TDS, total hardness, calcium, magnesium, sulfate, chloride, sodium, and potassium (Table 1). The varied ranges of all these parameters are considered as high, moderately high, moderate, moderately low, and low (Table 2).

About 60% of the area comprises low EC value (846.2 km^2) distributed partly in the districts of North, North West, North East, South, South West, East, Central, and New Delhi with EC value less than 1208 µS/cm. An area of 47.4 km^2 of North and North West districts recorded EC more than 3274.28 µS/cm (Figure 2(a)). According to Kazemi and Mohammadi (2012) EC values of more than 2000 µS/cm indicate salinization of aquifer (Gupta & Sarma, 2014). High electrical conductivity in groundwater may also be attributed to ion exchange and solubalization in the aquifer medium (Sanchez-Perez & Tremoliers, 2003; Sharma, Sarma, & Mahanta, 2012). According to Todd (2014), solution from rock materials is the primary source of soluble salts in groundwater. Similarly, TDS and hardness also show 59.8 and 58.8% of total area covered under low concentration values of less than 1163.38 mg/l and 382.35 mg/l, respectively (Figure 2(b) and 2(c)). Approximate areas under moderately low, moderate, moderately high, and high concentrations for EC, TDS, and hardness are also similar (Table 3). Similar trends for EC and TDS were also reported by Saka, Akiti, Osae, Appenteng, and Gibrilla (2013). Strong

Table 1. Descriptive statistics of groundwater parameters			
Parameters	**Minimum**	**Maximum**	**Mean**
E.C. (µS/cm)	322	5090	1437.89
TDS (mg/L)	284	5862	1323.44
Hardness (mg/L)	120	1594	428.38
Calcium (mg/L)	8.02	108.22	32.34
Magnesium (mg/L)	29.07	459.17	120.05
Sulfate (mg/L)	18.11	343.92	128.72
Chloride (mg/L)	25.56	1824.7	349.36
Sodium (mg/L)	23	1150	205.88
Potassium (mg/L)	2	21	7.65
Depth (mbgl)	0.83	60.85	14.94

Table 2. Classification of groundwater parameter values in five distribution classes					
Parameters	**Low**	**Moderately Low**	**Moderate**	**Moderately High**	**High**
E.C. (µS/cm)	<1208	1208–1896.85	1896.85–2585.56	2585.56–3274.28	>3274.28
TDS (mg/L)	<1163.38	1163.38–1888	1888–2613	2613–3339	>3339
Hardness (mg/L)	<382.35	382.35–582.86	582.86–783.37	783.37–983.87	>983.87
Calcium (mg/L)	<24.71	24.71–31.38	31.38–38.05	38.05–44.72	>44.72
Magnesium (mg/L)	<85.53	85.53–133.9	133.9–182.3	182.3–230.8	>230.8
Sulfate (mg/L)	<71.25	71.25–110.47	110.47–149.69	149.69–188.90	>188.90
Chloride (mg/L)	<343.52	343.52–634.81	634.81–928	928–1217.19	>1217.19
Sodium (mg/L)	<171.95	171.95–281.42	281.42–390.88	390.88–500.35	>500.35
Potassium (mg/L)	<5.41	5.41–7.99	7.99–10.56	10.56–13.14	>13.14
Depth (mbgl)	<12.9	12.9–23.04	23.04–33.14	33.14–43.24	>43.24

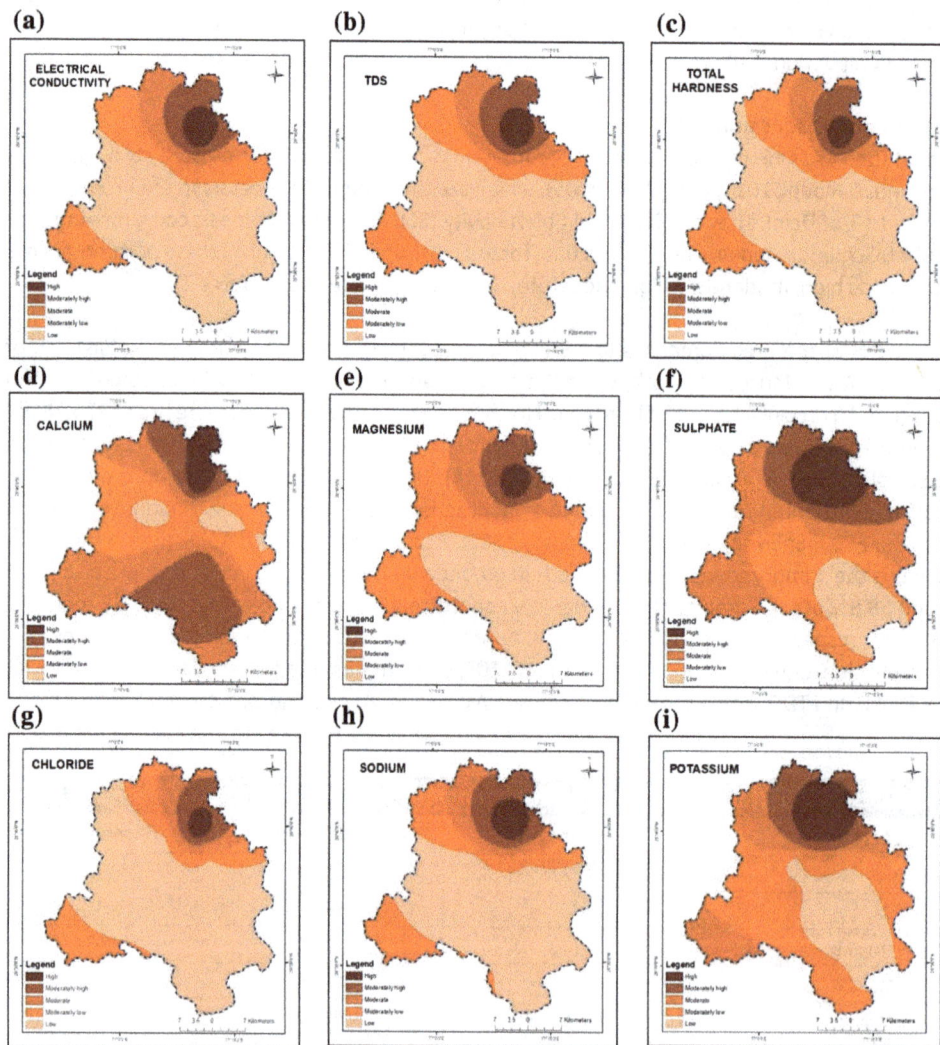

Figure 2. Spatial variations of different groundwater quality parameters.

positive correlation among EC, TDS, and total hardness justifies the distribution of these parameters in groundwater of Delhi (Table 4). Similar correlations were also reported by Srivastava and Ramanathan, (2008) and Gupta and Sarma (2013). High salinity values restrict the use of water for

Table 3. Area in sq. km under each distribution class of groundwater parameter values in Delhi										
Classes	**EC**	**TDS**	**TH**	**Ca**	**Mg**	**Sulfate**	**Chloride**	**Na**	**K**	**Depth**
Low	846.2	886.8	872	82.1	547.2	231.7	1010.6	855	347.3	528.5
Moderately Low	330.2	306	369.6	424.3	577.8	391.1	268.3	317.4	590.7	600.7
Moderate	140.2	130.5	116.8	527.2	201.5	467.2	96.5	129.1	261.1	150.5
Moderately High	119	114.4	96.1	371.2	120.3	224.2	82.2	127.2	145.4	157.8
High	47.4	45.3	28.5	78.2	36.2	168.8	25.4	54.3	138.5	45.5
Total Area	1483	1483	1483	1483	1483	1483	1483	1483	1483	1483

Table 4. Correlation table for groundwater parameters and depth										
	EC	**TDS**	**Hardness**	**Calcium**	**Magnesium**	**Sulfate**	**Chloride**	**Sodium**	**Potassium**	**Depth**
EC	1.00	0.98	0.90	0.33	0.66	0.75	0.95	0.83	0.79	−0.43
TDS	0.98	1.00	0.89	0.37	0.65	0.74	0.94	0.86	0.79	−0.41
Hardness	0.90	0.89	1.00	0.45	0.75	0.59	0.94	0.79	0.75	−0.44
Calcium	0.33	0.37	0.45	1.00	0.19	0.09	0.43	0.32	0.36	0.19
Magnesium	0.66	0.65	0.75	0.19	1.00	0.49	0.68	0.57	0.48	−0.45
Sulfate	0.75	0.74	0.59	0.09	0.49	1.00	0.63	0.61	0.76	−0.62
Chloride	0.95	0.94	0.94	0.43	0.68	0.63	1.00	0.77	0.71	−0.35
Sodium	0.83	0.86	0.79	0.32	0.57	0.61	0.77	1.00	0.73	−0.34
Potassium	0.79	0.79	0.75	0.36	0.48	0.76	0.71	0.73	1.00	−0.39
Depth	−0.43	−0.41	−0.44	0.19	−0.45	−0.62	−0.35	−0.34	−0.39	1.00

drinking purpose. Calcium concentration between 31.38 mg/l and 38.05 mg/l under moderate class has maximum coverage of 527.2 km^2 (35.5%) distributed in parts of North, North West, South, South West, and New Delhi districts. While for magnesium, maximum area of 577.8 km^2 covering 39% of total area reported moderately low range of 85.53 mg/l to 133.9 mg/l. Parts of South, South West, and New Delhi districts have low magnesium concentration of less than 85.53 mg/l (Figure 2(e)). Sulfate in Delhi groundwater was reported in the range 18 mg/l–343 mg/l, out of which 168.8 km^2 area is with more than 228.12 mg/l and 231.7 km^2 (15.6%) areas having sulfate less than 71.25 mg/l (low range). Maximum area of Delhi (31.5%) was reported to have moderate sulfate concentration of 110.47–149.69 mg/l in groundwater. Craig and Anderson (1979), Miller (1979), and Kumar, Sharma, Ramanathan, Rao, and Kumar (2009) reported breakdown of organic substances from soil and leachable sulfate from anthropogenic activities are sources of sulfate in groundwater. Chloride concentrations are found to be low throughout Delhi with the average value of 349.36 mg/l. Only 1.7% of total area is reported to have high chloride value (more than 1508.37 mg/l) while 1010.6 km^2 and 268.3 km^2 areas are reported under low and moderately low concentrations of chloride, respectively. Similarly for sodium, maximum parts of Delhi have low sodium contents (171.95 mg/l) except few small pockets in the districts of North and North West having high sodium concentration. Dissolution activity of clay, gravel, feldspar, and kankar enhances the sodium concentration in groundwater (Kumar et al., 2009). According to CPCB (Central Pollution Control Board, 2008) report, high sodium in Delhi can be due to base exchange phenomenon. Strong correlation between sodium and chloride shows the same source of origin for the ions (Belkhiri & Mouni, 2013; Saka et al., 2013). Saka et al. (2013) also reported correlation of TDS with sodium and chloride. Potassium content is found less in Delhi with average concentration of 7.65 mg/l. An area of 138.5 km^2 of North and North West districts has potassium more than 15.72 mg/l (Figure 2(i), Table 3).

Groundwater depth in Delhi varies from 0.83 meters below ground level (mbgl) to 60.85 mbgl (Figure 3). Low groundwater depth of less than 12.9 mbgl was recorded from about 35.6% of the total area and depth higher than 53.34 mbgl was recorded from only 3% of the area. Depth was

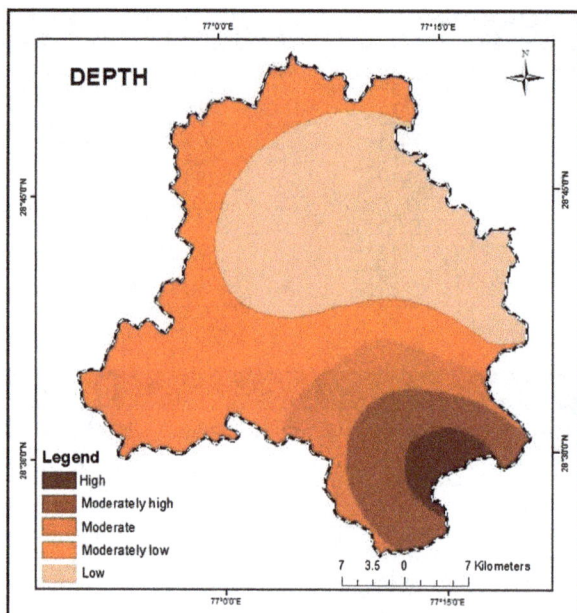

Figure 3. Spatial distribution of groundwater depth.

found to be improving toward north from the south, which may be attributed to River Yamuna as a source of recharge. Lowering of depth toward north Delhi may also be due to increase in hydraulic gradient toward northeast as reported by Dash et al. (2010). Kumar, Ramanathan, and Keshari (2009a) also supported the role of River Yamuna through western Yamuna canal as groundwater recharge source in parts of North, North West, East and North East districts. Moderately low levels of groundwater from 12.9 mbgl to 23.04 mbgl were reported from 600.7 km^2 covering parts of North West, West, South West, and Central districts. Areas of South Delhi include ridge parts which are considered as recharge zones for groundwater, but face uncontrolled groundwater extraction, leading to deeper water levels (Gupta & Sarma, 2014). Chatterjee et al. (2009) also found deep water table in the ridge area.

All the parameters reported highest values from north and northwest parts of the NCT of Delhi, while groundwater depth recorded maximum values from South Delhi. High nutrient concentrations in northern part of the study area were also recorded by Kumar et al. (2009). Groundwater depth was also found to be negatively correlated with all the physico-chemical parameters, thus justifying the good quality of water at sites with deep groundwater table. Hu et al. (2005), Dash et al. (2010) and Gupta and Sarma (2013) also reported high salinity in areas with minimal groundwater depth.

5. Conclusion

Being an important source of water supply, groundwater quality and depth are among the major issues of concern in an urban capital like NCT of Delhi, India. Knowledge of spatial distribution of physico-chemical parameters and depth of groundwater is a basic requirement for planning and management of resources. Kriging interpolation is one of the advanced geostatistical tools to study the spatial distribution of any parameter based on some spatial values available. The spatial variability maps of groundwater quality parameters and depth show the increase of all the physico-chemical parameters from north to south, while groundwater depth decreases from south to northern parts of Delhi. Thus, information obtained from the maps generated using kriging can be useful to delineate areas in NCT of Delhi under various categories of water availability or different water quality classes for planning, management, and policy-making.

Acknowledgment

The authors wish to acknowledge GGSIP University, New Delhi, for providing financial support to carry this study in the form of fellowship to the researcher. Authors are also thankful to Central Groundwater Board, Delhi, for necessary assistance during the field study.

Funding

The authors received no direct funding for this research.

Author details

Parul Gupta[1]
E-mail: parulgupta_87@yahoo.co.in
Kiranmay Sarma[1]
E-mail: kiranmay@ipu.ac.in
[1] University School of Environment Management, GGSIP University, Sector-16C, Dwarka 110078, New Delhi, India.

References

Belkhiri, L., & Mouni, L. (2013). Groundwater geochemistry of Ain Azel Area, Algeria. *Chemie Erde-Geochemistry*. doi:10.1016/j.chemer.2013.09.009

Belkhiri, L., Boudoukha, A., & Mouni, L. (2011). A multivariate statistical analysis of groundwater chemistry data. *International Journal of Environmental Research, 5*, 537–544.

Central Pollution Control Board. (2008). *Status of groundwater quality in India-Part-II* (Groundwater quality Series). GWQS/10/2007–2008. Delhi: Ministry of Environment and Forest, Government of India.

CGWB. (2006). *Groundwater year book of national capital of territory, Delhi*. New Delhi: Ministry of Water Resources, Government of India.

Chang, K.-T. (2012). *Introduction to Geographic Information System*. New Delhi: Tata McGraw Hill Education Private Limited.

Chatterjee, R., Gupta, B. K., Mohiddin, S. K., Singh, P. N., Shekhar, S., & Purohit, R. (2009). Dynamic groundwater resources of National Capital Territory, Delhi: Assessment, development and management options. *Environmental Earth Sciences, 59*, 669–686. http://dx.doi.org/10.1007/s12665-009-0064-y

Chaudhary, B. S., Kumar, M., Roy, A. K., & Ruhal, D. S. (1996). Application of remote sensing and GIS in groundwater investigations in Sohna block, Gurgaon district, Haryana, India. In K. Kraus & P. Waldhausl (Eds.), *International Archives of Photogrammetry and Remote Sensing 31, B-6* (pp. 18–23). Vienna: International Society for Photogrammetry and Remote Sensing.

Choi, B. -Y., Yun, S. -T., Yu, S. -Y., Lee, P.-K., Park, S. -S., Chae, G. -T., & Mayer, B. (2005). Hydrochemistry of urban groundwater in Seoul, South Korea: Effects of land-use and pollutant recharge. *Environmental Geology, 48*, 979–990. http://dx.doi.org/10.1007/s00254-004-1205-y

Craig, E., & Anderson, M. P. (1979). The effects of urbanization of ground water quality. A case study of ground water ecosystems. *Environmental Conservation, 30*, 104–130.

Das, S. (2011). *Groundwater Resources of India*. New Delhi: National Book Trust.

Dash, J. P., Sarangi, A., & Singh, D. K. (2010). Spatial variability of groundwater depth and quality parameters in the National Capital Territory of Delhi. *Environmental Management, 45*, 640–650. http://dx.doi.org/10.1007/s00267-010-9436-z

Datta, P. S., & Tyagi, S. K. (1996) Major ion chemistry of groundwater in Delhi area, India. *Journal of Geological Society of India, 47*, 179–188.

Ella, V. B., Melvin, S. W., & Kanwar, R. S. (2001) Spatial analysis of NO_3-N concentration in glacial till. *Transactions of American Society of Agricultural Engineering, 44*, 317–327.

Goovaerts, P. (2000). Geostatistical approaches for incorporating elevation into the spatial interpolation of rainfall. *Journal of Hydrology, 228*, 113–129. http://dx.doi.org/10.1016/S0022-1694(00)00144-X

Goyal, S. K., Chaudhary, B. S., Singh, O., Sethi, G. K., & Thakur, P. K. (2010) Variability analysis of groundwater levels — AGIS-based case study *Journal of the Indian Society of Remote Sensing, 38*, 355–364. http://dx.doi.org/10.1007/s12524-010-0024-8

Gupta, P., & Sarma, K. (2013). Evaluation of groundwater quality and depth with respect to different land covers in Delhi, India. *International Journal of Applied Sciences and Engineering Research, 2*, 630–643.

Gupta, P., & Sarma, K. (2014). Variations in groundwater quality under different land covers in Delhi, India. *Journal of Global Ecology and Environment, 1*(1), 1–10.

Hu, K., Huang, Y., Li, H., Li, B., Chen, D., & White, R. E. (2005) Spatial variability of shallow groundwater level, electrical conductivity and nitrate concentration, and risk assessment of nitrate contamination in North China Plain. *Environment International, 31*, 896–903.

Kazemi, G. A., & Mohammadi, A. (2012) Significance of hydrogeochemical analysis in the management of groundwater resources: A case study of northeastern Iran. In G. A. Kazemi (Ed.), *Hydrogeology-a global perspective* (pp. 141–159). InTech. ISBN: 978-953-51-0048-5. http://dx.doi.org/10.5772/1523

Kumar, M., Ramanathan, A. L., Rao, M. S., & Kumar, B. (2006). Identification and evaluation of hydrogeochemical processes in the groundwater environment of Delhi, India. *Environmental Geology, 50*, 1025–1039.

Kumar, M., Ramanathan, A. L., & Keshari, A. K. (2009a). Understanding the extent of interactions between groundwater and surface water through major ion chemistry and multivariate statistical techniques. *Hydrological Processes, 23*, 297–310. http://dx.doi.org/10.1002/hyp.v23:2

Kumar, M., Sharma, B., Ramanathan, A. L., Rao, M. S. and Kumar, B. (2009) "Nutrient chemistry and salinity mapping of the Delhi aquifer, India: Source identification perspective" *Environmental Geology*, Vol. 56, pp. 1171-1181. http://dx.doi.org/10.1007/s00254-008-1217-0

Miller, G. T. (1979) "Living in the environment" Belmonol California, Wordsworth Publishing company, CA, p. 470.

Moukana, J. A., & Koike, K. (2008) Geostatistical model for correlating declining groundwater levels with changes in land cover detected from analyses of satellite images. *Computers & Geosciences, 34*, 1527–1540. http://dx.doi.org/10.1016/j.cageo.2007.11.005

Saka, D., Akiti, T. T., Osae, H., Appenteng, M. K., & Gibrilla, A. (2013). Hydrogeochemistry and isotope studies of groundwater in the Ga West Municipal Area, Ghana. *Applied Water Science, 3*, 577–588. http://dx.doi.org/10.1007/s13201-013-0104-3

Sanchez-Perez, J. M., & Tremoliers, M. (2003). Change in groundwater chemistry as a consequence of suppression of floods: The case of the Rhine floodplain. *Journal of Hydrology, 270*, 89–104. http://dx.doi.org/10.1016/S0022-1694(02)00293-7

Sarangi, A., cox, C. A. and Madramootoo, C. A. (2005) "Geostatistical methods for prediction of spatial variability of rainfall in a mountainous region" *Transactions of the ASAE, 48*, 943–954. http://dx.doi.org/10.13031/2013.18507

Sarma, K., Sarma, R. K., & Barik, S. K. (2012). Soil Erosion Vulnerability Mapping of Nokrek Biosphere Reserve, Meghalaya Using Geographic Information System. *Disaster and Development, 6*, 19–32.

Sharma, P., Sarma, H. P., & Mahanta, C. (2012). Evaluation of groundwater quality with emphasis on fluoride concentration in Nalbari district, Assam, Northeast India. *Environmental Earth Sciences, 65*, 2147–2159. http://dx.doi.org/10.1007/s12665-011-1195-5

Singh, D. K., & Singh, A. K. (2002). Groundwater situation in india: Problems and perspective. *International Journal of Water Resources Development, 18*, 563–580. http://dx.doi.org/10.1080/0790062022000017400

Srivastava, S. K., & Ramanathan, A. L. (2008). Geochemical assessment of groundwater quality in vicinity of Bhalswa landfill, Delhi, India, using graphical and multivariate statistical methods. *Environmental Geology, 53*, 1509–1528. http://dx.doi.org/10.1007/s00254-007-0762-2

Standard Methods. (1998). *Standard methods for the examination of water and wastewater* (20th ed.). Washington, DC: American Public Health Association (APHA).

State of India's Environment. (2015). *Water and Sanitation.* Delhi: A Down to Earth Annual, Centre for Science and Environment.

Todd, D. K. (2014) "Groundwater Hydrology", 2nd edition, John Wiley and Sons, Inc.

Weise, U. (2001). *Kriging- a statistical interpolation method and its applications.* Department of Geography, The University of British Columbia. Retrieved from http://ibis.geog.ubc.ca/courses/geog570/talks_2001/kriging.

13

Water quality index development for groundwater quality assessment of Greater Noida sub-basin

Sajal Singh[1] and Athar Hussian[1*]

*Corresponding author: Athar Hussian, Civil Engineering Department, School of Engineering, Gautam Buddha University, Greater Noida 201310, Uttar Pradesh, India
E-mail: athariitr@gmail.com
Reviewing editor: Shashi Dubey, Hindustan College of Engineering, India

Abstract: The water quality index (WQI) is an important parameter for determining the drinking water quality for the end users. The study for the same has been carried on the groundwater by collecting 47 groundwater samples from 25 blocks of Greater Noida city, India. In order to develop WQI the samples were subjected to a comprehensive physicochemical and biological analysis of 11 parameters such as pH, calcium, magnesium, chloride, nitrate, sulphate, total dissolved solids, fluorides, bicarbonate, sodium and potassium. Geographical information system has been used to map the sampling area. The coordinates in terms of latitude and longitude of the sampling locations were recorded with the help of global positioning system. Piper plots and cation–anion correlation matrix were plotted from the values obtained by the analysis of various parameters. The WQI index for the same has been calculated and the values ranged from 53.69 to 267.85. The WQI values from present study indicate the very poor quality water in the area dominated by industrial

ABOUT THE AUTHORS

Sajal Singh holds M Tech in Environmental Science & Engineering and completed his MTech dissertation at National Institute of Hydrology (Ministry of water resources), Roorkee (IIT Campus). His research and development work includes monitoring of effluent treatment plants, sewage treatment plants, common effluent treatment plant and other water and wastewater sources. His research interest includes the treatment of the typically contaminated water and wastewater treatment.

Athar Hussain is currently working as an associate professor at Ch. B.P. Govt. Engineering College, Jaffarpur, Delhi, India. His research interest includes water and wastewater treatment, solid and hazardous waste management. He is specialized in biological wastewater treatment and solid waste management.

PUBLIC INTEREST STATEMENT

Groundwater resources are affected in principle by major activities such as excessive use of fertilizers and pesticides in agricultural areas, untreated/partially treated wastewater to the environment and excessive pumping. Also, the improper management of aquifers results in excessive water depletion in an area. Another important aspect is solid waste disposal in open unengineered landfill is the one of the factors that cause ground water pollution due to lack of pollution control interventions such as waterproof layer, leachate treatment pond and monitoring wells. The present study has been aimed to develop the Water Quality Index (WQI) of swift developing city Greater Noida located Uttar Pradesh, India. WQI is one of the most effective tools to communicate information on the quality of water to the concerned citizens and policy-makers. WQI is a mathematical equation used to transform large number of water quality data into a single number. The water quality of the study was determined for all samples considering the eleven important parameters such as pH, Total Dissolved Solids (TDS), Total Hardness, Calcium (Ca^{2+}), Magnesium (Mg^{2+}), Sulphates (SO_4^-), Chlorides (Cl^-), Nitrates (NO_3^-), Fluorides (F^-) Bicarbonate (HCO_3^-) , Sodium (Na^+) and Potassium (K^+) were taken for the assessment.

and construction activities. Poor water quality has been observed in commercial zone of the study area. The analysis reveals the fact that the ground water of the Greater Noida needs a degree of treatment before consumption and needs to be protected from further contamination.

Subjects: Pollution; Water Engineering; Water Science

Keywords: water quality index; ground water quality assessment; physico chemical analysis; biological analysis; GIS; environmental sustainability engineering

1. Introduction

Groundwater is an important source of water supply throughout the world. Groundwater occurs almost everywhere beneath the earth surface not in a single widespread aquifer but in thousands of local aquifer systems and compartments that have similar characters. Knowledge of the occurrence, replenishment and recovery of groundwater has special significance in arid and semi-arid regions due to discrepancy in monsoonal rainfall, insufficient surface waters and over drafting of groundwater resources.

The ground water quality is still important to the community, therefore it is important to ensure its high quality at all time so that the consumer health is not compromised. Groundwater resources are affected in principle by three major activities. First of these activities is excessive use of fertilizers and pesticides in agricultural areas. The second one is untreated/partially treated wastewater to the environment. Finally, excessive pumping and improper management of aquifers result. The activity of solid waste disposal in open un-engineered landfill is the one of the factor that cause the ground water pollution due to lack of pollution control interventions such as water proof layer, leachate treatment pond, monitoring wells, etc. (Girija et al., 2007).

According to WHO organization, about 80% of all the diseases in human beings are caused by water. High rates of mortality and morbidity due to water borne diseases are well known in India. Access to safe drinking water remains an urgent necessity, as 30% of urban and 90% of rural households still depend completely on untreated surface or groundwater (Palanisamy et al., 2007). The quality of water is defined in terms of its physical, chemical and biological parameters. Its development and management plays a vital role in agriculture production, poverty reduction, environmental sustenance and sustainable economic development. In India, most of the population is dependent on groundwater as the only source of drinking water supply. As per the latest estimate of Central Pollution Control Board, about 29,000 million litre/day of wastewater generated from class-I cities and class-II towns out of which about 45% is generated from 35 metro-cities alone (Mangukiya et al., 2012).

Water quality index (WQI) is defined as a rating reflecting the composite influence of different water quality parameters. Horton (1965) has firstly used the concept of WQI, which was further developed by Brown, Mc Clelland, Deininger, and Tozer (1970) and improved by Deininger (Scottish Development Department, 1975). WQI is one of the most effective tools to communicate information on the quality of any water body. WQI is a mathematical equation used to transform large number of water quality data into a single number. WQI is one of the most effective tools to communicate information on the quality of water to the concerned citizens and policy-makers. The advent of satellite technology and geographical information system (GIS) has made it very easy to map of the sampling area. GIS has wide application in water quality mapping using which informative and user-friendly maps can be obtained.

The water quality of the study was determined for all samples using the weighted arithmetic index method as per Tiwari and Mishra (1985). In this method, the eleven important parameters such as pH, total dissolved solid (TDS), total hardness, calcium (Ca^{2+}), magnesium (Mg^{2+}), sulphates (SO_4^-), chlorides (Cl^-), nitrates (NO_3^-), fluorides (F^-) bicarbonate (HCO_3^-), sodium (Na^+) and potassium (K^+) were taken for the assessment.

2. Study area

Greater Noida is a town with a population of 107,676 (till March 2014) approximately is a part of Gautam Budh Nagar district (Figure 3) of Uttar Pradesh (Figure 2), India. It comes under the purview of the National Capital Region of India (Figure 1). Greater Noida is 48 km distance from New Delhi, the capital of India. The city was developed based on Greater Noida Master Plan 2001, 2021 plan report (2013). The notified area of Greater Noida comprising of 124 villages and about 40,000 hectare of area is broadly bounded by National Highway 24 in the north-west. Also River Hindon lies in the western side of the city. Due to nearness to Delhi and both these towns being are being well developed. Due to the pressure for development on Greater Noida the number of industries during the last decade, has grown more than ten times. Accordingly the problems related to environmental degradation have increased many folds. In summer i.e. from March to June the weather remains hot and average temperature ranges from minimum of 23 C to maximum of 45 C. Monsoon season prevails during mid-June to mid-September with an average rainfall of 93.2 cm (36.7 inches). Average temperature falls substantially down to as low as 3 to 4 C at the peak of winter. Total land use is 13,570 hectares with the total institutional area around 1,970 hectares along with 30 hectares of commercial area. The area is divided into different zones for water supply such as tube wells, overhead tanks and trunk and other supply lines. At present approximately 500 km length of water supply lines with approximately 460 km length of sewerage network and approximately 500 km length of drainage exists. The general slope of the ground water movement is from eastern side towards river Hindon in the west.

3. Materials and methods

A total number of 47 samples during winter period were collected from different selected locations of the study area. The coordinates of the sampling locations in terms of latitudes and longitudes were taken with the GPS. The samples were collected from various sources such as private hand pumps, government hand pumps, bore wells during February 2014. The samples were preserved as per the method prescribed in American Public Health Association manual (APHA, 1999). The depth of less than 30 m was considered as shallow aquifer while the depth greater than 30 m is considered as deep aquifer.

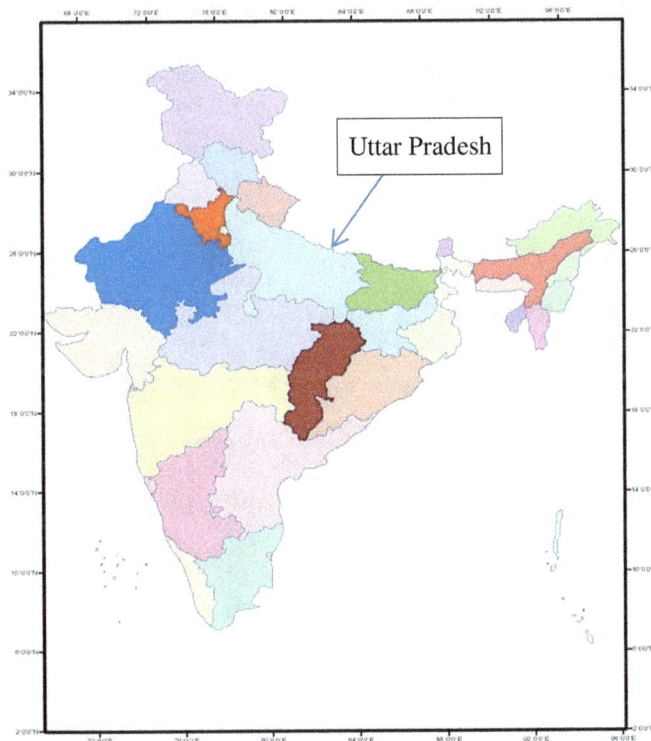

Figure 1. Map of India.

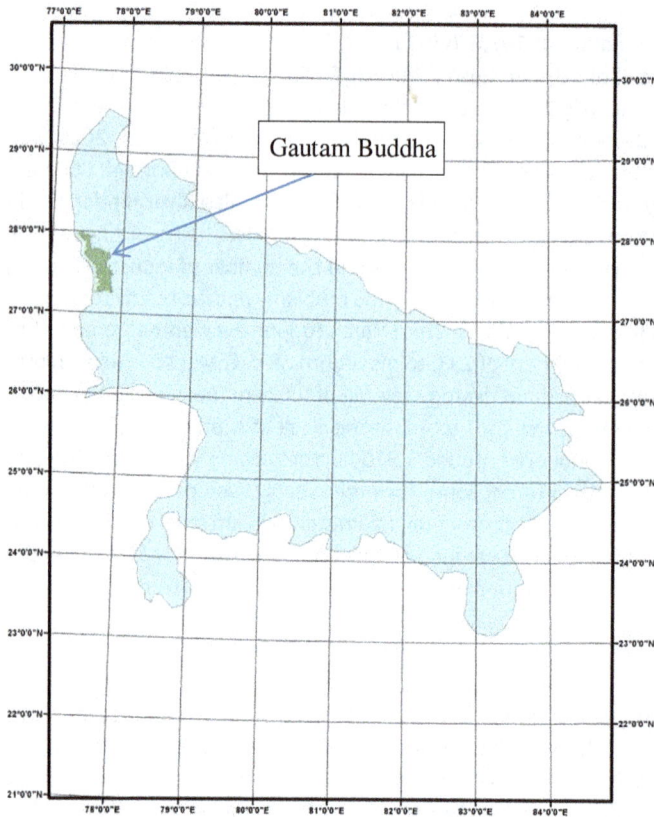

Figure 2. Map of Uttar Pradesh.

The samples were analysed for Ca, Mg, Na, K, HCO_3, SO_4, Cl, SO_4, F, NO_3 pH, EC, TDS, TSS, TH, turbidity, taste and odour, colour and total coliform. All the analyses were carried out as per the standard procedures prescribed in American Public Health Association manual (APHA, 1999). The pH, TDS and EC were measured by Systronics Water Quality Analyzer 371 F using portable Consort electrochemical analyzer model. Na and K were determined by using flame photometer (Systronics mk-1/mk-III). Parameters Ca, Mg, HCO_3 and Cl were analysed by titrimetric method. Total coliform were measured

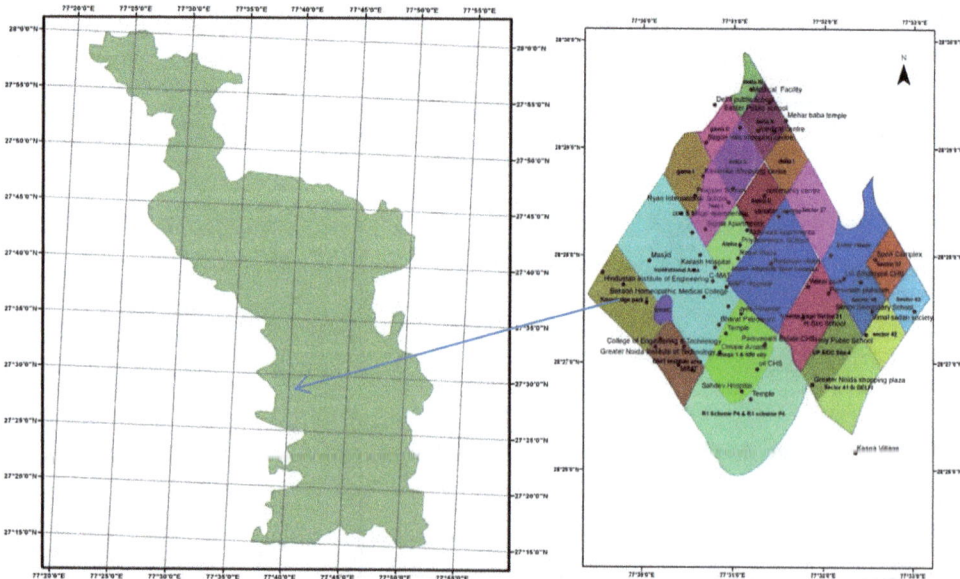

Figure 3. (a) Map of district Gautam Buddha Nagar (b) Sampling point locations in Greater Noida city.

by most probable number (MPN) method as per standard method prescribed in APHA 1992. The ionic charge balance was also calculated and it lies < 10% confirming to the reliability of the analytical results. Geochemical facies type, WQI and correlation coefficient matrix of parameters were also determined.

3.1. Piper diagram and cation–anion correlation matrix
In present study, piper plots and cation–anion correlation matrix has been plotted using Aquachem software (Version 14.1) for showing trends of major ions in all the samples. In the piper plot, major ions are plotted as cation and anion percentages in terms of milliequivalents in two base triangles. The piper plot allows comparisons of six parameters between a large numbers of samples. Like all trilinear plots, it does not portray absolute ionic concentrations. Determination of hydrochemical facies has been extensively used in the chemical assessment of groundwater and surface water for several decades. This method is able to provide sufficient information on the chemical quality of water, particularly the origin. Correlation coefficient (r) values have been determined using correlation matrix to identify the highly correlated and interrelated water quality parameters.

3.2. Water quality index
WQI is one of the most effective tool to monitor the surface as well as ground water pollution and can be used efficiently in the implementation of water quality upgrading programmes. WQI provide information on a rating scale from zero to hundred. Eleven parameters have been selected for developing the water quality index.

In the present study the WQI has been calculated in three steps. In the first step, each of the 11 parameters (PH, TDS, HCO_3, Cl, SO_4, NO_3, F, Ca, Mg, Na and K) has been assigned a weight (w_i) according to its relative importance in the overall quality of water for drinking purposes in Table 1.

The maximum weight of five has been assigned to the parameter nitrate due to its major importance in water quality assessment (Srinivasamoorthy et al., 2008). Bicarbonate is given the minimum weight of 1 as it plays an insignificant role in the water quality assessment. Other parameters like calcium, magnesium, sodium and potassium were assigned weight between 1 and 5 depending on their importance in water quality determination. In the second step, the relative weight (W_i) is computed from the following equations (Equation 1–4).

$$W_i = \frac{w_i}{\sum_{i=1}^{n} w_i} \tag{1}$$

Wi and w_i is the relative weight and weight of each parameter, respectively, and n is the number of parameters.

In the third step, a quality rating scale (Q_i) for each parameter is assigned by dividing its concentration in each water sample by its respective standard according to the guidelines laid down in the BIS and the result for the same is multiplied by 100 (Equation 2)

$$Q_i = \frac{C_i \times 100}{S_i} \tag{2}$$

where, Q_i is the quality rating, C_i is the concentration of each chemical parameter in each water sample in mg/L. Also S_i is the Indian drinking water standard for each chemical parameter in mg/L according to the guidelines of the BIS.

For computing the WQI, the SI is first determined for each chemical parameter, which is then used to determine the WQI as per the following Equations (3 and 4)

$$SI_i = W_i \times Q_i \tag{3}$$

Table 1. Details of chemical parameters with their relative weight and assigned weight with drinking water standards as per BIS (2012) and WHO (2012)				
S. No.	Chemical parameters	Drinking water standard (7 and WHO)	Weight (w₍ᵢ₎)	Relative weight (W₍ᵢ₎)
1	TDS (mg/l)	500	5	0.1190
2	Bicarbonate (mg/l)	244	1	0.0238
3	Chloride (mg/l)	250	5	0.1190
4	Sulphate (mg/l)	200	5	0.1190
5	Nitrate (mg/l)	45	5	0.1190
6	Fluoride (mg/l)	1.0	5	0.1190
7	Calcium (mg/l)	75	3	0.0714
8	Magnesium (mg/l)	30	3	0.0714
9	Sodium (mg/l)	200	4	0.0952
10	Potassium (mg/l)	–	2	0.0476
11	pH	7.5	4	0.0952

Table 2. Range of water quality index specified for drinking water used in India		
S. No.	WQI range	Water quality
1	<50	Excellent water
2	50–100	Good water
3	100–200	Poor water
4	200–300	Very poor water
5	>300	Water unsuitable for drinking purpose

$$WQI \sum\nolimits_{i=1}^{n} SI_i \tag{4}$$

where, SI_i is the sub index of ith parameter, Q_i is the rating based on concentration of ith parameter, n is the number of parameters.

The computed WQI values are categorized into five types as "excellent water" to "water, unsuitable for drinking". The range for WQI for drinking purpose is tabulated in Table 2.

4. Results and discussion

Greater Noida is an industrial city of Uttar Pradesh. Most of the wastewater from the city finds its way directly to the natural water bodies such as river, pond, etc. due to the insufficient treatment capacity of treatment plants. The contaminants also reach the ground water aquifers and making it unfit for human consumption.

4.1. Physicochemical parameters

4.1.1. pH
It plays an important role in clarification process and disinfection of drinking water. For effective disinfection with chlorine, the pH should preferably be less than eight, however, lower-pH water (<7) is more likely to be corrosive. Failure to minimize corrosion can result in the contamination of drinking water and adverse effect on its taste and appearance. Bureau of Indian Standard (BIS) has prescribed permissible limit of pH to be 6.5–8.5. The pH value of groundwater samples in the present study has been analysed and it lies in the range 7.22–8.46 (Figure 4).

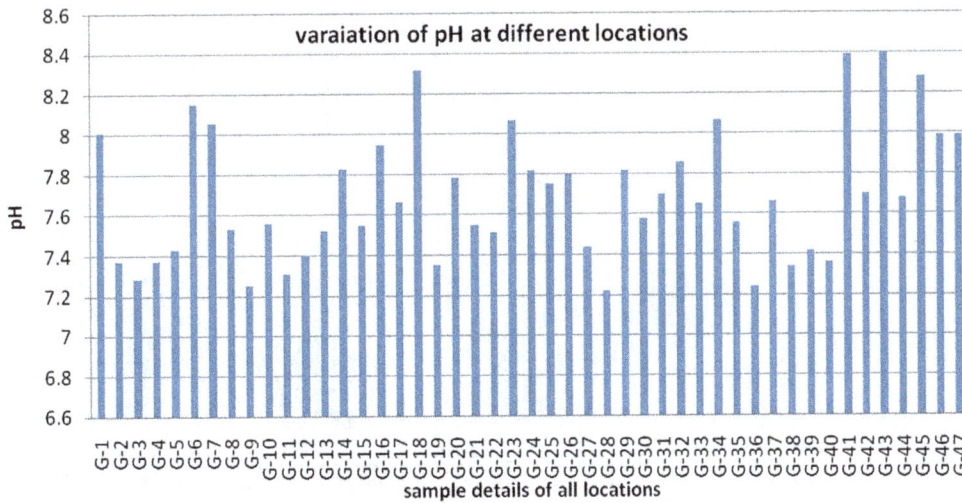

Figure 4. Variation of pH in samples of all the locations.

4.1.2. Conductivity

Conductivity is a parameter in water affected by the presence of dissolved ions. Organic compounds do not conduct electric current very well and hence their contribution to conductivity is very low. Significant changes in conductivity could then be an indicator that a discharge or some other source of pollution has entered in a stream. Conductivity of collected samples varies in the range of 710–3,410 µS/cm (Figure 5).

4.1.3. Turbidity

Recent research establishes a correlation between gastro-intestinal infections with high turbidity and turbidity events in distribution. The turbidity of the collected samples has been observed in the range 0.28–8.33 NTU. BIS has prescribed 1 NTU as the acceptable limit and 5 NTU as the permissible limit in absence of alternate source of drinking water. In the present study, 87% of the samples were well within the acceptable limit and rest were within the permissible limit except one sample collected from location G-26.

4.1.4. TDS

The presence of dissolved solids in water may affect its taste. The palatability of drinking water has been rated by panels of tasters in relation to its TDS level as follows: excellent (less than 300 mg/l), good (300–600 mg/l); fair (600–900 mg/l), poor (900–1,200 mg/l) and unacceptable (>1,200 mg/l). BIS has prescribed 500 mg/L as the acceptable limit and 2,000 mg/L as the permissible limit for TDS

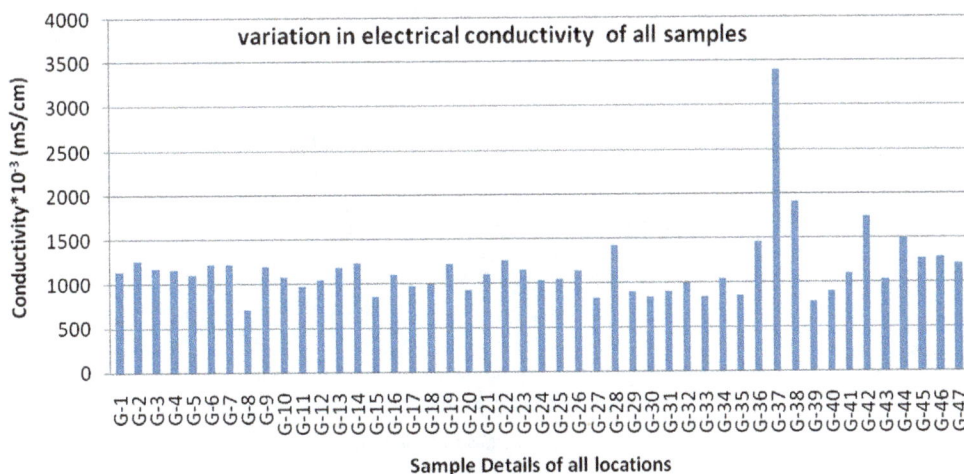

Figure 5. Variation of electrical conductivity in all samples of different locations.

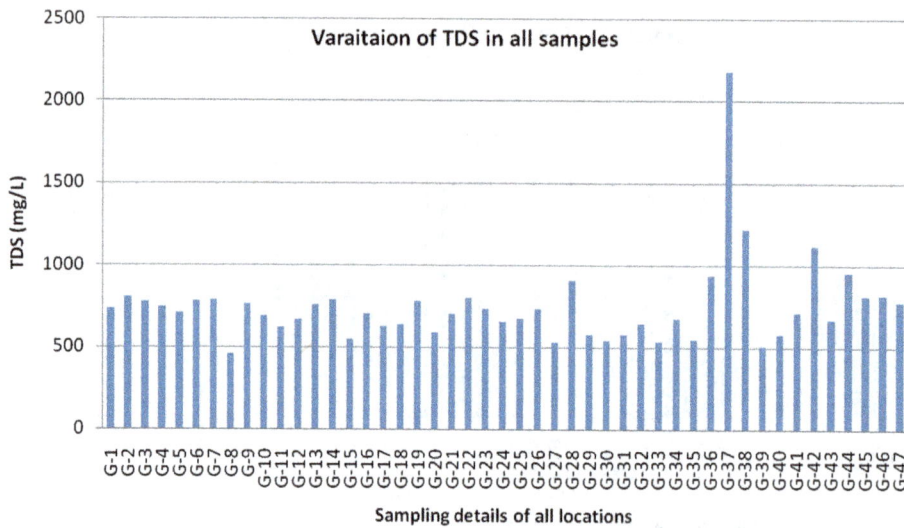

Figure 6. Variation of total dissolved solids in all the samples.

for the water to be used for drinking purpose. In present study, the TDS concentration of analysed samples lies in the range of 454–2,182 mg/L (Figure 6).

It is inferred that TDS of all the samples is well within the permissible limit prescribed by BIS except the sample No. G-38, where the TDS concentration is found to be 2,182 mg/L.

4.1.5. Total hardness

In fresh water sources, hardness is mainly due to presence of calcium and magnesium salts. Temporary hardness more than 200 mg/L as $CaCO_3$ may cause scale deposition in the treatment works, distribution system and pipe work and tanks within buildings. Water with hardness less than 100 mg/l may, in contrast, have a low buffering capacity and will be more corrosive for water pipes. BIS has prescribed 200 mg/l as the acceptable limit and 600 mg/l as the permissible limit for total hardness in absence of alternate source of drinking water. The hardness of groundwater samples in the study area is found to be in the range 91–678 mg/L as $CaCO_3$ (Figure 7). The hardness value of sample no is found to be 678 mg/L as Ca CO3 which is exceeding the permissible limit of 600 mg/L, while 23% of total samples were found to be within the desirable limit of 200 mg/l.

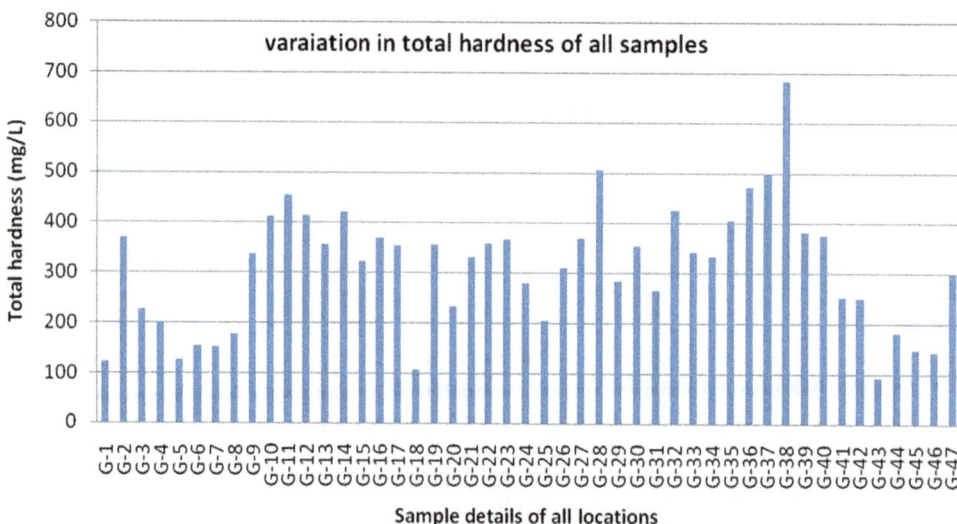

Figure 7. Variation of total hardness in all the samples from different locations.

4.1.6. Calcium

Analysis of calcium has also been carried out in all the samples in present study. The BIS limit for calcium is 75 mg/l as acceptable limit and 200 mg/L as permissible limit for drinking water. The calcium concentration in all groundwater samples of the study area lies in the range of 10.91–135.69 mg/l (Figure 8). The concentration of all the samples is well within the permissible limit of 200 mg/l.

4.1.7. Magnesium

Magnesium is another important parameter that has been analysed in all the samples taken in the present study. The BIS limits the magnesium concentration of 30 mg/l as acceptable value and 100 mg/L as a permissible value for drinking water. The magnesium concentration of all the samples has been found to be in the range of 7.5–120.5 mg/l (Figure 9). Only 25% of the total samples has been found to be within the acceptable limit. However, the concentration of magnesium in sample number G-38 is found to be 120.5 mg/L which is greater than the permissible limit prescribed by BIS.

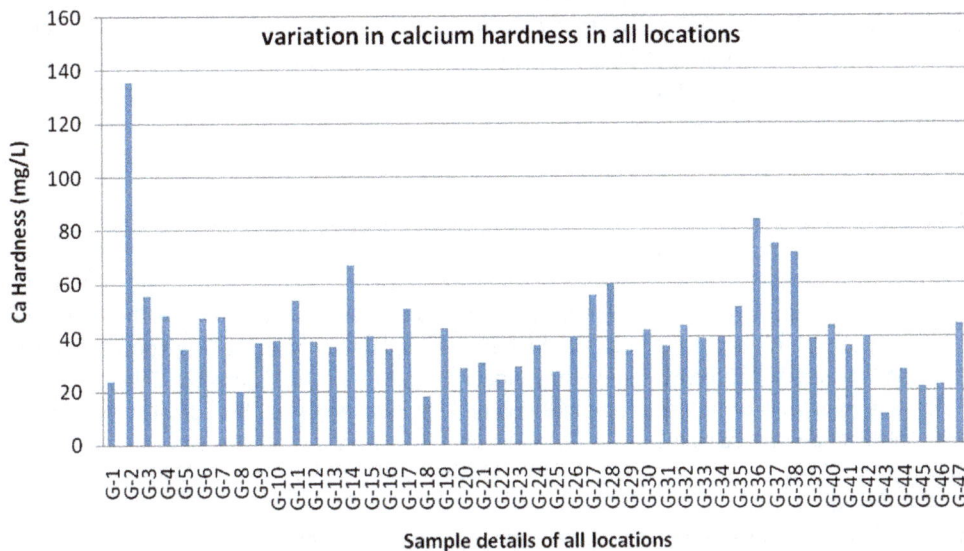

Figure 8. Variation of calcium hardness in all the samples.

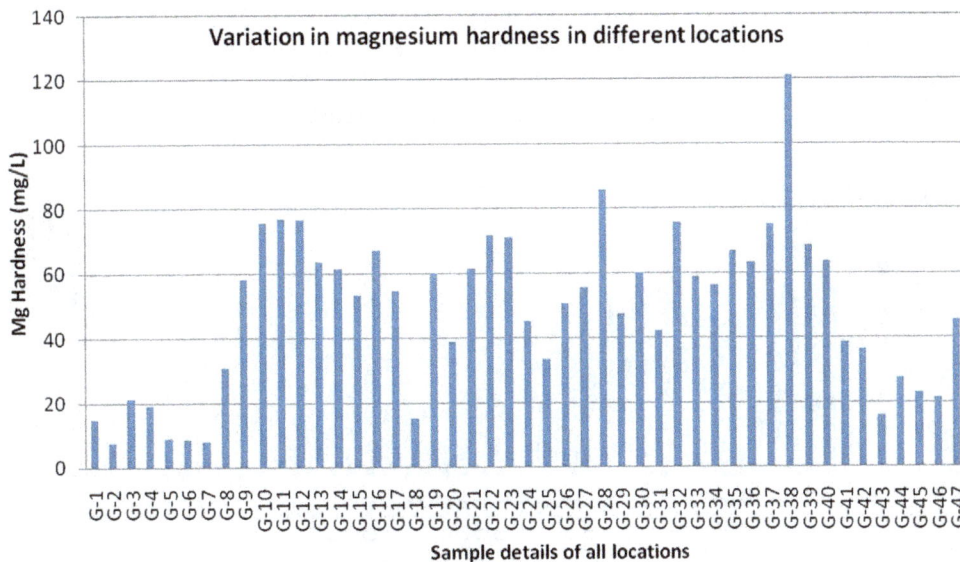

Figure 9. Variation of magnesium hardness in samples from different locations.

4.1.8. Alkalinity

Alkalinity in the water may be due to hydroxides, carbonates and bicarbonates. The relationship between pH, alkalinity and water stability is shown in Figure 10. BIS has prescribed 200 mg/l as the acceptable limit and 600 mg/l as the permissible limit for total alkalinity as $CaCO_3$ in absence of alternate source of drinking water. In present study, the alkalinity of all samples lies between 137 and 287 mg/l (Figure 10) and is well within the permissible limit prescribed by BIS for water used for drinking purpose.

Bicarbonate ions are the major contributor of alkalinity in ground water. In present study, the concentration of bicarbonate ions lies in the range of 168–350 mg/L in all the collected water samples (Figure 10).

4.1.9. Sodium

Sodium is a very reactive metal, and therefore does not occur in its free form in nature. High sodium intake can have adverse effects on humans with high blood pressure or pregnant women suffering from toxaemia. At room temperature, the average taste threshold for sodium is about 200 mg/l. Based on this, WHO has prescribed 200 mg/l as a limit for sodium in drinking water and BIS has not prescribed any limit. The concentration of sodium in groundwater samples of the study area ranges between 2.5 and 995.5 mg/l during study period (Figure 11.). 89% samples were well within the prescribed limit for sodium.

4.1.10. Chlorides

Some common chloride compounds found in natural water are sodium chloride (NaCl), potassium chloride (KCl), calcium chloride ($CaCl_2$) and magnesium chloride ($MgCl_2$). Taste thresholds for the chloride anion depend on the associated cations and the concentration ranges from 200 to 300 mg/L for sodium, potassium and calcium chloride. Based on taste threshold, BIS has prescribed 250 mg/l as the acceptable limit and 1,000 mg/l as the permissible limit for chloride. The concentrated of chloride in the collected samples were in the range of 33.2–675.6 mg/l (Figure 12). Chloride level in 89% samples was well within the acceptable limit and chloride level in all the samples were within permissible limit prescribed by BIS.

4.1.11. Potassium

Potassium is an essential element in humans and is seldom, if ever, found in drinking water at levels that could be a concern for humans health. Adverse health effects due to potassium consumption from drinking water are unlikely to occur in healthy individuals. The concentration of potassium in all groundwater samples of the study area ranges between 3.9 and 29.7 mg/l (Figure 13). The potassium concentration of 10 mg/l is permissible limit being prescribed by BIS (2012). The maximum concentration is found to be in the sample G-38. Result analysis indicates that the concentration of

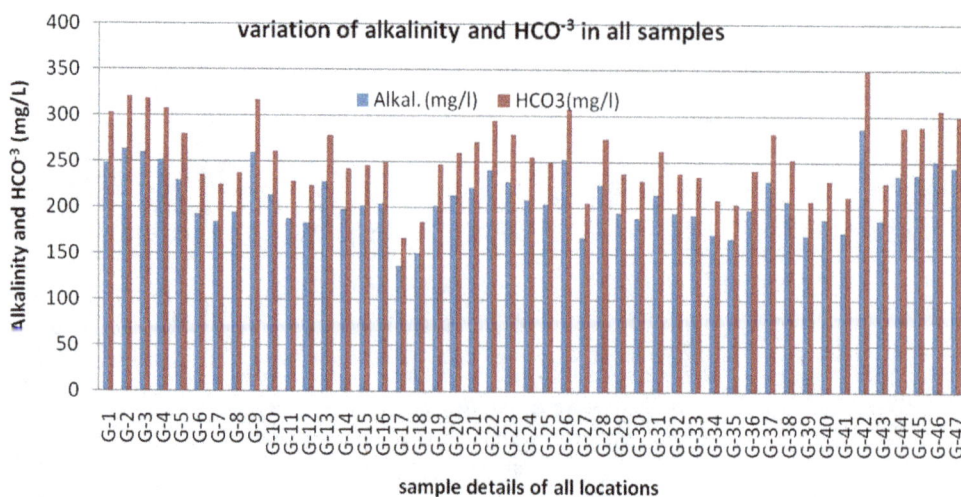

Figure 10. Comparison of alkalinity and bicarbonate variations in all the analysed samples.

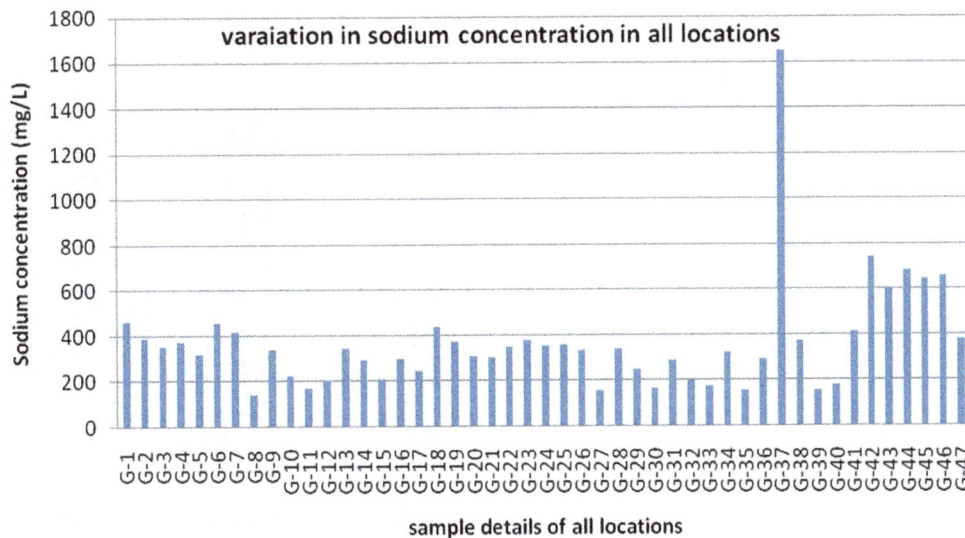

Figure 11. Variation of sodium concentration in all the samples.

potassium in 87% of the total samples is found to be well within the permissible limit prescribed by BIS for drinking water.

4.1.12. Sulphate

The most common form of sulphur in well-oxygenated water is sulphate. The presence of sulphate in drinking water can cause noticeable taste, and very high levels might cause a laxative effect in unaccustomed consumers. Taste thresholds have been found to range from 250 mg/l for sodium sulphate to 1,000 mg/L for calcium sulphate. BIS has prescribed 200 mg/l as the acceptable limit and 400 mg/L as the permissible limit. Sulphate concentration of groundwater samples in the study area lies in the range of 2 mg/L–1,300 mg/L (Figure 13). Sulphate concentration in all the samples is found to be within the acceptable limit prescribed by BIS for drinking water.

4.1.13. Nitrate

Nitrate (NO_3) is found naturally in the environment and is an important plant nutrient. Some ground waters may also have nitrate contamination as a consequence of leaching from natural vegetation. The presence of nitrate in drinking water is a potential health hazard when present in large

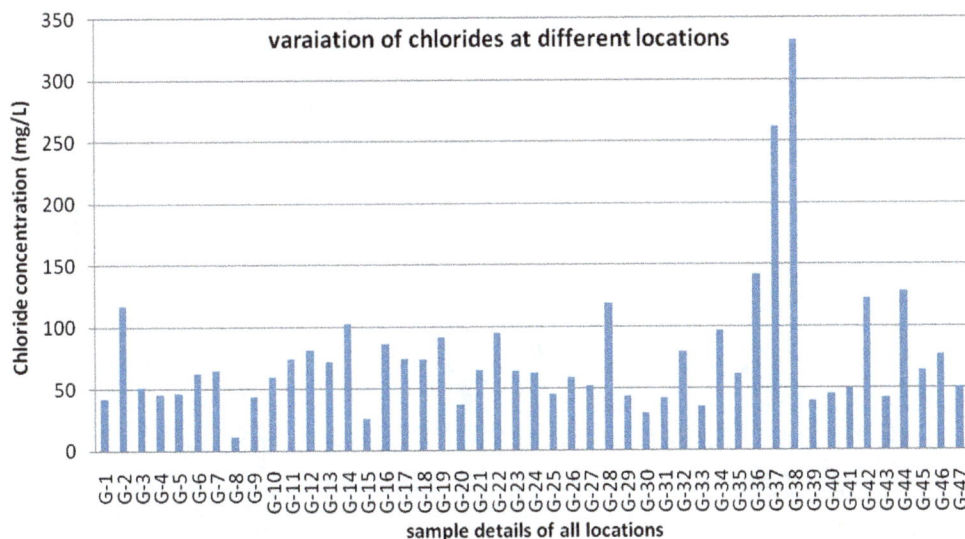

Figure 12. Variation of chlorides in all the samples from different locations.

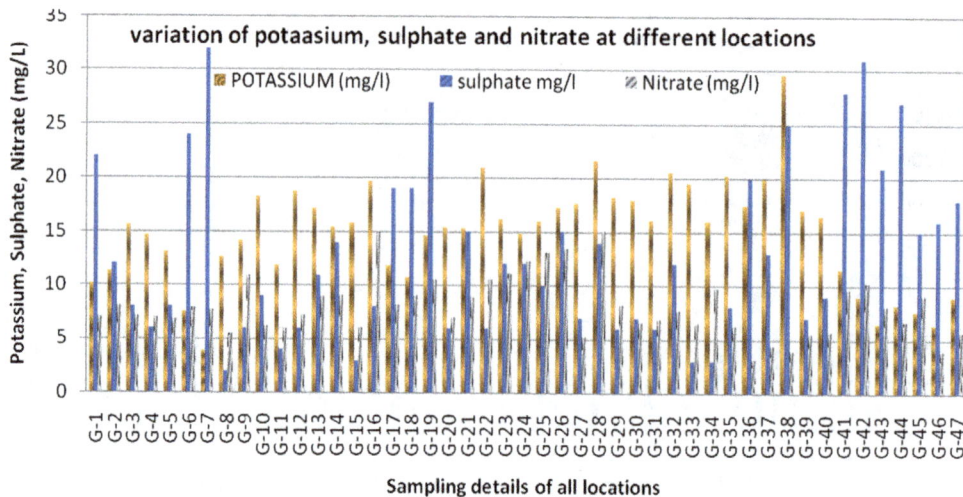

Figure 13. Comparison of potassium, sulphate and nitrate variations in different samples of all the locations.

quantities. The combination of nitrates with amines, amides, or other nitrogenous compounds through the action of bacteria in the digestive tract results in the formation of nitrosamines, which are potentially carcinogenic. The maximum allowable nitrate concentration as per BIS for drinking water is 45 mg/L as NO_3. The concentration of nitrate in groundwater samples of the study area ranges between 3.2 and 15.1 mg/l (Figure 13) and is found to be well within the desirable limit prescribed by BIS.

4.1.14. Fluoride

Fluoride is found in all natural type of waters at different concentrations. The fluoride concentration in water is limited by fluorite solubility, so that in the presence of 40 mg/L calcium it should be limited to 3.1 mg/L. It is the absence of calcium in solution which allows higher concentrations to be stable. Excess fluoride intake causes different types of fluorosis, primarily dental and skeletal fluorosis. BIS has prescribed 1 mg/l as the acceptable limit and 1.5 mg/l as the permissible limit for fluoride. The fluoride concentration of all groundwater samples in present study is in the range 0–1.70 mg/l (Figure 14). It is found that Fluoride concentration in 89% of total samples is well within the acceptable limit prescribed by BIS. However, it can also be inferred that 98% of total samples were with within permissible limit prescribed by BIS expect one sample (GW-1) exceeded the permissible limit.

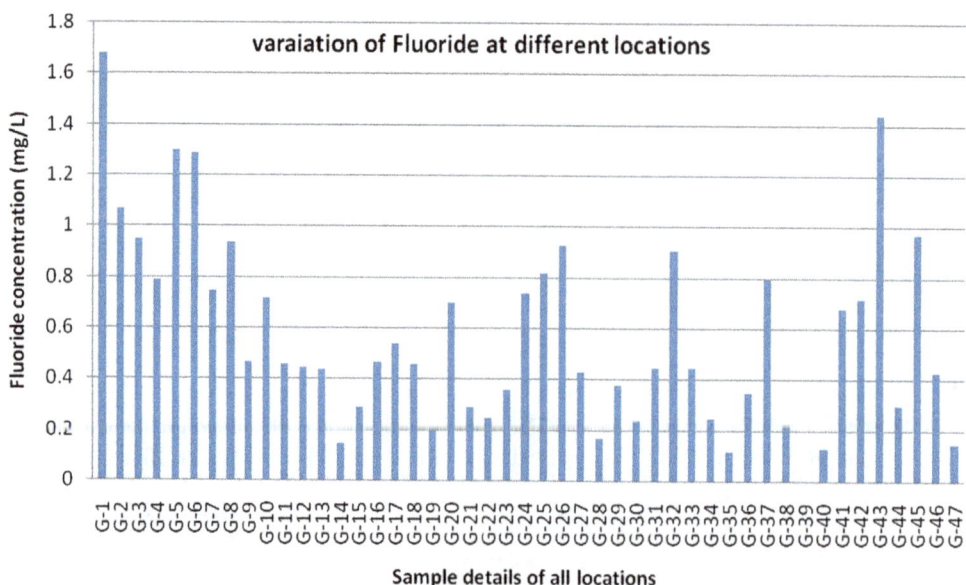

Figure 14. Fluoride variation in all the samples of different locations.

4.1.15. MPN

Total coliform bacteria include a wide range of aerobic and facultative anaerobic, Gram-negative, non-spore-forming bacilli capable of growing in the presence of relatively high concentrations of bile salts with the fermentation of lactose and production of acid or aldehyde within 24 h at 35–37 C. This test is first in line to micro-biological analysis. Negative results of the analysis test indicate the absence of any pathogens. The total coliform count in all the collected samples of the present study is found to be negative.

4.2. Hydrogeochemical facies of groundwater

Hydrochemical concepts can help to elucidate mechanisms of flow and transport in groundwater systems, and unlock an archive of paleoenvironmental information (Hem, 1992; Ophori & Toth, 1989; Pierre, Glynn, & Plummer, 2005). Every sample is represented by three data points; one in each triangle and one in the projection diamond grid. Also the pH, alkalinity and water stability standard relationship has been shown through Figure 15.

The main purpose of piper plots is to indicate water type of an area. All cations and anions designate a decrease trend in results as shown in Table 3. It is very useful for indication of the sample of

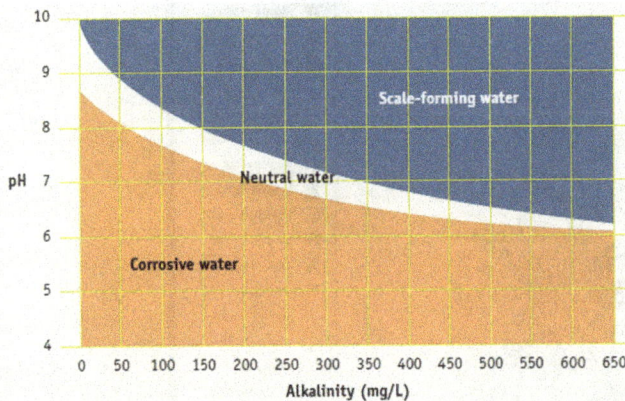

Figure 15. pH, alkalinity and water stability standard relationship.

Table 3. Sample analysis of the mixed concentrations of cations and anions		
Group No.	**Group sample ID**	**Cation and anion mixed concentration**
Group 1.	G-39, G-30, G-40, G-15, G-10, G-33	Mg-Ca-HCO$_3$-Cl
Group 2.	G-35, G-32	Mg-Ca-Cl-HCO$_3$
Group 3.	G-12	Mg-Cl-HCO3
Group 4.	G-17	Mg-Ca-Na-Cl-HCO$_3$
Group 5.	G-28, G-14	Mg-Na-Ca-Cl-HCO$_3$
Group 6.	G-23, G-13, G-21	Mg-Na-HCO$_3$-Cl
Group 7.	G-2, G-7	Na-Ca-Cl-HCO$_3$
Group 8.	G-37	Na-SO$_4$-Cl
Group 9.	G-2, G-20, G-25	Na-Mg-HCO$_3$-Cl
Group 10.	G-3, G-6, G-4, G-5	Na-Ca-HCO$_3$-Cl
Group 11.	G-42, G-44, G-18	Na-Cl-HCO$_3$
Group 12.	G-45, G-46, G-43, G-1	Na-HCO$_3$-Cl
Group 13.	G-38	Na-Mg-Cl
Group 14.	G-36	Na-Mg-Ca-Cl-HCO$_3$
Group 15.	G-34, G-16, G-19	Mg-Na-Cl-HCO$_3$
Group 16.	G-29, G-26, G-31, G-9	Mg-Na-Ca-HCO$_3$-Cl

mixed concentrations such as magnesium (Mg^{2+}), bicarbonate (HCO_3^-), sodium (Na^+), sulphate (SO_4^-) and chloride (Cl^-), etc. that varies from area to area.

Most of the samples indicate that the cations and anions are in mixed concentration in piper triangle. The plot shows majority of water samples fall in mixed Mg-Ca-HCO_3-Cl, Na-Ca-HCO_3-Cl, Na-HCO_3-Cl and Mg-Na-Ca-HCO_3-Cl concentration type (Figure 16) with minor representations from mixed types Mg-Ca-Cl-CO_3, Mg-Cl-HCO_3, Mg-Ca-Na-Cl-HCO_3, Mg-Na-Ca-Cl-HCO_3, etc. Only two samples indicate the higher mixed concentration Na-SO_4-Cl and Na-Mg-Cl sample G-37 (ID) and G-38 (ID), respectively. Alkali metal (Ca^{2+} + Mg^{2+}) exceeds over the alkaline earth metal (Na^+ + K^+) and the temporary hardness prevails over permanent hardness (Figure 16)

4.3. Correlation coefficient matrix of water parameters

A correlation coefficient (nearly 1 or −1) values lies between −1 and +1 and a correlation coefficient around zero means no relationship (Muthulakshmi et al., 2013). Positive values indicate a positive relationship while negative values of r indicate an inverse relationship. The values of correlation coefficients (r) are given in Table 4.

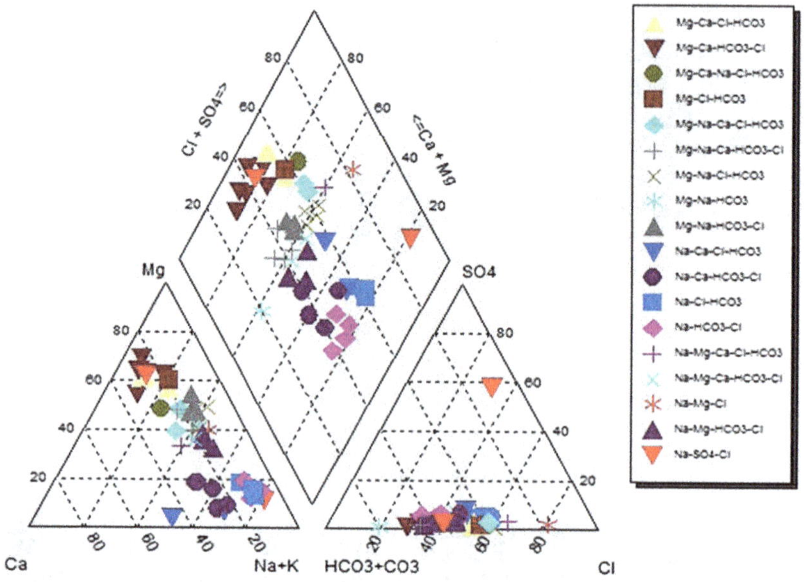

Figure 16. Piper triangle diagram for comparison of cations and anions in all the samples.

Table 4. Correlation coefficient matrix of analysed water quality parameters in all samples											
		pH	Cond	TDS	Mg	Ca	Na	K	Cl	SO$_4$	HCO$_3$	NO$_3$
pH		1	0	0	0	0	0	0	0	0	0	0
Cond	µS/cm		1	1	0.178	0.348	0.942	0.117	0.803	0.837	0.369	−0.101
TDS	mg/l			1	0.178	0.348	0.942	0.116	0.803	0.837	0.37	−0.101
Mg	mg/l				1	0.176	−0.026	0.814	0.46	0.145	−0.208	0.0064
Ca	mg/l					1	0.231	0.246	0.485	0.237	0.1	−0.195
Na	mg/l						1	−0.028	0.664	0.911	0.296	−0.164
K	mg/l							1	0.393	0.137	−0.127	0.044
Cl	mg/l								1	0.514	0.127	−0.15
SO$_4$	mg/l									1	0.093	−0.197
HCO$_3$	mg/l										1	0.154
NO$_3$	mg/l											1

Table 5. Details of water quality and Index rate of analysed samples

Sample ID	Index rate	Water quality
G-1	77.43	Good water
G-2	81.14	Good water
G-3	70.11	Good water
G-4	67.02	Good water
G-5	68.68	Good water
G-6	72.14	Good water
G-7	65.5	Good water
G-8	53.69	Good water
G-9	71.12	Good water
G-10	77.46	Good water
G-11	70.72	Good water
G-12	77.46	Good water
G-13	77.46	Good water
G-14	78.75	Good water
G-15	55.5	Good water
G-16	82.86	Good water
G-17	68.92	Good water
G-18	64.98	Good water
G-19	78.37	Good water
G-20	63.64	Good water
G-21	72.06	Good water
G-22	84.76	Good water
G-23	77.04	Good water
G-24	73.72	Good water
G-25	69.68	Good water
G-26	79.91	Good water
G-27	62.15	Good water
G-28	94.36	Good water
G-29	62.66	Good water
G-30	58.0	Good water
G-31	61.79	Good water
G-32	84.00	Good water
G-33	61.84	Good water
G-34	75.06	Good water
G-35	64.53	Good water
G-36	92.05	Good water
G-37	267.85	Very poor water
G-38	152.24	Poor water
G-39	56.69	Good water
G-40	60.23	Good water
G-41	68.82	Good water
G-42	97.06	Good water
G-43	70.39	Good water
G-44	84.97	Good water
G-45	74.29	Good water
G-46	70.35	Good water
G-47	63.91	Good water

Electric conductivity (EC) has a strong positive and signification correlation with TDS, Na^+, Cl^-, SO_4^- weak correlation with magnesium, calcium, potassium and bicarbonate and negative correlation with nitrate. The TDS showed negative correlation with nitrate. The TDS showed max correlation with sodium, chloride and sulphate and min with magnesium, calcium, potassium and bicarbonate.

The magnesium showed positive correlation with Ca^{2+}, Cl^-, SO_4^- and NO_3^-. A strong positive correlation of magnesium vs. potassium in the water was observed and negative with sodium and bicarbonate. Calcium has a weak positive correlation with Na^+, K^+, Cl^-, SO_4^-, and HCO_3^- and negative with nitrate. Sodium showed negative correlation with K^+ and NO_3^-. The sodium showed max correlation with sulphate and minimum with chloride and bicarbonate. Potassium showed negative correlation with bicarbonate and weak positive with chloride, sulphate and nitrate. Chloride has weak positive correlation with sulphate and bicarbonate and negative with nitrate. Sodium showed weak positive correlation with bicarbonate and negative with nitrate. Bicarbonate showed weak positive correlation with nitrate.

4.4. Developed index for ground water quality

During study period, very poor quality water has been observed in sample (G-37) from CHI-3 or Temple (R1 Scheme P5). This may be due to the location of the study area which is dominated by industrial and construction activities. Poor water quality is noted in (G-38) location from Radisson Hotel (Jaypee Integrated complex) of the study area, dominated by the influence of direct dump of waste on land in back side of location and domestic activities. Water quality was observed good in locations with Sample ID as G-1, G-2, G-3, G-4, G-5, G-6, G-7, G-8, G-9, G-10, G-11, G-12, G-13, G-14, G-15, G-16, G-17, G-18, G-19, G-20, G-21, G-22, G-23, G-24, G-25, G-26, G-27, G-28, G-29, G-30, G-31, G-32, G-33, G-34, G-35, G-36, G-39, G-40, G-41, G-42, G-43, G-44, G-45, G-46 and G-47.

However the study area is of granulite gneiss region, where industrial activities are dominant, and human activities such as domestic and construction, etc. are in progress (Vasanthavigar et al., 2010). Excellent water quality has not been observed in any study area. The index rate and type of water of ground water sample of the study area were calculated and the values are given in Table 5. Perusal of the data from Table 5 indicates that 96% of total water samples are observed to be of good quality of water while 2% of water sample shows the poor quality of water. However the remaining 2% of water sample shows the very poor quality of water.

5. Conclusion

In the present study, the computed WQI values ranges from 53.69 to 267.85. Very poor quality water has been observed in sample G-37 taken from location CHI-3 or Temple (R1 Scheme P5) of the study area. Poor water quality has been observed in sample G-38 from location Radisson Hotel (Jaypee Integrated complex) of the study area. It has been observed that 97% of samples indicate a good water quality around the study area. Analysis of results reveals the fact that WQI pertaining to the groundwater of the area needs some degree of treatment before consumption. Only one sample of location G-8 from HGVA (Echar Village) has been observed to be best in quality for drinking and all the parameters are well within the acceptable limit prescribed by BIS (2012) for drinking water. Piper plot of the physicochemical plot indicates alkali metals exceeding over the alkaline earth metals and the temporary hardness prevails over permanent hardness.

The results of the correlation analysis indicate a strong positive correlation of magnesium with potassium ions and negative correlation between sodium and bicarbonate ions. Correlation of calcium with Na^+, K^+, Cl^-, SO_4^- is found to be weakly positive and negative for nitrate. Sodium indicates max correlation with sulphate, minimum with chloride and bicarbonate, and negative with bicarbonate. Chloride showed a weak positive correlation with sulphate and bicarbonate and negative correlation with nitrate.

Acknowledgements

The authors of this paper are thankful to Noida authority and Gautam Buddha University for using lab facilities.

Funding.

This research was carried out at Gautam Buddha University, Greater Noida. The research has not been funded by government or private funding agency.

Author details

Sajal Singh[1]
E-mail: choudharysajal5@gmail.com
Athar Hussian[1]
E-mail: athariitr@gmail.com
[1] Civil Engineering Department, School of Engineering, Gautam Buddha University, Greater Noida 201310, Uttar Pradesh, India.

References

APHA. (1999). *Standard methods for the examination of water and waste waters* (18th ed.). Washington, DC: Author.

BIS. (2012). *Drinking water specification IS: 10500:2012*. New Delhi: Author.

Brown, R. M., Mc Clelland, N., Deininger, R. A., & Tozer, R. G. (1970). A water quality index - do we dare. *Water Sewage Works, 117*, 339–343.

Girija, T. R., Mahanta, C., & Chandramouli, V. (2007). Water quality assessment of an untreated effluent impacted urban stream: The Bharalu tributary of the Brahmaputra River, India. *Environmental Monitoring and Assessment, 130*, 221–236. http://dx.doi.org/10.1007/s10661-006-9391-6

Greater Noida Master Plan, 2001, 2021 plan report. (2013). Greater Noida Authority.

Hem, J. D. (1992). *Study and interpretation of chemical characteristics of natural water* (3rd ed., U.S. Geological Survey Water-Supply Paper 2254, 263 p.). Alexandria, VA: U.S. Geological Survey.

Horton, R. K. (1965). An index number system for rating water quality. *Journal of Water Pollution Control Federation, 37*, 300–305.

Mangukiya, R., Bhattacharya, T., & Chakraborty, S. (2012). Quality characterization of groundwater using water quality index in Surat city, Gujarat, India. *International Research Journal of Environment Sciences, 1*, 14–23.

Muthulakshmi, L., Ramu, A., Kannan, N., & Murugan, A. (2013). Application of correlation and regression analysis in assessing ground water quality, Virudhunagar, India. *International Journal of ChemTech Research, 5*, 353–361.

Ophori, D. U., & Toth, J. (1989). Characteristics of ground water flow by field mapping and numerical simulation, Ross Creek Basin, Alberta, Canada. *Ground Water, 27*, 193–201.

Palanisamy, P. N., Geetha, A., Sujatha, M., Sivakumar, P., & Karunakaran, K. (2007). Assessment of ground water quality in and around Gobichettipalayam town Erode District, Tamil Nadu, India. *E-Journal of Chemistry, 4*, 434–439.

Pierre, D., Glynn, L., & Plummer, N. (2005). Geochemistry and the understandings of the groundwater systems. *Hydrogeol Journal, 13*, 263–287.

Srinivasamoorthy, K., Chidambaram, S., Prasanna, M. V., Vasanthavihar, M., John peter, A., & Anandhan, P. (2008). Identification of major sources controlling groundwater chemistry from a hard rock terrain—A case study from Mettur taluk, Salem district, Tamil Nadu, India. *Journal of Earth System Science, 117*, 49–58. http://dx.doi.org/10.1007/s12040-008-0012-3

Tiwari, T. N., & Mishra, M. A. (1985). A preliminary assignment of water quality index of major Indian river and ground water. *Indian Journal of Environmental Protection, 5*, 276–279.

Vasanthavigar, M., Srinivasamoorthy, K., Vijayaragavan, K., Ganthi, R. R., Chidambaram, S., Anandhan, P., ... Vasudevan, S. (2010). Application of water quality index for groundwater quality assessment: Thirumanimuttar sub-basin, Tamil Nadu, India. *Environmental Monitoring and Assessment, 171*, 595–609.

WHO. (2012). *Guidelines for drinking water, recommendations*. Author.

PERMISSIONS

The contributors of this book come from diverse backgrounds, making this book a truly international effort. This book will bring forth new frontiers with its revolutionizing research information and detailed analysis of the nascent developments around the world.

We would like to thank all the contributing authors for lending their expertise to make the book truly unique. They have played a crucial role in the development of this book. Without their invaluable contributions this book wouldn't have been possible. They have made vital efforts to compile up to date information on the varied aspects of this subject to make this book a valuable addition to the collection of many professionals and students.

This book was conceptualized with the vision of imparting up-to-date information and advanced data in this field. To ensure the same, a matchless editorial board was set up. Every individual on the board went through rigorous rounds of assessment to prove their worth. After which they invested a large part of their time researching and compiling the most relevant data for our readers.

The editorial board has been involved in producing this book since its inception. They have spent rigorous hours researching and exploring the diverse topics which have resulted in the successful publishing of this book. They have passed on their knowledge of decades through this book. To expedite this challenging task, the publisher supported the team at every step. A small team of assistant editors was also appointed to further simplify the editing procedure and attain best results for the readers.

Apart from the editorial board, the designing team has also invested a significant amount of their time in understanding the subject and creating the most relevant covers. They scrutinized every image to scout for the most suitable representation of the subject and create an appropriate cover for the book.

The publishing team has been an ardent support to the editorial, designing and production team. Their endless efforts to recruit the best for this project, has resulted in the accomplishment of this book. They are a veteran in the field of academics and their pool of knowledge is as vast as their experience in printing. Their expertise and guidance has proved useful at every step. Their uncompromising quality standards have made this book an exceptional effort. Their encouragement from time to time has been an inspiration for everyone.

The publisher and the editorial board hope that this book will prove to be a valuable piece of knowledge for researchers, students, practitioners and scholars across the globe.

LIST OF CONTRIBUTORS

Samuel Asumadu-Sarkodie, Çağlan Sevinç, Herath M.P.C. Jayaweera, Phebe Asantewaa Owusu, Samuel Asumadu-Sarkodie, Samuel Asumadu-Sarkodie Phebe Asantewaa Owusu, Phebe Asantewaa Owusu, Samuel Asumadu-Sarkodie, Obaidullah Mohiuddin, Abdullah Mohiuddin, Madina Obaidullah, Humayun Ahmed and Samuel Asumadu-Sarkodie
Sustainable Environment and Energy Systems, Middle East Technical University Northern Cyprus Campus, Mersin, Kalkanli- Guzelyurt, 99738, TRNC, Turkey

Polycarp Ameyo
Department of Civil Engineering, Middle East Technical University, Northern Cyprus Campus, Morphou, 99738 TRNC, Turkey

Sajal Singh
School of Engineering, Gautam Buddha University, Greater Noida 201310, Uttar Pradesh, India

Athar Hussain
Civil Engineering Department, Ch. Brahm Prakash Government Engineering College, Jaffarpur, Delhi, India

S.D. Khobragade
Hydrological Investigations Division, National Institute of Hydrology, Roorkee 247667, Uttarakhand, India

N. Vivekanandan
Central Water and Power Research Station, Pune, Maharashtra 411024, India

Sujay Raghavendra. N and Paresh Chandra Deka
Department of Applied Mechanics and Hydraulics, National Institute of Technology Karnataka, Surathkal, Mangalore 575025, India

Pushpendra Kumar Sharma
Department of Civil Engineering, Hindustan College of Science and Technology, Farah, Mathura, Uttar Pradesh, India

Sohail Ayub
Department of Civil Engineering, Aligarh Muslim University, Aligarh, Uttar Pradesh, India

Chandra Nath Tripathi
Environmental Engineering, Hindustan College of Science and Technology, Farah, Mathura, Uttar Pradesh, India

K.N. Patil, S.C. Kaushik and S.N. Garg
Centre for Energy Studies, Indian Institute of Technology Delhi, Hauz Khas, New Delhi 110016, India

Dattatraya R. Mahajan and Basavanand M. Dodamani
Department of Applied Mechanics and Hydraulics, National Institute of Technology Karnataka, Surathkal, Mangalore, India

Parul Gupta and Kiranmay Sarma
University School of Environment Management, GGSIP University, Sector-16C, Dwarka 110078, New Delhi, India

Sajal Singh and Athar Hussian
Civil Engineering Department, School of Engineering, Gautam Buddha University, Greater Noida 201310, Uttar Pradesh, India

Index

9 781682 865422